Progress in
Colloid & Polymer Science

Editors: H.-G. Kilian (Ulm) and A. Weiss (Munich)

Surfactants, Micelles, Microemulsions and Liquid Crystals

Editor:
A. Weiss (Munich)

Springer-Verlag Berlin Heidelberg GmbH 1984

ISBN 978-3-662-16060-2 ISBN 978-3-7985-1702-8 (eBook)
DOI 10.1007/978-3-7985-1702-8
ISSN 0340-255 X

Originally published by Dr. Dietrich Steinkopff-Verlag GmbH & Co. kG in 1984
Softcover reprint of the hardcover 1st edition 1984

Contents

Progress in Colloid & Polymer Science Progr Colloid & Polymer Sci 69:1–11 (1984)

Effect of the molecular structure of components on micellar interactions in microemulsions*)

D. Roux, A. M. Bellocq, and P. Bothorel

Centre de Recherche Paul Pascal (CNRS), Talence, France

Abstract: The effect of the chemical nature of oil, alcohol and surfactant on interactions in W/O microemulsions has been studied by light scattering and photon correlation spectroscopy. The influence of the chain length and of the branching of oil has been investigated. For this purpose the following oils have been used: dodecane, octane, cyclohexane and trimethyl 2-2-4, pentane. The alcohol chain length has also been varied from 5 to 7 carbons. In order to examine the influence of the surfactant structure we have used SDS and α methyl SDS.

All these molecular changes lead to very different behavior in the interactions. Indeed the second virial coefficients vary in a large range from positive to very negative values indicating that interactions in the studied systems change from hard spheres to largely attractive. It seems that the important parameters are the length for alcohol, the molecular volume for oil and the polar head area for surfactant. A very simple intermicellar potential is proposed. It allows to account for all the obtained light scattering results. This potential is due to Van Der Walls interactions and the interpenetration between micelles is taken into account. The most important values of the attractive interactions appear in the overlapping region. It is shown that the proposed potential is proportional to the penetrated volume.

Key words: Microemulsion, micelles, micellar interactions.

1. Introduction

The scattering properties of microemulsions were first investigated by Hoar and Schulman [1]. Since that time, these systems have attracted considerable attention. In some part of the phase diagram their structure can be pictured as dispersions of water droplets surrounded by a film of surfactant and alcohol molecules in a continuous medium mainly made of oil. Light scattering techniques have been extensively applied to these systems, since these methods provide information on micellar sizes and interactions between droplets [2, 3, 4, 5].

Previous results have shown that in water in oil(W/O)microemulsions a large range of interaction forces can be obtained by varying chemical composition. Modern theories of fluids were used by several authors to explain the light scattering experimental results [3, 4]. The difficulty in evaluating the interaction potential for inverted micelles did not permit in most cases to give a complete analysis of data. Recently two of us have calculated the intermicellar energy potential between W/O microemulsions [6]. In this calculation interactions have been evaluated for penetrable particles formed by spherical aqueous core and concentric spherical layers. Attractive interactions are calculated through integration of semi-empirical interatomic potential over the various regions of interacting micelles. This potential allows the interpretation of the effects of size and alcohol on the behavior of microemulsions [5].

Our interest in this paper lies mainly devoted in investigating the effect of the molecular structure of the various components which form a microemulsion – alcohol, oil and surfactant – on the interactions between W/O micelles. For this purpose we have studied a large number of W/O microemulsions by static and dynamic light scattering. Indeed, the static experiments give access to the second virial coefficient B of the osmotic pressure which is directly related to the

*) Lecture given at the 31th Conference of the Kolloid Gesellschaft, Bayreuth October 11–24, 1983.

KT-W 800

interaction potential. Moreover, study of the concentration dependence of the diffusion coefficient D at moderate concentrations allows the determination of the virial coefficient α of D. Several recent theories relate this virial coefficient α to the interaction potential [7, 8, 9]. So in this paper, we have compared the experimental B and α values with those calculated by using a simplified interaction potential.

In the following, we will first recall some theoretical results (sect. II), then we describe the experimental procedure (sect. III), and intensity and diffusion coefficient measurements (sect. IV). In section V we will present an analytical interaction potential between W/O micelles, then we will give the results of the calculation of virial coefficients B and α. In the concluding section we will discuss the experimental and theoretical results.

2. Theoretical background

For unpolarized light the excess scattering of the particles assumed to be of a constant size over that of the continuous phase is [10]:

$$I(\theta) = (1 + \cos^2\theta)\, K v_m\, \phi\, S(q)\, P(q)$$

where $q = 4\Pi n/\lambda_o \sin \theta/2$ is the scattering wave vector, θ is the scattering angle, v_m is the volume of the micelle, ϕ the micellar volume fraction and

$$K = 2\Pi^2 n^2 \left(\frac{dn}{d\phi}\right)^8 (\lambda_o^4)^{-1} \tag{1}$$

with n the refractive index and λ_o the light wavelength in vacuo. $P(q)$ is the intraparticle form factor: we set $P(q) \sim 1$ for particles under study since their radius is less than 100 Å. $S(q)$ is the structure factor. In the limit $q \to o$, $S(q)$ is related to the osmotic pressure Π by the compressibility relation [11]

$$S(O) = \frac{k_B T}{v_m} \left(\frac{\partial \Pi}{\partial \phi}\right)^{-1}$$

where k_B is the Boltzman constant and T the absolute temperature. The relation between compressibility and interaction between particles is not easy. One of the simplest ways to have an idea of the interactions is to develop the osmotic pressure according to the virial formula. Π can be written as a function of different powers of ϕ:

$$\Pi = \frac{\phi k_B T}{v_m}\left(1 + \frac{B}{2}\phi + \frac{C}{3}\phi^2 + \ldots\right)$$

where B and C are virial coefficients. B is directly related to the interaction potential $U(r)$ by:

$$B = -\frac{4\Pi}{v_m}\int \left(e^{-\frac{U(r)}{k_B T}} - 1\right) r^2\, dr$$

In the limit of very small volume fractions, the osmotic pressure can be approximated by

$$\Pi \sim \frac{k_B T}{v_m}\, \phi \left(1 + \frac{B}{2}\phi\right)$$

From the above equations one can deduce:

$$\frac{\phi}{I} = \frac{1}{K v_m}(1 + B\phi) \tag{2}$$

It appears that droplet size and the second virial coefficient B can be extracted from a study of ϕ/I in the low concentration range. In addition, results obtained from more concentrated solutions allow one to determine the variation of the osmotic compressibility.

Modern theories of fluids have been used by several authors to explain microemulsion experimental results. One of the main conclusions of these theories is that in dense liquids the spatial structure, which can be represented either by the function of pair distribution $g(r)$ or by the thermodynamic properties, is to a large extent determined by steric repulsions between close particles. The attractive or repulsive effects are treated as perturbation of hard spheres. Vrij et al. consider that the intermicellar interaction potential in W/O microemulsions can be expressed as the sum of two terms U_{HS} and U_A [3]. The hard sphere contribution to the osmotic pressure Π_{HS} is described by the equation of state proposed by Carnahan and Starling [12]. Only binary interactions are considered in the term of perturbation Π_A. The total osmotic pressure of the solution can be written as $\Pi = \Pi_{HS} + \Pi_A$ with

$$\Pi_{HS} = \frac{k_B T}{v_{HS}} \frac{1 + \phi_{HS} + \phi_{HS}^2 - \phi_{HS}^3}{(1 - \phi_{HS})^3}$$

$$\Pi_A = \frac{k_B T}{v_m} \frac{A}{2} \phi^2$$

$$A = \frac{4\Pi}{k_B T} \frac{1}{v_m} \int_{2R_{HS}}^{\infty} U_A(r)\, r^2\, dr$$

$U_A(r)$ is the perturbation to the hard sphere potential $U_{HS}(r)$. ϕ_{HS} is the volumic fraction of hard spheres of radius R_{HS}. ϕ_{HS} is related to ϕ by $\phi_{HS}/\phi = a$. The second virial coefficient deduced from the whole expression of Π is written $B' = 8a + A$. In this model, the scattered intensity is:

$$I(\phi) = Kv_m \frac{\phi(1-a\phi)^4}{1+4a\phi+4a^2\phi^2-4a^3\phi^3+a^4\phi^4+A\phi(1-a\phi)^4} \qquad (3)$$

The fit of the experimental $I(\phi)$ curve by a least square method allows the determination of three parameters, Kv_m, a and A from which we can deduce the micellar radius and the second virial coefficient B'.

Moreover, the autocorrelation function of the scattered light is given by [13]:

$$g^{(2)}(\tau) = 1 + e^{-2Dq^2\tau}$$

where D is the translational diffusion and q the wave vector. D is related to osmotic pressure by $D = v_m/f\ \partial\Pi/\partial\phi$ where v_m is the volume of the micelle and f the friction coefficient between micelle and continuous phase.

In the low concentration range, D can be written as [7]

$$D \sim D_o\,(1+\alpha\phi) \text{ with } D_o = \frac{k_B T}{6\,\Pi\,\eta\,R_H} \qquad (4)$$

η is the viscosity of the continuous phase, R_H the hydrodynamical radius of the micelle. The virial coefficient α is related to that of the osmotic pressure by the equation

$$\alpha = B - \beta$$

β represents the dynamic part which takes into account the volume fraction dependence of the friction coefficient f. Both the static and dynamic contributions of α are related to the interaction potential of particles $U(r)$. The expression for B is well established in the case of rigid spherical particles of radius R_{HS} with a pair interaction potential $U(x)$ where $x = r/R_{HS}$ and r is the distance between the centers of the two particles.

$$B = 8 + \frac{24}{R_{HS}^3}\int_{R_{HS}}^{\infty} (1-e^{-\frac{U(r)}{k_BT}})\, r^2\, dr$$

where 8 is the hard sphere contribution.

In opposition, there is still some discrepancy amongst the calculations for β presented by different authors [7, 8, 9]. However, in the case where $U(r) = U_{HS}(r) + U_A(r)$, it is possible to write down a relation for β which has the same structure as the preceeding one, that is

$$\beta = \beta_o + \int_{R_{HS}}^{\infty} F(x)\,(1-e^{-\frac{U(x)}{kT}})\, dx$$

where β_o is the hard sphere contribution.

A complete treatment has been given by Feldherof [7] who obtains:

$$\beta_o = 6.44$$

and

$$F(x) = 12x - 15/(8x^2) + 27/(64x^4) + 75/(64x^5) \qquad (6)$$

Others treatments have been proposed by Batchelor [8] and Goldstein and Zimm [9]. The α values derived from these three calculations being very close, in the following we only give the result obtained with the Feldherof equation (eq. 6).

3. Experimental part

a) Sample preparation

The studied samples are quaternary mixtures of water, oil, alcohol and surfactant. N-dodecane (D), N-octane (O), isooctane (trimethyl 2-2-4-pentane) (I) and cyclohexane (C) have been used as oil; 1-pentanol (C5), 1-hexanol (C6), 1-heptanol (C7) as alcohol; sodium dodecyl sulfate (S) and sodium methyl-1 dodecyl sulfate (M) as surfactant. For the systems containing SDS and dodecane several microemulsions of water/surfactant ratio (W/S expressed in volume) ranging from 1.74 to 3.50 were studied in order to obtain size variation. All the studied systems are located on the surface of demixion of the one-phase volume. They are designated in an abridged form by the surfactant followed by the alcohol and the oil, then as an example a microemulsion formed with SDS, pentanol and dodecane is named S-C5-D. SDS of quality puriss was purchased from Touzart and Matignon; the sodium α-methyl dodecyl sulfate was synthetized in the laboratory according to the method of ref. [14]. The other compounds are Fluka products. The overall composition of the studied systems are given in table 1 and ref. [5].

A schematic description of the W/O studied micelles is given in figure 1. The water cores are surrounded by a mixed film of surfactant and alcohol molecules. These micelles are dispersed in a com-

Table 1. Volumic compositions of microemulsions made with different oils, and with α-methyll-SDS. The compositions of the S-Alcohol-D microemulsions are given in ref. [5].

	S-C_6-O	S-C_6-C	S-C_6-I	S-C_6-D	M-C_6^1-D	M-C_6^2-D	M-C_7^1-D	M-C_7^2-D
W/S	2.55	2.55	2.55	2.55	1.75	2.55	2.75	2.55
Water	19.25	21.36	19.33	19.56	13.78	20.51	15.93	20.43
Oil	55.19	55.90	56.46	52.94	65.59	55.47	63.73	55.24
Surfactant	7.51	8.33	7.54	7.66	7.49	8.05	5.80	8.02
Alcohol	18.05	14.41	16.67	19.83	13.14	15.96	14.53	16.31

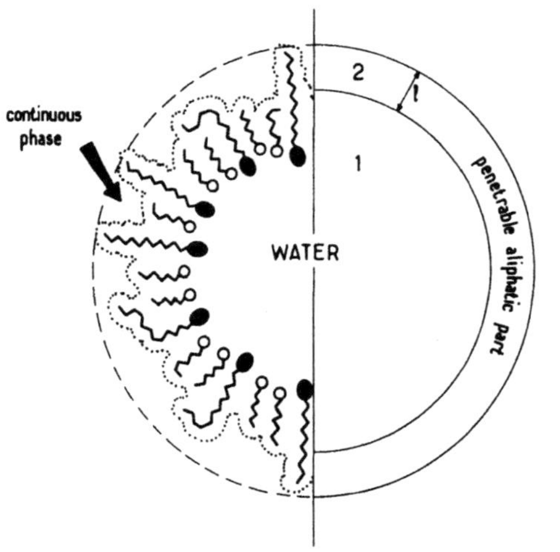

Fig. 1. Schematic picture of a W/O microemulsion

plex solvent named „continuous phase". The continuous phase contains primarily oil and alcohol but also a small amount of water. Analysis of light scattering data in terms of size and interaction requires an extrapolation of the results to zero concentration, therefore it is necessary to use a dilution procedure which keeps constant both the size and composition of micelles. The composition of the continuous and dispersed phases of the studied microemulsions has been determined by using the dilution procedure described by Graciaa et al. [2]. The validity of such a method has been checked by Taupin et al. [15] by means of neutron scattering. Indeed variable contrast gives information about the internal structure of the object. It has been shown that structure and composition of the elementary droplet is unchanged up to a water volume fraction equal to 0.3 in microemulsions where attractive forces are not very strong.

The volume fraction of the micelles has been defined as:

$$\phi = \frac{V_A^M + V_W^M + V_S}{V}$$

Where V is the total volume, V_S the volume of surfactant and V_A^M and V_W^M are respectively the volumes of alcohol and water contained in the micelles. Straight dilution lines are obtained in a very large concentration range. They allow the dilution of the studied microemulsions approximately 30 times. Solutions of volumic micellar fractions ranging from 0.01 to 0.30 have been prepared by this method. One has observed a kinetic effect in systems for which the second virial coefficient is positive (this corresponds to the less attractive interactions, tables 2 and 3). Indeed in these cases, the solution becomes clear only a few hours after mixing of the components. Compositions of the continuous and dispersed phases are reported in tables 2 and 3. The composition of the continuous phase is characterized by the molecular alcohol oil ratio A^C/oil. It is known that the continuous phase penetrates into the micelle [16]; usually one assumes that its composition is not changed. In this assumption it is possible to calculate the alcohol volume contained in the micelle: the used alcohols being very slightly soluble in water, we consider that the whole alcohol contained in the micelle A^M is only located at the interface. Hence the composition of the interface is defined by the molecular alcohol surfactant ratio (tables 2 and 3). For a given alcohol, as the W/S ratio increases the alcohol concentration in the continuous phase increases, whereas in the interface this concentration decreases. For a given W/S ratio, the alcohol concentration in the continuous phase depends on alcohol and oil. The A^C/oil ratio decreases as the alcohol chain length increases and as the oil chain length decreases. Microemulsions formed with the branched surfactant contained less alcohol than those formed with SDS; in these systems the decrease of the alcohol concentration is much more marked in the continuous phase than in the interface. This means that the quantity of surfactant molecules necessary to solubilize a given amount of oil and water is less with Me-SDS than with SDS.

b) Method

The various liquid components of the microemulsions (oil, water and alcohol) were first filtered on fine sintered glass ~ 1.4 µm) before preparation of the samples and then the solutions were centrifugated at 5000 rpm for 30 mn. Refractive indexes have been measured using a Pulfrich refractometer. The usual sin θ correction was made to allow for the angular variation of the size of the scattering volume. Correction of solid angle was also carried out. The angular range studied was $30° < \theta < 150°$.

The intensity and the correlation function of the scattered light were successively measured with a laser beam (Argon ion laser, Spectra Physics Model 165, $\lambda_o = 5145$ A). All the static and dynamic measurements were made at 21.5 ± 0.5 °C.

Measurements of the viscosities of the continuous phases were carried out using an improved Oswald-like viscosimeter.

Table 2. Effect of alcohol and W/S ratio on the sizes and virial coefficients of microemulsions formed with SDS and dodecane. R and B are determined by the extrapolation method R' and B' by the Vrij method. a) volumic ratio of total water to surfactant; b) molecular ratio of alcohol to SDS in the micelle; c) molecular ratio of alcohol to dodecane in the continuous phase

		S-C_5^1-D	S-C_5^2-D	S-C_6^1-D	S-C_6^2-D	S-C_6^3-D	S-C_6^4-D	S-C_7^1-D	S-C_7^2-D	S-C_7^3-D
W/S	a	2.55	1.74	1.75	2.32	2.90	3.50	2.55	2.74	3.48
A^M/S	b	2.99	2.71	2.58	2.8	2.61	1.9	2.69	2.7	2.82
A^c/C_{12}	c	0.4	0.29	0.24	0.263	0.34	0.454	0.22	0.25	0.34
R (Å)		50 ± 6	62 ± 7	47.6 ± 1	64 ± 4	66 ± 1	70 ± 2	54	62 ± 2	66 ± 3
R' (Å)				48 ± 1	63 ± 5	66 ± 1	74 ± 2	56	61 ± 2	65 ± 3
B		-23 ± 3	-27 ± 3	-0.5 ± 0.5	-4.4 ± 1	-6.1 ± 1	-8.8 ± 2	$+6$	4 ± 2	3 ± 3
B'				-1.5 ± 0.5	-3.8 ± 1	-6 ± 1	-9.2 ± 2	$+5$	3 ± 2	2 ± 3
R_H (Å)		55 ± 10	65 ± 10	52 ± 5	67 ± 10	69 ± 10		61	73 ± 2	75 ± 10
α		-18 ± 3	-21 ± 3	-5.8 ± 1	-9.7 ± 1	-12 ± 1		$+1$	$\sim 0 \pm 1$	$\sim 0 \pm 1$

Table 3. Effect of oil and surfactant on the virial coefficients

		S-C_6-O	S-C_6-C	S-C_6-I	S-C_6-D	M-C_6^1-D	M-C_6^2-D	M-C_7^1-D	M-C_7^2-D
W/S	a	2.55	2.55	2.55	2.55	1.75	2.55	2.55	2.751
A^M/S	b	2.45	1.68	2.46	2.77	2.25	2.45	2.4	2.44
A^C/O	c	0.306	0.114	0.17	0.36	0.14	0.22	0.16	0.17
R			55.6	68.	62.	54.	75.	61.	57.
R'			54.	68.	65.	56.		60.	58.
R_H		61.5	61.	61.	65.		71.	64.	
B		~ 0.	10.	4.	− 10.	− 13.	− 16.	− 4.	− 9.
B'			4.5	5.	− 5.	− 9.		− 3.	− 4.
α		~ 2.	~ 3.	4.	− 8.		− 20.	− 7.	

In the determination of the dilution line, the transition from turbid polyphasic state to transparent one-phase system is visually observed. However this observation becomes very difficult in the low concentration range. Moreover we have observed that in some cases, scattered intensity varies largely in the vicinity of the demixion line. Therefore, all the investigated samples have been prepared in the photometer according to the procedure described in the previous paper [5].

4. Light scattering results

We have measured the scattered intensity I_{90} (which is expressed as the Rayleigh ratio in cm^{-1}) and the diffusion coefficient at $\theta = 90°$ by the various microemulsions as a function of volumic fraction. Figures 2 and 3 show examples of intensity and diffusion coefficient

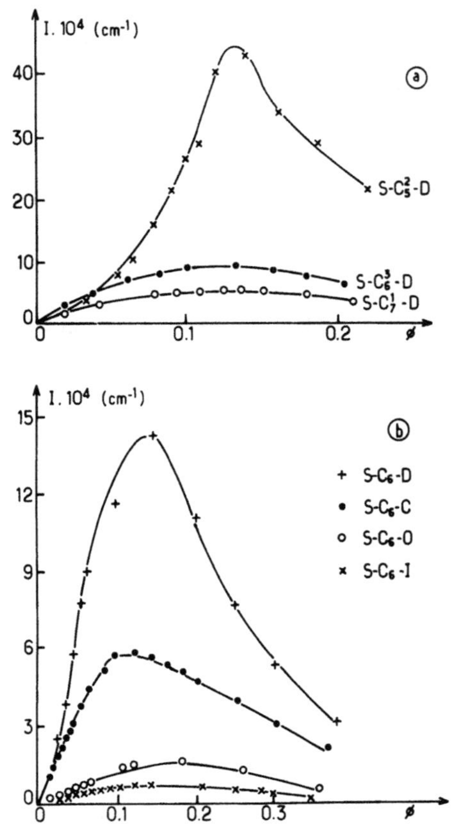

Fig. 2. Scattered intensities at $\theta = 90°$ versus the micellar volume fraction. a) influence of alcohol and surfactant; b) influence of oil

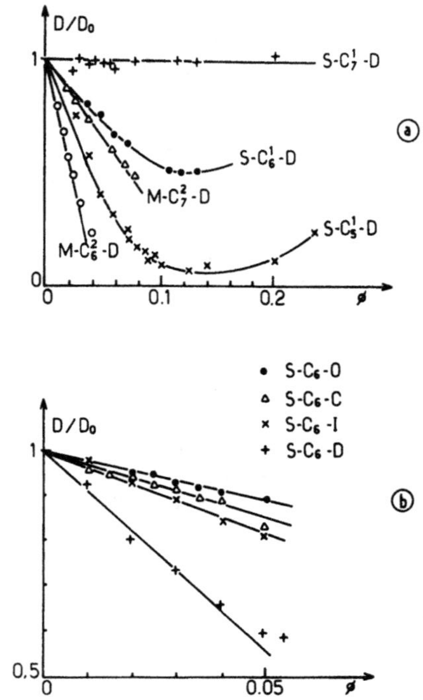

Fig. 3. a) influence of alcohol and surfactant; b) influence of oil on diffusion coefficient

variation versus ϕ for various microemulsions. Both variations $I(\phi)$ and $D(\phi)$ are strongly dependent on the alcohol, the surfactant and the oil. All the curves $I(\phi)$ show in the studied concentration range a maximum for a certain ϕ_{max} value. $D(\phi)$ curves relative to most of the studied microemulsions presents a minimum. However for the microemulsions S–C7–D the translational diffusion coefficient is found independent of ϕ.

Data analysis shows that the observed differences of intensity are due to various causes:

i) variation of size and interactions
ii) vicinity of a critical point
iii) variation of the increment of refractive index, this latter varies between 0.077 in the pentanol-dodecane system and 0.009 in the hexanol-isooctane microemulsion.

Besides this we have measured the angular dependence of the V_v component of the scattered light and of the D coefficient at different volume fractions. It appears that for all the studied systems I and D are independent of the wave vector in the low concentration range ($\phi < 0.04$). For most of the concentrated microemulsions, I and D remain independent of q. However in the case of the S–C5–D and M–C6–D microemulsions which exhibit a strong variation of I and D versus ϕ, an angular dependence is observed in the concentration range around the extremum of I and D. These variations are related to a critical behavior of these microemulsions [18]. Similar behavior has been found recently in the three-phase microemulsion systems [19].

Figure 4 shows plots of ϕ/I versus ϕ in the low concentration range ($\phi < 0.1$) for different microemulsions. Analysis of these data allows one to derive by extrapolation at infinite dilution (eq. 2) the apparent radius R of the micelle and the second virial coefficient B (tables 2 and 3). As well as the microemulsions for which I is independent of q, these two values R and B have also been determined by analysis of the whole $I(\phi)$ curve by using equation 3 proposed by Vrij [3]. The corresponding values R' and B' are reported in tables 2 and 3. In most cases values of R and B obtained by the two methods are in the limit of experimental accuracy in good agreement. However this agreement is only obtained as one takes volumic frac-

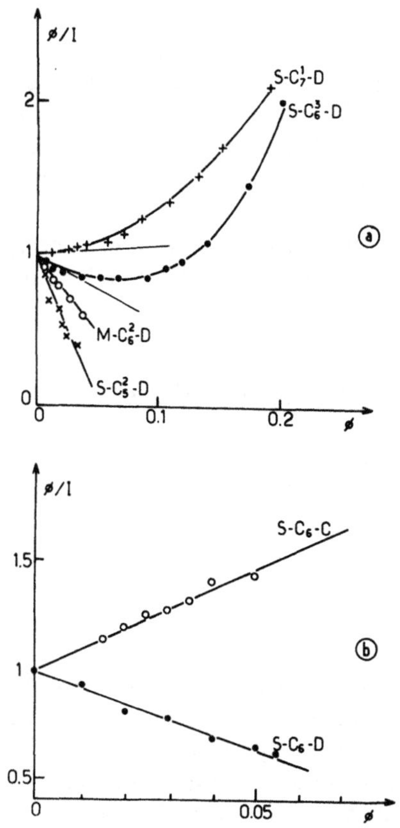

Fig. 4. Variation of ϕ/I versus ϕ for various microemulsions a) influence of alcohol and surfactant; b) influence of oil, ϕ/I has been normalized to 1 for $\phi = O$

tions $\phi < 0.3$. The hydrodynamic radius of the micelle R_H and the coefficient α are extracted from analysis of the variation of D versus ϕ. The obtained values are given in tables 2 and 3.

A comparison of the results given in tables 2 and 3 shows that the droplet size is mainly dependent on the W/S water to soap ratio. The radius increases with this ratio. Also, tables 2 and 3 show that the apparent radius is not very much affected neither by the alcohol nort by the oil, but seems to depend on the surfactant. The B and α values vary similarly, they are strongly dependent on the molecular structure of the components. Experimental results obtained for the various studied alcohols in the S-D series exhibit very different behavior, the obtained values are largely negative for pentanol, negative for hexanol and positive for heptanol. It follows from these data that interactions are strongly attractive in microemulsions containing pentanol and much less in hexanol and heptanol systems. Indeed, a larger attractive potential leads to a more negative second virial coefficient. Also for a given alcohol, an increase of the droplet size seems to induce a decrease of the second virial coefficient. Thus, for example for the S-C_6 series, the B values become more negative as the W/S ratio increases; these two effects are acompanied by an increase of the alcohol content in the continuous phase. Interactions are also strongly dependent upon the molecular structure of oil. The obtained values are positive for cyclohexane, close to zero for octane and isooctane and negative for dodecane. Finally our results evidence that the introduction of a methyl group close to the polar head of SDS leads to a strong decrease of the B and α values.

5. Intermicellar potential

Modern theories of liquids have been successfully used by several authors to explain the first light scattering results on inverse microemulsions. In particular Vrij et al. [3] have shown that the obtained data can be interpreted by a model of spherical particles in interaction. Interactions in microemulsions can vary in a very large range; indeed our results evidence it is possible to go from very strong attractive interactions to much weaker ones in changing the molecular structure of the microemulsion components (alcohol, oil or surfactant). Then the question arises as to what is the origin of these interactions. Vrij et al. [3] have proposed to interpret these interactions as the result of the difference in the molecular compositions between the micelles and the continuous medium. The calculation for homogeneous spheres dispersed in a solvent has been carried out by Hamaker [20]. Application of this calculation to microemulsions leads to a very weak intermicellar potential when realistic values are given to the Hamaker constant [3]. Therefore in order to explain this incoherence Vrij suggested that a possible interpenetration of the micelles be taken into account. This assumption is supported by the structure of the interface; indeed the presence of alcohol molecules allows the penetration either of the continuous phase when the micelle is isolated or of the surfactant chain of another micelle during a collision (fig. 5). A previous calculation has shown that the most important values of the attractive interactions are obtained in the overlapping region [6]. Therefore in the following we first evaluate the interaction potential caused by the interpenetration of two micelles and in a second step we examine if such a potential accounts satisfactorily for the experimental values of B and α.

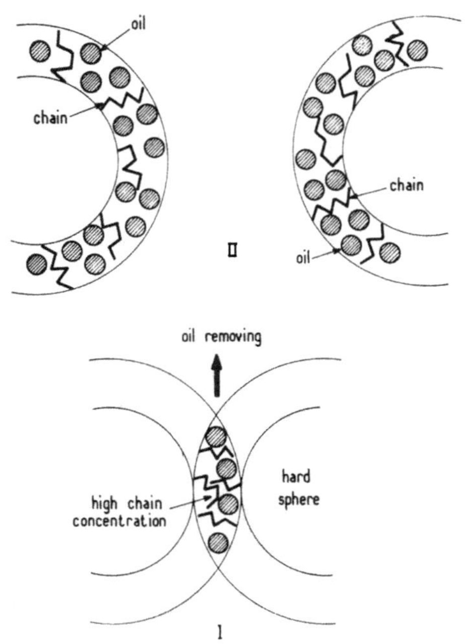

Fig. 5. Schematic representation of: I two interpenetrated micelles II two isolated micelles

Calculation of the interaction potential between W/O micelles

We will consider the case where two micelles overlap; the case of non-penetrated micelles can be calculated according to the Hamaker formula. Following the principle of the Hamaker calculation one has to estimate the internal energy difference between the states I and II:

- In state I the two micelles are located at a distance r ($r < 2R$).
- In state II the two micelles are located at an infinite distance one from the other. In this case the oil of the continuous phase replaces the surfactant chains (fig. 5).

The internal energy of each state X (X is I or II) can be written

$$U^X = \sum_i \int_v \varrho_i \, dv' \sum_j \int_v U_{ij}(d) \, g_{ij}(d) \, \varrho_j \, dv$$

Where i and j are relative to the different present molecules U_{ij} is the interaction potential between two molecules i and j distant by r, $g_{ij}(r)$ is the radial correlation function between these two molecules and ϱ_i the number of molecules i by volume unit. Although calculation of this integral is difficult, an attempt has been recently carried out [6]. However a certain number of simplifications allows one to obtain a very simple result.

We only calculate the part of the internal energy relative to the penetration volume; let $V_o(r)$ be this volume then:

$$U(r) = V_o(r) \Big[\sum_i \varrho_i^c \sum_j \varrho_j^c S_{ij} + \sum_i \varrho_i^p \sum_j \varrho_j^p S_{ij} - 2 \sum_i \varrho_i^{is} \sum_j \varrho_j^{is} S_{ij} \Big]$$

$$S_{ij} = \int_{v_o(r)} U_{ij}(d) \, g_{ij}(d) \, dv$$

ϱ^{is} is the density number of the molecules i in the penetrable part of an isolated micelle (state II); ϱ^p is the density number in the overlapped part (state I) and ϱ_i^c is the density number in the continuous phase.

If one assumes that the potential $U_{ij}(d)$ is short ranged relative to the size of the considered volume (this potential is typically in $1/d^6$) then one can consider that the integrals S_{ij} are independent of r. Consequently $U(r)$ is proportional to the volume of penetration $V_o(r)$:

$$U(r) = V_o(r) \cdot \Delta\varrho \cdot kT$$

$\Delta\varrho$ is a constant depending only upon the composition of the interface and the continuous phase. $V_o(r)$ can be calculated as a function of the radius of the micelle and the distance r between the micelles.

$$V_o(r) = \frac{\Pi}{6} (2R - r)^2 \left(2R + \frac{r}{2} \right)$$

The presence of alcohol in the micellar interface limits the interpenetration of the micelles; therefore we assume that penetration occurs in a thickness equal to the difference between the lengths of the entirely stretched alcohol and surfactant molecules $l = 1.26 (n_s - n_A)$ Å; n_S and n_A are the numbers of carbon in the alcohol and surfactant molecules; so in the systems SDS/pentanol, SDS/hexanol and SDS/heptanol l respectively equals 8.82, 7.56 and 6.30 Å.

Then one can write:

$$U(r) = 0 \qquad r > 2\,R_H$$

$$U(r) = -\Delta\varrho\, kT \frac{\Pi}{6}(2R-r)^2\left(2R+\frac{r}{2}\right)$$

$$2\,R_H - l < r < 2\,R_H$$

$$U(r) = +\infty \qquad r < 2\,R_H - l$$

Because of the approximations made in this calculation, this potential is realistic if $\Delta\varrho$ varies only with very strong variations of composition. In practice, we will consider that $\Delta\varrho$ is constant for a given oil whatever the micellar size or the used alcohol. $\Delta\varrho$ will be used as a parameter, i. e. for a series of measurements with the same oil, we look for the $\Delta\varrho$ value which allows an agreement between the experimental second virial coefficient values and those calculated according to equation 7. Then using this $\Delta\varrho$ value and the Feldherof formula (eq. 6), we calculate α. In both cases we have carried out numerical integration.

Table 4. Comparison of the experimental (B, B', α) and calculated values (B_{th}, α_{th}) of the virial coefficients α and B.

	R_H	B	B'	α	B_{Th}	α_{Th}
S-C_5^1-D	55	−23.		−18.	−15.	−19.
S-C_5^2-D	65	−27.		−21.	−29.	−33.
S-C_6^1-D	52	− 0.5	−1.5	− 5.8	− 1.3	− 6.2
S-C_6^2-D	67	− 4.4	−3.8	− 9.7	− 5.1	−10.3
S-C_6^3-D	69	− 6.1	−6.	−12.	− 6.	−11.4
S-C_6^4-D	76	− 8.8	−9.2		− 8.6	−13.
S-C_7^1-D	61	− 7.	−5.	− 1.	− 3.5	− 1.8
S-C_7^2-D	73	− 4.	−3.	~ 0.	2.5	− 2.5
S-C_7^3-D	76	3.	2.	~ 0.	2.	− 3.

In a first step we have compared the experimental and calculated values obtained with the 9 microemulsions formed with SDS and dodecane. The $\Delta\varrho$ value has been found equal to 7.1 10^{-4} Å^{-3}, the experimental and calculated values are given in Table IV; these results are plotted in figure 6. An excellent agreement

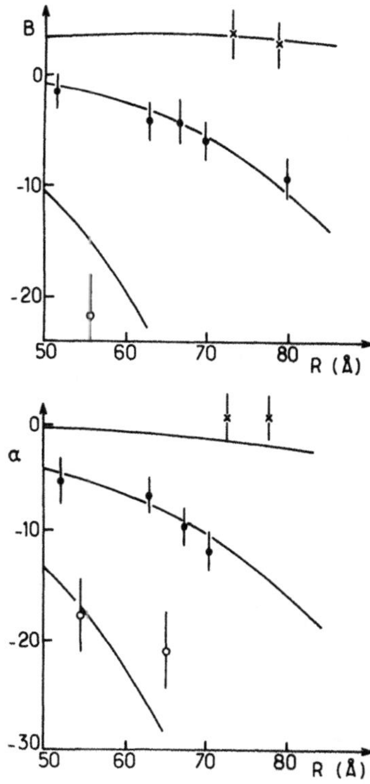

Fig. 6. Comparison of the calculated (full line) and experimental B and α values for the microemulsions S-Alcohol-D.
O pentanol, ● hexanol, x heptanol. The value of $\Delta\varrho$ is kept equal to 7.1 10^{-3} Å^{-3}

is obtained for both the α and B values. Therefore the proposed potential allows to interpret very satisfactorily both the static and dynamic results. Also the hypothesis that $\Delta\varrho$ is constant for a given oil is checked. In order to interpret results obtained in part II of this work with different oils and also some data published in literature for various systems containing cyclohexane and toluene [4] we have applied the described intermicellar potential. Figure 7 shows the results of the comparison and table 5 gives the determined

Table 5. Comparison between values of $\Delta\varrho$ for different oils and their molecular volume

	$\Delta\varrho$ Å^{-3}	Molecular Volume Å^3
Dodecane	7.1 10^{-4}	377.
Octane	5.5 10^{-4}	270.
Isooctane	5.5 10^{-4}	273.
Cyclohexane	4.6 10^{-4}	179.
Toluene	4.2	10^{-4}176.

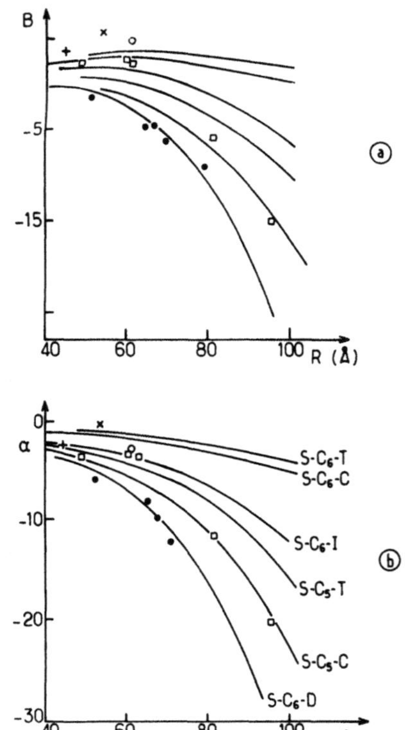

Fig. 7. Comparison of the calculated (full line) and experimental B and α values for microemulsions made with different oils: ● dodecane-hexanol; ○ cyclohexane-hexanol; □ cyclohexane-pentanol; ■ isooctane-hexanol and octane hexanol; + toluene-pentanol, x toluene-hexanol (from ref. [4])

values for $\Delta\varrho$. Let us notice that numerous experimental data for the cyclohexane and dodecane systems are available, this permits us to obtain $\Delta\varrho$ with a good accuracy; this is not the case for the other oils.

The parameter $\Delta\varrho$ appears very sensitive to the oil molecule. A very interesting comparison can be made between the values of $\Delta\varrho$ and the molecular volume of oil (table V). These two values are strongly correlated, indeed an increase in the interactions corresponds to an increase of the molecular volumes.

The preceding results show that the branching of SDS produces two main effects: first a decrease of the amount of alcohol in the interface and secondly an increase of the attractive interactions. One can suppose that the polar head area of the branched surfactant is larger than that of the SDS molecule; the accuracy of our data is not sufficient to measure this effect. However this assumption is consistent with the decrease of the alcohol content in the interface and with what is generally observed in monolayers. Therefore the micelles formed with the methyl-SDS molecule are probably less compact than those formed with SDS and the strengthening of the attractive forces could result from an increase of the volume of penetration. It is possible to explain this behavior with the interaction potential as an increase of the value of the length of penetration l.

6. Conclusion

The results presented in this study clearly show that interaction forces in W/O microemulsion are very sensitive to the molecular structure of the components and to the chemical compositions of the continuous and dispersed phases. Indeed the experimental values of the second virial coefficient B are ranging from -27 to $+10$. Our data evidence that very important variations of B are obtained by changing each component. In order to analyse the intensity data in the low concentration range we propose an intermicellar potential. In this potential, the most important values of the attractive interactions appear in the overlapping region and result from short interatomic attractive interactions. The attractive energy is proportional to the penetrated volume and to one parameter $\Delta\varrho$ which only depends on oil and surfactant. The contribution for the nonpenetrated volumes corresponds to the classical Hamaker contribution, which is negligible in these systems. The behavior of calculated and experimental virial coefficients are quite similar; therefore the proposed potential allows one to account for the experimental data obtained in the low concentration range. In particular the effect of alcohol chain length and micellar size are well represented. The calculation provides an approach for the understanding of the variations of the interactions. Indeed variations of the penetrated volume can be obtained by two different methods: either in changing the alcohol chain length or in changing the micellar size. The alcohol effect corresponds to the principal effect; for a given micellar radius, as the alcohol chain length decreases, the thickness of the penetrated layer is increased and the interaction is stronger. The second effect relative to the micellar size is a secondary effect and can be explained in the same way. An increase of the micellar radius leads to an increase of the volume of interaction.

The calculation of the potential indicates also that interactions can be changed by varying the $\Delta\varrho$ parameter. Our results show that this parameter is very sensitive to the molecular volume of oil. The effect of the molecular structure of the surfactant can be interpreted as a larger penetration of the aliphatic layer.

Acknowledgements

The authors wish to thank Mrs Dubien and Mrs Dupart for the synthesis of the branched surfactant and Mrs Maugey for technical assistance. The authors gratefully acknowledge valuable discussions with *B.* Lemaire. This work was supported by a DGRST contract.

References

1. Shulman JH, Hoar TP (1943) Nature 152:102
2. Graciaa A et al. (1976) CR Acad Sci Paris B 282:547
3. Calje AA, Agterof WGM, Vrij A (1977) Micellization, solubilisation, microemulsion (Proc Int Symp) 77(2):779
4. Cazabat AM, Langevin D (1981) J Chem Phys 74(6):3148
5. Brunetti S, Roux D, Bellocq AM, Fourche G, Bothorel P, J Phys Chem
6. Lemaire B, Bothorel P, Roux D, Part I, J Phys Chem
7. Feldherof BU (1978) J Phys A 11:929
8. Batchelor GK (1976) J Fluid Mech 74:1
9. Goldstein B, Zimm BH (1971) J Chem Phys 54:4408
10. Riley DP, Oster G (1951) Disc Faraday Soc 11:107
11. Ornstein LS, Zernike F (1914) Proc Acad Sci, Amsterdam 17:793; (1918) Physik Z 19:134; (1926) ibid 27:761
12. Carnahan NF, Starling KE (1969) J Chem Phys 51:635
13. Berne BJ, Pecora R (1976) Dynamic light scattering Wiley, New York
14. Dreger EE et al. (1944) Ind Eng Chem 36:316
15. Lagues M, Ober R, Taupin C (1978) J Phys, Paris Lett 39:L-487
16. Dvolaitzky M, Guyot M, Lagues M, Lepesant JP, Ober R, Sauterey C, Taupin C (1978) J Chem Phys 69:3279
17. Aniansson EA (1978) J Phys Chem 82:2805
18. Fourche G, Bellocq AM, Brunetti S, J Coll Int Sci, in press
19. Bellocq AM, Bourbon D, Lemanceau B, Fourche G, J Coll Int Sci, in press
    Cazabat AM, Langevin D, Meunier J, Pouchelon A, Adv in Coll Int Sci, in press
20. Hamaker HC (1937) Physica IV, 1058

Authors' address:

D. Roux
Centre de Recherche Paul Pascal
Domaine Universitaire
33405 Talence Cedex, France

From rodlike micelles to lyotropic liquid crystals*)

M. Angel, H. Hoffmann, M. Löbl, K. Reizlein, H. Thurn, and I. Wunderlich

Lehrstuhl für Physikalische Chemie der Universität Bayreuth

Abstract: Ionic surfactants with strongly binding counterions begin to form rods at rather low concentrations. In extreme cases rods form from the CMC on. The lengths of the rods increase rapidly with the surfactant concentration, with the growth depending on the ionic strength of the system. As soon as the rotational volumes of the micelles begin to overlap the growth slows down and finally reverses itself. For surfactants with alkyl chains consisting of more than 12 carbon atoms the surfactant concentration is still very low at the point of overlap and can be of the order 0.1 % by weight. Several parameters are very sensitive to the overlap concentration. The viscosity starts to increase sharply, the rotational diffusion coefficient and the light scattering decrease rapidly. The rotational diffusion coefficient controls the viscosity and the elasticity of the system. It can become so low that the solution takes on elastic properties for small shear fields. The overall intermicellar interaction is repulsive under these conditions. As a consequence the light scattering intensity decreases, and the effective translational diffusion coefficient increases with concentration. Upon a further increase of concentration the light scattering increases again and the effective translational diffusion coefficient slows down toward the phase boundary. On the basis of the experimental evidence we can thus distinguish four concentration regions Co.

region I: $0 < c_o <$ CMC monomers
region II: CMC $< c_o < c^*$ non-overlapping rods
region III: $c^* < c_o < c_a$ overlapping rods, repulsive interaction
region IV: $c_a < c_o < C_{l.c.}$ phase boundary for liquid crystals, overlapping rods with attractive interaction

Key words: Micelles, micellar interaction.

Introduction

Many surfactants with a single straight hydrocarbon chain form a hexagonal phase as a first liquid crystalline phase. This phase is built up from long cylindrical micelles. Examples of surfactants having such a phase are the classical soaps of the hydrocarbon carboxylic acids. Typically, these phases form at a soap concentration of around 40 % by weight [1]. In special cases the liquid crystalline phase is built up at 10 % of surfactant. Usually at concentrations considerably below the two phase region rodlike micelles can be observed in the isotropic solution. Most systems form globular micelles at the critical micelle concentration (CMC) and there is an extended concentration range for the existence of globular micelles. The rods are formed above a second well-defined transition concentration which again is characteristic for the system [2]. In extreme situations this transition concentration can be very low and it may even coincide with the CMC. In such cases the surfactant begins to form rods from the CMC on. Rodlike micelles can have very small and very large axial ratios depending on the equilibrium conditions. Their lengths are determined by a subtle balance of counteracting forces on the individual micelles and by intermicellar forces. Electrostatic charge on the micellar surface tends to increase the size of the micelles while the surface tension of the micellar interface tends to keep the micelles as small as possible. In the same direction as the electrostatic repulsive forces act steric effects which are due to bulky head groups. The interfacial tension at the micellar interface

*) Lecture given at the 31th Conference of the Kolloid Gesellschaft, Bayreuth October 11–24, 1983.

KT-W 806

is indirectly also partially controlled by the charge density and the electrostatic repulsion. The free surface of a micelle that is the area of the micelle which is not covered by a hydrophilic head group is determined by the repulsion of the head groups. Any effect which lowers this repulsion allows for a closer packing of the headgroups and hence reduces the free surface which in turn lowers the surface tension and favors the tendency for growth. It is clear from these outlines that the situation is indeed very delicate. Any small change in the force balance of such a system is bound to have major consequences as far as the sizes of the micelles are concerned. A theoretical treatment of the aggregation process which takes into account some of the mentioned effects was given by Ninham et al. and by Ruckenstein [3, 4]. These theories at least qualitatively give a good description of the aggregation process. They do not take into account however the intermicellar forces which strongly affect the system at higher concentration. At present there seems to be no quantitative theory available which treats all effects present in such a system. Light scattering measurements and SANS-measurements clearly show that the intermicellar energies are quite large in solution for rodlike systems of ionic micelles. Since the free energy in the system always adjusts itself to a minimum the intermicellar energies must play a role in the control of the lengths of the rodlike micelle. However at present it is only possible to make calculations for interacting spheres but not for rods [5]. Thus it is not possible to take such a theory into consideration. Several surfactant systems which form rodlike micelles at rather low concentrations have been studied recently [6]. Usually rodlike micelles are formed in solutions of ionic surfactant systems with strongly coordinating counterions or in solutions which contain both ionic and anionic surfactants. Growth of globular micelles to rodlike micelles is also induced in many systems when the salt concentration in the solution is increased [7–10]. Typical examples are the Na-alkylsulphates which were studied quite extensively.

A considerable number of questions are linked to the presence of rodlike micelles which are both of theoretical and practicall interest. First there is the question of the size dependence on concentration, temperature and ionic strength. Related to this problem is the size distribution of the rods. A more delicate question is the one after the flexibility of the rod or the radius of curvature of the rod. Both of these parameters are determined by the persistence length of the rods [11]. Surfactant solutions with rodlike micelles usually have a very complicated flow behaviour and we can pose the question of how the mentioned parameters affect the flow behaviour. Some solutions have viscoelastic properties [12]. If the elastic properties of the solutions be explained on the basis of the presence of the rods is another question that is difficult to answer. In this publication we will mainly be concerned with the change of the length of the rods with the increase of the concentration when we go from very dilute to semidilute and to concentrated solution and approach the phase boundary for liquid crystals.

The determination of the lengths of the rods in the semidilute and concentrated solutions presents a problem of considerable complexity because most measurable parameters depend on the size of the micelles and their interaction and it is no simple matter to separate the two contributions. For this reason we will discuss and use several techniques for the determination of the sizes and see whether the data are self-consistent.

Results

In this paper we will mainly present data from the alkyltrimethylammoniumsalicylates and alkylpyridiniumsalicylates. Data on the cetylsystems have already been published. The data on the shorter chain length surfactants are presented for the first time. Some of the older data are included again in this publication in order to show the general behaviour of the aggregation of the systems which seems to be independent of the chain length. The systems were prepared according to a described procedure from the alkylpyridinium- or trimethylammoniumchlorides and sodiumsalicylate which were commercially available products. Some samples were also prepared via ion exchange of the chloride solution in an ion exchange column which was charged with salicylate ions. The resulting solutions were freeze dried in order to obtain the solid material. The compounds were re-cristallized from ethylacetate and dried.

The critical micelle concentrations were determined from conductivity and surface tension measurements. The conductivity-concentration plots for the systems with chain length equal to or larger than 14 showed two breaks while the shorter chain length systems showed only one break. The first break was identical with that obtained from the surface tension measurements and corresponded to the CMC of the systems where they begin to form micelles. The second break in the conductivity plot corresponds to the transition concentration c_t above which the system form rodlike micelles (fig. 1). For the shorter chain length systems c_t coincides with the CMC. Anisometric

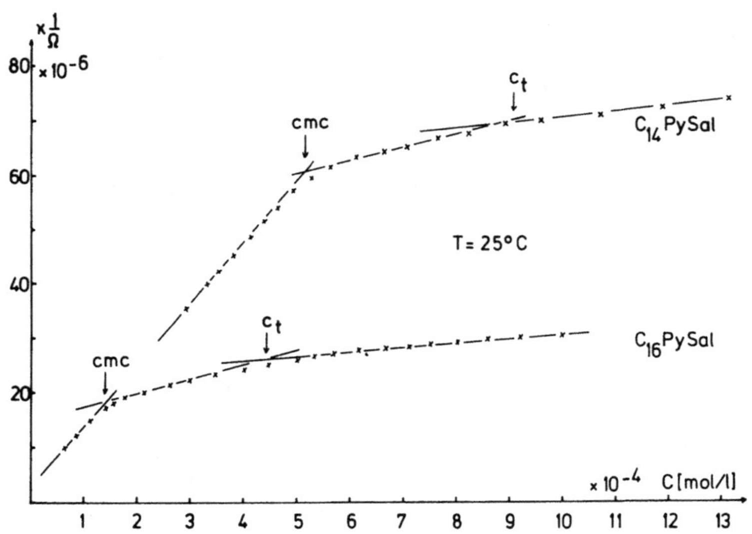

Fig. 1. Concentration dependence of the conductivity for aqueous solutoins of C_{14}PySal and C_{16}PySal

micelles in these systems are present for concentrations above CMC while for the long chain surfactants such micelles are observed only for concentrations above c_t. This different behaviour of the system is a consequence of their different CMC which controls the ionic strength. When small amounts of excess NaCl and Na-salicylate are added to the solutions the c_t values shift more rapidly to lower concentrations than the CMC values, and as a consequence the two characteristic concentrations finally coincide when enough salt is added. This is already the case in the presence of 0.01 M NaCl. When the surfactant concentration is increased in 0.01 M NaCl solution all systems begin to form rods at the CMC. The values for the CMC, c_t and dissociation degrees α of the micelles which were obtained from the slopes of the conductivity plots are given in table 1. For comparison values are included for the alkylpyridinium chloride. These systems form globular micelles in the whole concentration range of the micellar solution. Note the strong decrease of the CMC of the salicylate systems in comparison to the chloride systems. This shift of the CMC is a consequence of the strong binding of the counterions to the micelles. Furthermore the strong binding is also expressed in the very low dissociation degrees of the micelles. It is this strong binding of the counterion to the micelles which causes the micelles to grow. Rodlike micelles can therefore also be found for other systems with different counterions which also bind strongly to the micellar interface.

The molecular weight of micelles is often determined from light scattering measurements. Such data

Table 1. Values for the CMC, the transition concentration c_t and the corresponding apparent dissociation degrees α and α_t for n-alkylpyridiniumsalicylates and n-alkyltrimethylammoniumsalicylates

System	CMC (mole l^{-1})	c_t (mole l^{-1})	α	α_t	T (°C)
C_{12} Py Sal	$2.15 \cdot 10^{-3}$	–	–	0.06ª)	25
C_{14} Py Sal	$5.15 \cdot 10^{-4}$	$9 \cdot 10^{4}$	0.21	0.095	25
C_{16} Py Sal	$1.4 \cdot 10^{-4}$	$4.5 \cdot 10^{4}$	0.19	0.09	25
C_{12} $N(CH_3)_3$ Sal	$2.85 \cdot 10^{-3}$	–	–	0.06ª)	25
C_{14} $N(CH_3)_3$ Sal	$6.25 \cdot 10^{-4}$	–	–	0.09ª)	25
C_{14} $N(CH_3)_3$ Sal	$9.4 \cdot 10^{-4}$	$2.2 \cdot 10^{-3}$	0.18	0.11	50
C_{16} $N(CH_3)_3$ Sal	$1.5 \cdot 10^{-4}$	–	0.41	–	25
C_{16} Py Cl	$9.9 \cdot 10^{-4}$	–	0.43	–	25

ª) Rodlike micelles are formed above the CMC.

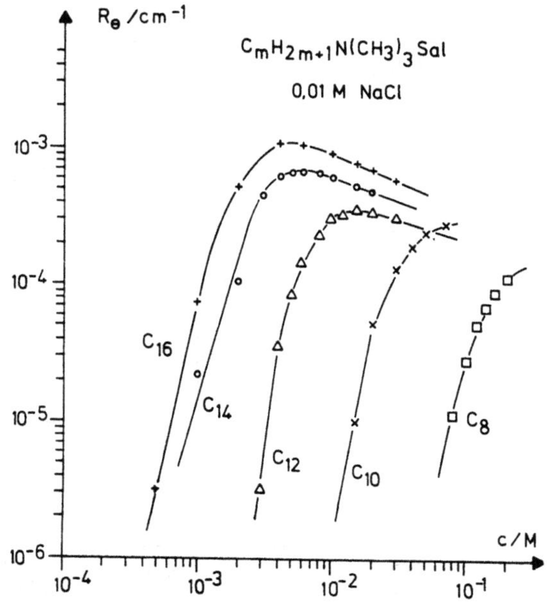

Fig. 2. Concentration dependence of the Rayleigh ratios for several alkyltrimethylammoniumsalicylates in presence of .01 M NaCl

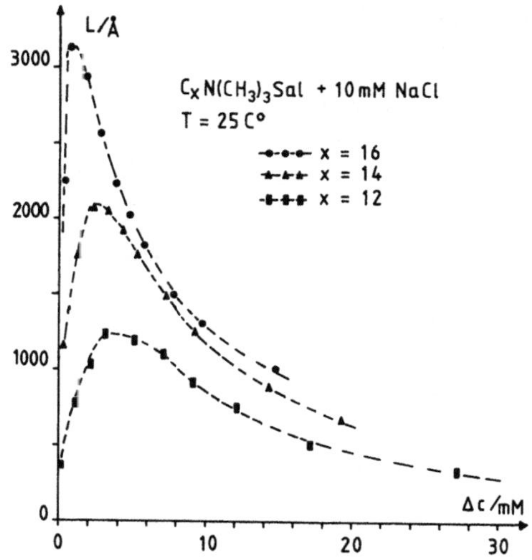

Fig. 3. Concentration dependence of the micellar length for some alkyltrimethylammoniumsalicylates in presence of .01 M NaCl

are presented in figure 2 for the alkyltrimethylammoniumsalicylates in the presence of 0.01 M NaCl. The NaCl was added in order to shield the charge density of the micelles to obtain a more ideal situation, at least in the dilute range, as far as the evaluation of the data is concerned. For all systems, the scattering intensity increases rapidly with concentration and then passes over a maximum. In previous publications we have observed a similar behaviour for other systems [12, 13]. In these publications we have shown that the maximum in the scattering intensity corresponds to a situation in which the rods have grown to a length at which the rotational volumes of the rods begin to overlap. For smaller concentrations the rods can freely rotate without much hindrance, while for larger concentrations the mean distance of separation is smaller then their length. If the lengths of the micelles are determined from the difference of forward and backward scattering the values were obtained that are given in figure 3.

For all systems the length passes over a maximum. The largest micelles are obtained for the surfactants with the longest chains. It is also noteworthy that very similar L-values are obtained from the absolute scattering rates. The data are given in table 2 for comparison.

Table 2. Values for the effective molecular weight M_{eff} and the corresponding length L_{eff} of the rodlike micelles evaluated from the light scattering intensity. Values for the radius of gyration R_G and the corresponding length $L(R_G)$ from the angular dependence of the scattered light. Intermicellar distane $\langle r \rangle$ calculated from $L(R_G)$. Structure factor $S(O)$ calculated from $L_{eff}/L(R_G)$.

2 a) The system $C_{12}N(CH_3)_3$ Sal in the presence of 10 mM NaCl at 25 °C

c / mmole l^{-1}	M_{eff} / $10^5 \cdot$ g mole^{-1}	L_{eff} / Å	R_G / Å	$L(R_G)$ / Å	$\langle r \rangle (R_G)$ / Å	$S(0)$
3	2.1	376	–	–	–	–
4	4.3	789	–	–	–	–
5	5.7	1038	216	746	954	1.39
6	6.8	1235	277	958	913	1.29
8	6.6	1195	286	991	784	1.20
10	6.1	1109	197	682	620	1.63
12	5.0	911	270	932	634	0.98
15	4.2	764	223	772	542	0.99
20	2.8	513	143	492	416	1.04
30	1.7	303	96	333	313	0.91

2b) The system $C_{14}N(CH_3)_3$ Sal + 10 mM NaCl

$\frac{c}{\text{mmole l}^{-1}}$	$\frac{M_{eff}}{10^5 \cdot \text{g mole}^{-1}}$	$\frac{L_{eff}}{\text{Å}}$	$\frac{R_G}{\text{Å}}$	$\frac{L(R_G)}{\text{Å}}$	$\frac{<r>(R_G)}{\text{Å}}$	$S(0)$
1	7.9	1163	–	–	–	–
2	19.2	2825	506	1755	1551	1.60
3	24.3	3580	595	2063	1361	1.73
4	23.2	3415	587	2034	1204	1.67
5	19.6	2886	551	1910	1080	1.51
6	16.1	2371	507	1756	980	1.34
8	11.1	1644	428	1483	833	1.10
10	8.2	1212	359	1243	725	0.97
15	4.7	693	248	860	556	0.80
20	3.2	465	188	649	458	0.71

2c) The system $C_{16}N(CH_3)_3$ Sal + 10 mM NaCl

$\frac{c}{\text{mmole l}^{-1}}$	$\frac{M_{eff}}{10^5 \cdot \text{g mole}^{-1}}$	$\frac{L_{eff}}{\text{Å}}$	$\frac{R_G}{\text{Å}}$	$\frac{L(R_G)}{\text{Å}}$	$\frac{<r>(R_G)}{\text{Å}}$	$S(0)$
0.5	26.4	3211	648	2244	2895	1.43
1	88.7	10800	905	3136	2334	3.44
2	54.7	6672	844	2926	1740	2.28
3	34.1	4159	743	2573	1439	1.61
4	22.4	2724	644	2230	1239	1.22
5	17.5	2137	583	2019	1109	1.05
6	13.5	1644	526	1823	1006	0.90
8	9.4	1144	431	1492	853	0.76
10	7.4	903	379	1313	757	0.68
15	4.3	522	290	1004	603	0.52

The agreement is satisfactory if one considers the many assumptions which are involved in obtaining the data. Table 3 gives data for the different chain length systems when no excess salt was added. Under these conditions the rods are considerably smaller and the maximum of the effective rodlength should be shifted to higher concentrations.

In small angle neutron scattering measurements we have tried to measure the persistence length of the rods by fitting these scattering curves to the scattering curves of straight rods or coils. The data which were obtained when increasing amounts of NaCl were added to a 5 mM CPySal solution are given in figure 4. The scattering curves give a very good fit with the theoretical curves for rigid rods [14]. The results are summarized in table 4.

For the CPySal we have also determined the L-values in salt-free solutions over a large concentration range. The scattering curves are given in figure 5 and the evaluated parameters are given in table 5. The

Table 3. Length L of the rodlike micelles in aqueous solutions of $C_{12}N(CH_3)_3Sal$ and $C_{14}N(CH_3)_3Sal$ without added electrolyte obtained from static light scattering measurements at $T = 25\,°C$. Values for the structure factor $S(0)$ are included.

$C_{12}N(CH_3)_3$ Sal

$\frac{c}{\text{mM}}$	10	12	15	20	30
L	592	523	454	382	304
$S(O)$	0.56	0.48	0.40	0.29	0.181

$C_{14}N(CH_3)_3$ Sal

$\frac{c}{\text{mM}}$	6	8	10	15	19.5	30
L	733	625	554	448	390	313
$S(0)$	0.17	0.13	0.1	0.07	0.06	0.05

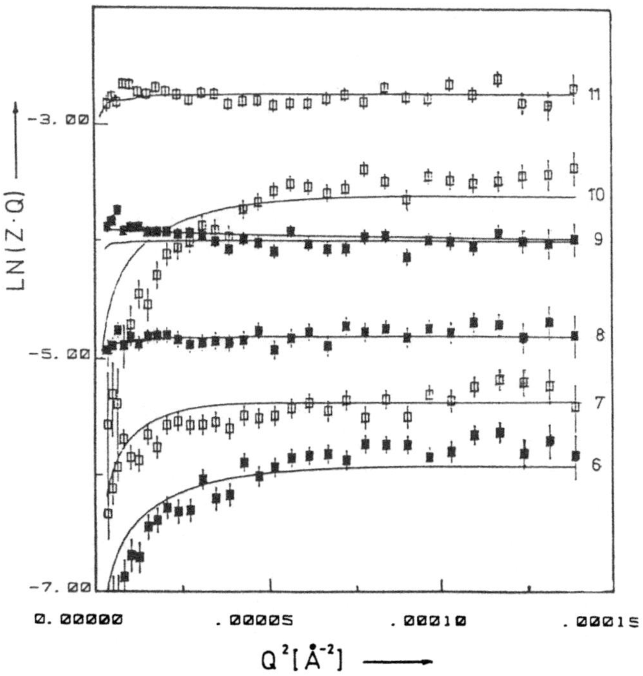

Fig. 4. Semilogarithmic plot of ln(ZQ) against Q^2 for low Q-values for 5 mM C_{16}PySal with different amount of salt added. The data are fitted by the model of stiff rods. The parameters are given in table 4 (6 = .5 mM NaCl, 7 = 2 mM NaCl, 8 = 8 mM NaCl, 9 = 32 mM NaCl, 10 = .2 mM NaSal, 11 = 1 mM NaSal). For convenience the countrates Z of the curves are multiplied by a factor of exp (.5) respectively. The upper curve of sample 9 is a fit for the persistence length model

Table 4. The range of length L and persistence length a for the 5 mM solution of C_{16}Py Sal for different concentrations c of added salt (NaCl and NaSal). In the third column the first number is L, the second a. The dots indicate that values between the pairs of numbers are also possible. – Indicates that we are not able to extract an a value.

Salt	c[mM]	L[Å]	a[Å]
NaCl	0.5	550,	–
NaCl	2	750,	–
NaCl	8	1000,	10000 ... 1800, ∞ ... 3000, ∞
NaCl	32	3000,	8000 ... 5000, 12000 ... 20000, 12000
NaSal	0.2	550,	–
NaSal	1	2500,	–

L-values were determined from the scattering maximum while the fit values were obtained from a curve fit of the whole scattering curves.

In previous publications it has already been pointed out that the overlap concentration of the rods is also noticeable in electric birefringence measurements [12]. As long as the rods can rotate freely the electric birefringence decay curve is single exponential, sometimes somewhat distorted. In the overlap region the single process splits into two processes, the fast one being due to the free rotation while the slow one is due to a

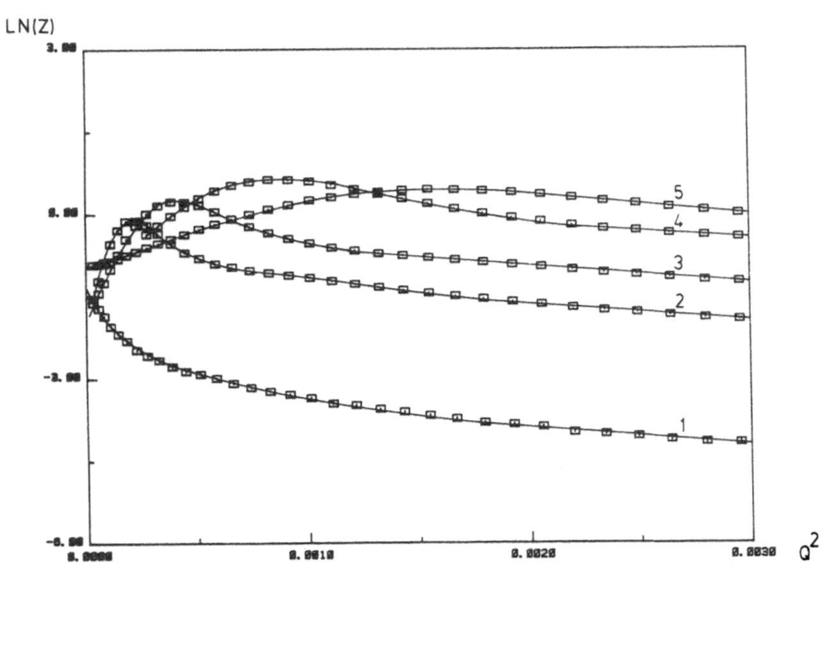

Fig. 5. Semilogarithmic plot of the scattering rates against Q^2 for different concentrations of C_{16}PySal in an intermediate Q-range (1 = 5 mM, 2 = 20 mM, 3 = 40 mM, 4 = 80 mM, 5 = 160 mM)

Table 5. SANS-data for C_{16}PySal at 25 °C. Mean distances $<r>$ gained from the maximum of the structure factor $S(Q)$. L-values used to fit the data (L^{fit}). Minimum value S_{min} of the $S(Q)$ data at a Q-value of about $3 \cdot 10^{-3}$ Å^{-1}

c_o/mM	$<r>$/Å	L/Å	L^{fit}/Å	S_{min}
5	–	–	530	–
10	573	–	650	–
20	339	426	557	0.05
40	286	312	550	0.046
80	196	205	395	0.048
160	140	152	300	0.049

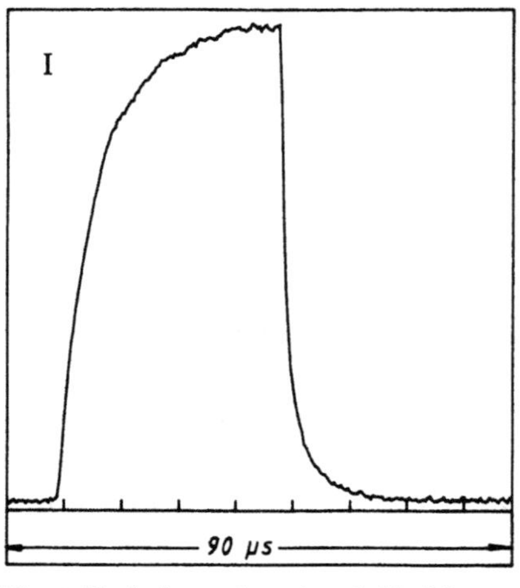

Fig. 6. Typical transient electric birefringence signal obtained for a 20 mM solution of C_{12}PySal at 25 °C, I = Intensity

coupled rotation-translation process. The time constant for the slow one depends very much on the overlap ratio and the length of the field pulse. Typical results are given for the C_{12}Py Sal in figure 6.

These measurements were obtained with a single electric pulse of high field of the order of 10^3 V/cm. Electric birefringence measurements can also be carried out with an AC technique using lock-in amplification for the birefringence detection [15]. With this technique the field strength can be two to three orders of magnitude lower than in the transient case. In these measurements the frequency is swept over a large range in order to obtain the relaxation frequency and hence the rotational diffusion coefficient. Such measurements are given in figure 7. It is noteworthy that with this technique we see only the slow relaxation process of the orientation but not the fast one. This shows that the two techniques, the transient method and the AC method, complement each other and it is helpful for the interpretation of the data to use both techniques.

Rotational diffusion coefficients can also be determined using the flow birefringence method [16, 17]. With this method the extinction angle of the solution is determined in a shear gradient. The extinction angle is the small angle between the direction of flow and the direction of the oriented rods. Typical data from such measurements are presented in figure 8 in which the angle of extinction is plotted against the shear rate. The data show that the micelles can be the easier aligned the higher the concentration is. The rotational diffusion

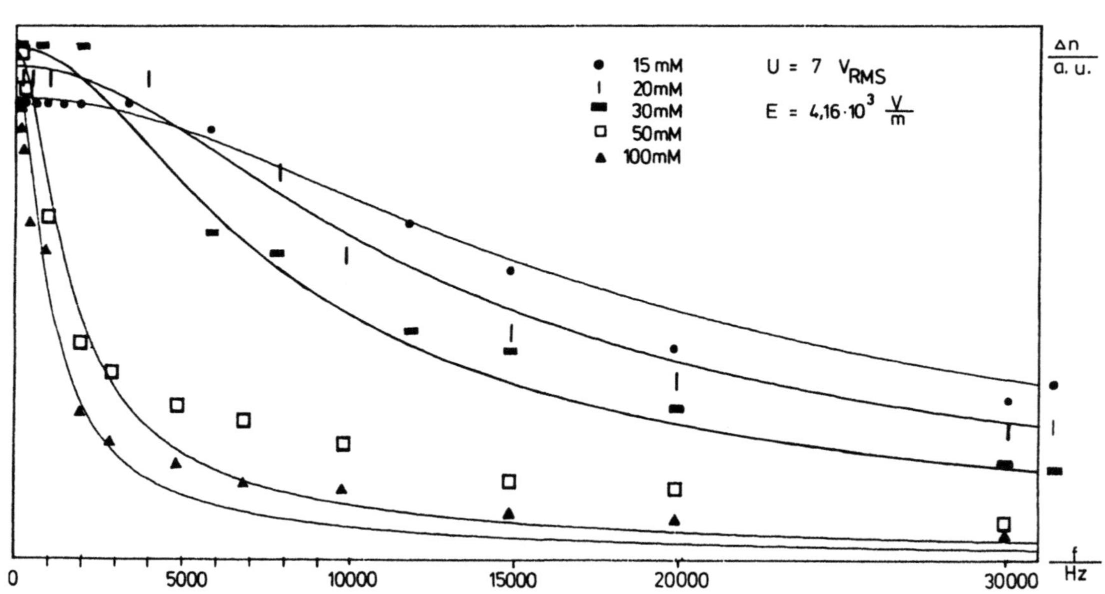

Fig. 7. Plot of the birefringence (in arbitrary units) versus the AC frequency

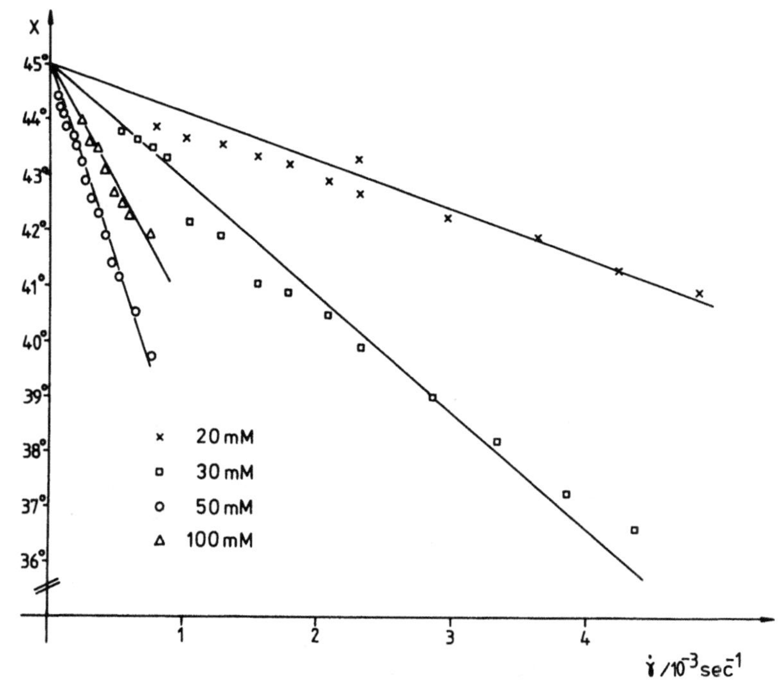

Fig. 8. Dependence of the optical extinction angle χ on the shear rate $\dot{\gamma}$ for various concentrations of C_{12}PySal at 25 °C

coefficient decreases rapidly with increasing concentration.

NMR measurements can also be used to monitor the size of the micelles [18]. While it is very complicated to calculate absolute sizes from the line shape, it should be possible to detect changes of the micelles with the concentration. Such line width measurements of the protons of the hydrocarbon chain are presented in figure 9. It is interesting to note that the line width changes very little over a concentration decade in the range from where the overlap begins to higher concentrations. There is a small tendency of the lines to

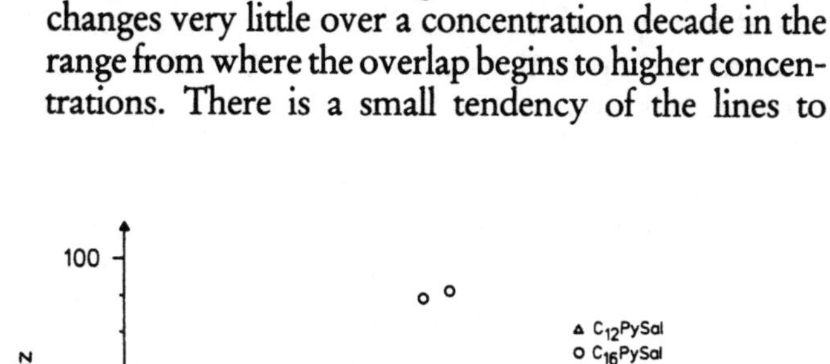

Fig. 9. Concentration dependence of the linewidth of the alkyl chain proton NMR signal for C_{12}PySal and C_{16}PySal at 25 °C

become narrower again for higher concentrations. This is very remarkable because in this concentration range parameters like the hydrodynamic radius and the rotation diffusion coefficient change strongly. Figure 9 contains data both for the C_{16} and the C_{12}-system. The line widths for the C_{16}-system are about 3 times as broad in the plateau region as for the C_{12}-system.

Most dramatic changes occur for the viscosity at the overlap concentration. From this concentration on the viscosity increases rapidly by several orders of magnitude within a small concentration region. Since the solutions in this region behave as non-Newtonian liquids it is necessary to make the viscosity measurements under well defined conditions. This can best be done by measuring the complex viscosity with an oscillating viscometer. The measurements in figure 10 were obtained in this manner. Measurements of the complex viscosity permit the determination of the storage and the loss modulus as a function of the oscillating frequency. Typical data for a 50 mM solution of C_{16}PySal are given in figure 11. The measurements indicate that the storage modulus reaches a plateau value. In this sense the surfactant solutions behave in the same way as entanglement networks from polymer solutions. This makes it likely that the rods in the overlap region form a kind of temporary three dimensional network in which part of the rods, not necessarily all of them, are energetically coupled. The rheologi-

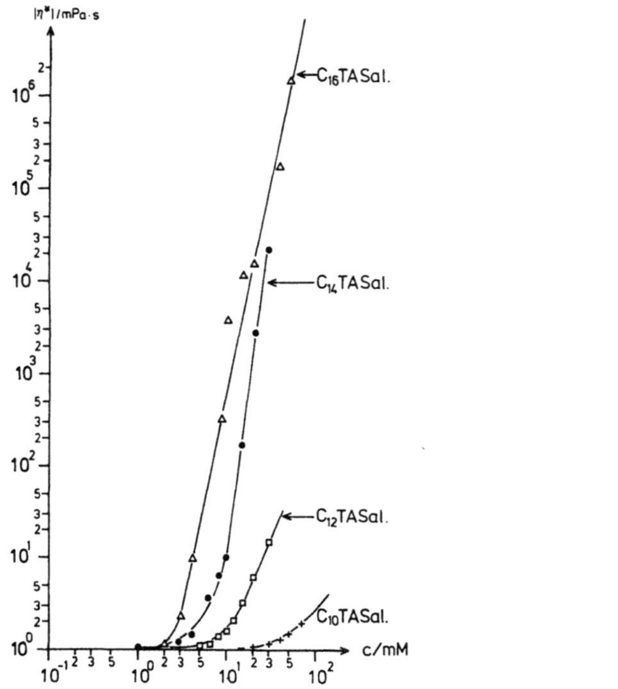

Fig. 10. Dependence of the complex viscosity η^* on the surfactant concentration for several alkyltrimethylammonium salicylates

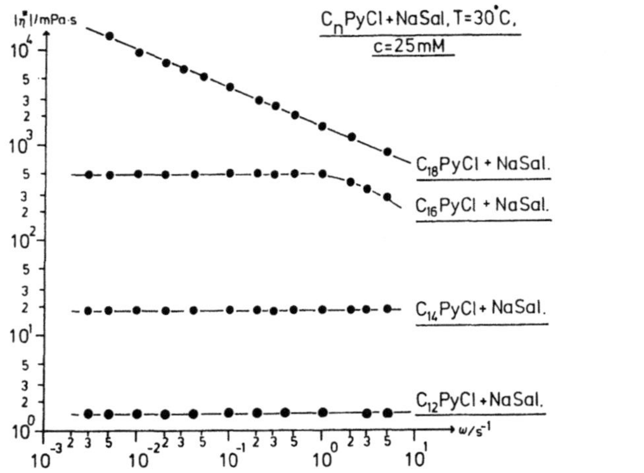

Fig. 12. Frequency dependence of the complex viscosity η^* for equimolar alkylpyridiniumchloride/sodiumsalicylate mixtures (C = 25 mM, T = 30 °C)

cal properties of the solution are then determined by the dynamic behaviour of this network. The dynamic properties in these networks seem to depend much on the chain length of the surfactant. The light scattering in figure 2, for different chain length solutions of the same concentration well above the overlap, was very similar, which makes it likely that the sizes of the rods in these solutions and the structures that are built up from these solutions should be very similar. The light scattering results in solutions of surfactants having a pyridinium head group instead of the trimethylammonium were very similar to the results in the trimethylammonium surfactants. The complex viscosities of solutions under indentical conditions but with different chain length surfactant vary many orders of magnitude as is shown in figure 12. This is likely to be a consequence of the difference in the dynamic behaviour of the network. It is likely however that the number of couplings of the effective elastically chains are very similar because the plateau values for the storage modulus which can be extrapolated from the measured storage and loss moduli are in the same order of magnitude. From the known G' and G'' values at low frequencies it is possible to extrapolate the angular frequency at which the two parameters should cross. From this extrapolated value and the measured complex viscosity η^* a value for G^o can be determined.

Unfortunately, at present we could only carry out measurements up to an angular frequency of $\omega = 10\ s^{-1}$ and could not directly reach the plateau value for the shorter chain length systems. However we have some other evidence for the conclusion that the difference in the rheological behaviour in the overlap region is controlled by differences in the dynamic behaviour of the structures and is not caused by different types of structures. Solutions of $C_{14}PySal$ and $C_{14}N(CH_3)_3$ Sal for the same concentration again give large differences in their complex viscosities and in their frequency dependence. Typical results are shown in figure 13 for the two surfactant solutions and for mixtures of the two compounds. For these systems we have been able to

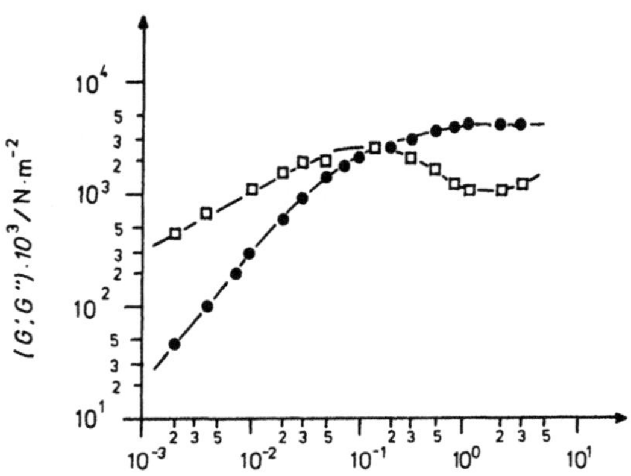

Fig. 11. Storage modulus G' and loss modulus G'' as a function of the oscillation frequency ω ($C_{16}PySal$, T = 20 °C, c = 50 mM, ● G', □ G'')

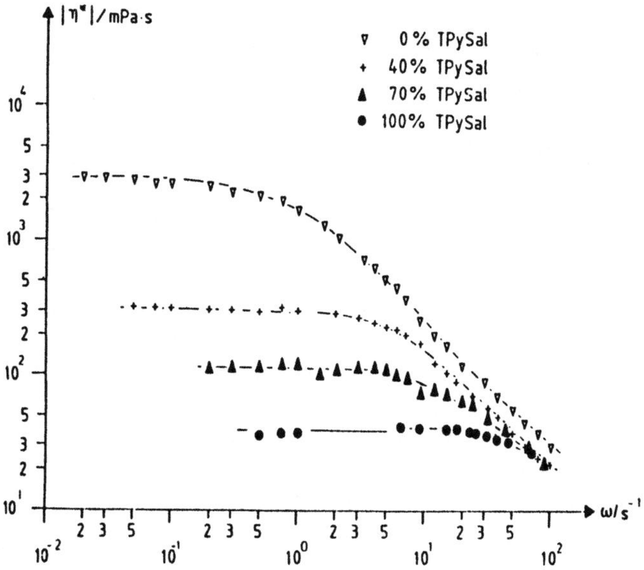

Fig. 13. Dependence of the complex viscosity η^* on the oscillation frequency for mixtures of C_{14}PySal (TPySal) and $C_{14}N(CH_3)_3$Sal and NaBr (total surfactant concentration 25 mM, NaBr concentration 25 mM) at 20 °C

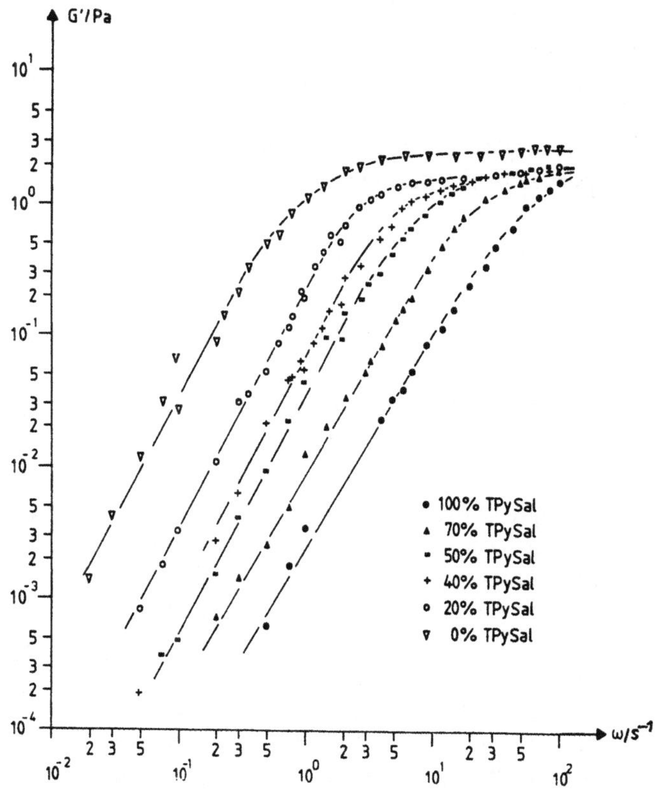

Fig. 14. Dependence of the storage modulus calculated from flow birefringence data on the oscillation frequency for solutions of the same surfactants as in figure 13 (total surfactant and salt concentrations were the same as in figure 13) at 20 °C

extend the frequency range of the oscillating measurements by making use of stress-optical law and by calculating the stress from the flow birefringence. The results of these measurements are shown in figure 14. For nearly all mixtures as well as the pure surfactants we could extend the frequency range into the plateau region of the storage modulus. These data clearly show that the plateau region of G' is practically the same for all solutions even though their complex viscosities in the low frequency range are almost two orders of magnitude different. These data clearly show that the large differences in the viscosities of surfactant systems which form rodlike micelles are due to different relaxation times for the network structures and not due to the difference in sizes. The question which then of course remains to be answered is why are these relaxation times so much different?

Theory

In the experimental part very different techniques were used to collect information on the system. One set of information comes from scattering experiments which permit the evaluation of sizes for the micelles. The length of the micelles in the case of the light scattering measurements can be determined from the absolute scattering rates, from the ratio of forward and backward scattering if the particles are large enough, and from the effective translation diffusion coefficient which comes from dynamic light scattering measurements. There is no room here to go through the detailed theory of the methods and we refer the reader to detailed articles in the literature where the methods are treated in detail [19, 20]. We shall however give the essential equations which were used for the evalutation of the data. The same applies to the SANS-data. Again, from these measurements we can obtain information from the structure factor, the absolute scattering rates and from the position of the scattering maximum which is due to the correlation of the micelles and from which it is possible to calculate a mean distance and hence also a size for the micelles.

For the light scattering measurements the Rayleigh factor R_Θ is defined by equation 1:

$$R_\Theta = I_{sc} \cdot r^2 / I_o \cdot V_{sc} \qquad (1)$$

where I_{sc} and I_o are the intensities of the scattered and the incident light, respectively, r is the distance between the scattering volume and the detector and V_{sc} is the scattering volume.

The Rayleigh factor R_Θ is related to the molecular properties of the particles:

$$R_\Theta = K \cdot c \cdot M_w \cdot P(q) \cdot S(q). \tag{2}$$

The scattering vector q is defined by equation (3)

$$q = (4 \cdot \pi \cdot n/\lambda_o) \cdot \sin(\Theta/2). \tag{3}$$

Here n means the refractive inde, λ_o the vacuum wavelength of the incident light, Θ the scattering angle, M_w the molecular weight of the scattering aggregates and c their weight concentration which is given by the expression $c = (c_o - c_{CMC}) * M_o$ where M_o is the molecular weight of the surfactant monomers and K is a constant. $P(q)$ is the scattering factor or form factor of a single micelle and $S(q)$ is the structure factor which contains the interaction between the aggregates [13]. In the Rayleigh-region where the size of the micelles is smaller than $\lambda_o/20$ the scattering factor $P(q)$ becomes equal to 1. Equation (2) can be re-written as equation (4):

$$K \cdot c/R_{\Theta \to 0} = \frac{1}{M_w S(0)}. \tag{4}$$

For micelles which have dimensions larger than $\lambda_o/20$ the scattering intensity depends on the q-value and hence on the scattering angle Θ:

$$R_\Theta = R_{\Theta \to O} \cdot (S(\Theta)/S(\Theta = 0)) \cdot \exp(-(1/3) \cdot R_G^2 \cdot q^2). \tag{5}$$

R_G is the radius of gyration which can be obtained if the q-dependence of $S(q)$ is known. With the R_G-values the lengths of the micelles can be calculated if their form is known [14]. One obtains for rigid rods:

$$L_{\text{rod}}^2 = 12 \cdot R_G^2. \tag{6}$$

The other experimental data come from completely different techniques and are based on the orientation of the rods in electric fields and shear gradients. The decisive parameter which is obtained here is the rotational diffusion coefficient. This parameter controls, as we will see in the semidilute region of the micelles, the macroscopic flow behaviour of the solution and hence such properties as the viscosity and the elasticity as a function of the shear rate.

The rotational diffusion coefficient can be determined from the transient electric birefringence effect, from the extinction angle of the flow birefringence measurements and from the viscosity of the solution.

For the dilute region in which the rods can rotate freely theories were developed a long time ago for stiff non-interacting particles. Rotational diffusion coefficients can again be translated into dimensions of the particles. For the overlapping region, the situation is not so clear. Progress however has recently been made in this area and Doi and Edwards have given equations for stiff non-interacting rods which relate the effective rotational diffusion coefficient to the dimensions and the concentration of the rods [21]. Since no other equations are available we will use these theories and compare the dimensions of the obtained micelles with the data from the scattering experiments. Deviation of the different sets of data must then be due to the interaction of the rods which can be both repulsive and attractive depending on the conditions. It is also possible now to relate the rotational diffusion coefficient to the viscoelastic properties of the systems with a theory which was developed recently by S. Hess [22].

In a dilute solution the rotational diffusion coefficient D_{rot}^o can be determined from the orientation time τ of a transient electric birefringence experiment

$$\tau = \frac{1}{6\,D_{\text{rot}}^o} = \frac{\pi \eta L^3}{18\,kT\,(\ln 2p - 0.5)} \tag{7}$$

where L is the length of the rod and p the axial ratio. The rotational diffusion coefficient can also be determined from the extinction angle χ according to the equation

$$\chi = \frac{\pi}{4} - \frac{\dot{\gamma}}{12\,D_{rot}^o} \text{ with } \dot{\gamma} \to O. \tag{8}$$

According to Doi and Edwards, the viscosity is expressed by

$$\eta_o = \eta_s\,(1 + \hat{c}\,L^3) = \eta_s\left[1 + \left(\frac{L}{d}\right)^3\right] \tag{9}$$

where d is the mean distance of separation between the rods and L/d consequently is the overlap. For non-overlapping rods $L < d$ and $(L/d)^3 \ll 1$ which means the viscosity is little affected by the presence of the micelles. The situation is very much different in the semidilute region where $L > d$. For this region Doi and Edwards have derived

$$\eta = \eta_L (\hat{c} \cdot L^3)^3 = \eta_L \left(\frac{L}{d}\right)^9 . \tag{10}$$

For this situation the rotational diffusion coefficient

$$D_{\text{rot}} = D^o_{\text{rot}}/(\hat{c} \cdot L^3)^2 = D^o_{\text{rot}}/(L/d)^6 . \tag{11}$$

The consequence of this is that the time constant for the decay of the tail of the birefringence signal τ_{n2} is given by

$$\tau_{n2} = \tau_{n1} \left(\frac{L}{d}\right)^3 \tag{12}$$

where τ_{n1} is the time constant of the free rotation of the rods. The overlap ratio of L/d can therefore easily be determined from the two time constants of the orientation of overlapping rods. As already mentioned attractive and repulsive forces between the rods are not considered in these equations. Furthermore, the emphasis is on stiff rods. The effect of flexibility of the rods has not been considered.

A different approach is taken by S. Hess [22, 28]. He does not try to come up with an expression for how the rotation diffusion coefficient depends on the number and dimension of the rods, but introduces an interaction energy ε. The rotation time τ_c is then given by

$$\tau_c = \tau_a/(1 - \varepsilon/kT) \tag{13}$$

when τ_a is the rotation time of the non-interacting free rod. If τ_c is determined experimentally from the extinction angle of the flow birefringence then the loss and storage modulus can be calculated from the following equations

$$G' = (\eta - \eta_\infty) \frac{\omega^2 \tau_c}{1 + \omega^2 \tau_c^2} \tag{14}$$

$$G'' = \eta_\infty \omega + (\eta - \eta_\infty) \frac{\omega}{1 + \omega^2 \tau_c^2} \tag{15}$$

in which η is the viscosity at $\omega = 0$ and η_∞ is the viscosity in the high frequency limit. The complex viscosity $|\eta^*|$ finally is given by

$$|\eta^*| = \frac{\sqrt{G'^2 + G''^2}}{\omega} . \tag{16}$$

For $\omega\tau < 1$ the magnitude of the complex viscosity should be the same as the viscosity which is determined from the shear rate $\dot{\gamma} \rightarrow O$.

Discussion

1. The length of the rods and their flexibility

The lengths of the rods which have been determined from the light scattering data in the presence of excess salt and without it pass over a maximum with increasing surfactant concentration. We have reported this tendency in several systems. We see now that it is a general phenomenon for rod forming systems and is independent of the chain length of the surfactant. While the increase of the rods with concentration can be explained – theories predict the L values grow with concentration with a power law with an exponent ranging from 1/2 to 1 depending on the theory – the decrease is not clear at present [3, 23]. Recently we proposed an explanation in which the freezing out of the rotational entropy of the rods with increasing crowding of the system would be responsible for the decrease of the L-values [14]. The system therefore tries to avoid this by adjusting the lengths of the rods accordingly.

In the presence of 0.01 M NaCl the rods are very long at the maximum as the values from Tables 2–4 indicate. The L-values which are determined from the light scattering data are contour lengths because they are calculated from the molecular weight. However these values agree fairly well with the L-values which have been determined from the radius of gyration, assuming stiff rods. Different L-values would result if we assumed coiled rods with a persistence length very much smaller than the contour length. Under these conditions the dimensions of the coil however would be considerably smaller than the mean distance of separation between the coil.

A solution with coils which do not overlap would have a viscosity which is only slightly higher than the viscosity of water. The viscosities of the solutions are several times higher than that of water. This can only be the case if the rods are fairly stiff. From geometrical considerations we can calculate a minimum size which the rods must have for a given surfactant concentration in order to overlap. If the mean intermicellar distance $<r>$ is equal to or smaller than the micellar length L_m one obtains

$$L_m \geq \sqrt{\frac{1000 \cdot \varrho\pi}{m_o^c}} <r>. \quad (17)$$

With $r = 20$ Å, $m_o = 450$ g/mol and $c = 5$ mM, neglecting the CMC, assuming $\varrho = 0.9$ g/cm^3 we obtain $L \geq 710$ Å. This simple estimation tells us already that the dimensions of the rods for these conditions must at least be of this size. The rods would not overlap if they were coiled. We can furthermore conclude that the persistence length must at least be of the same size or larger. This conclusion is contrary to what has been found by Porte on C_{16}PyBr in the presence of excess NaBr [24]. These solutions are of rather low viscosity. It is conceivable that the reason for this different behaviour lies in the different persistence lengths of the rods. The persistence length of a rod is determined both by interfacial forces and by electrostatic repulsion between the charged groups. The latter is controlled by the ionic strength of the system. It is probably the increase of the flexibility of a rod which causes the viscosity of a surfactant solution to decrease again with increasing salt concentration, a phenomenon which is quite general for all rod forming systems. The forward light scattering intensity and the radius of gyration tend to level off with increasing salt concentration, indicating that the micelles do not become smaller again. The determination of the L-values before the maximum is straightforward because under these conditions the intermicellar interaction is small and can practically be neglected in the presence of excess salt. The determination of the L-values in the overlap region is a more difficult problem. So far we have shown that the experimental results from different techniques are consistent with the rods becoming smaller again. Evidence comes from electric birefringence, light scattering, viscosity and from SANS-measurements. In this publication we show additional evidence from the NMR data. The line width from methylene-protons decreases again with increasing crowding of the system. This at least qualitatively seems only to make sense when indeed the rods decrease in size. Further support for this conclusion can be based on the phase behaviour of the systems. The data show that the long chain systems in the presence of salt already reach the state in which the rods begin to overlap at a rather small concentration of around 0.1 % by volume. At this concentration the system is still more than two orders of magnitude away from the phase boundary of the hexagonal phase. According to theories by Onsager and by Flory on phase transitions in solutions with rodlike particles, the system should undergo a phase transition if the two dimensional rotational volume of the rods begins to overlap [25, 26]. Even under the condition that the rods stop growing after the overlap concentration is reached, we would soon attain the condition where according to Flory a nematic-like liquid crystalline phase should separate out. This is clearly not the case and we have to assume therefore that we never reach the condition for liquid crystalline phase formation until we get to the real phase boundary. The only way of explaining this is by assuming that the rods become smaller again after the overlap.

Assuming constant size for the rods after the overlap concentration c^* we can calculate a hypothetical concentration c_2 for the liquid crystalline phase. From simple geometrical consideration the result is $c_2 = c^* \cdot \frac{L}{3r}$.

2. Effects which are connected to the rotational diffusion coefficient

The data show that the rotational diffusion coefficient of the rods in the overlap region becomes very low, that means orders of magnitude lower than the value the rod would have in the non-interacting state. This is very remarkable in view of the fact that the translational diffusion coefficient of the rods becomes even faster in the overlap region than in the dilute region. Qualitatively this can easily be explained. The overlap region is – concentrationwise – still a very dilute system. Consequently, the Brownian motion along the axis of the rods can proceed completely unhindered. In contrast the rotational motion is restricted because of the hindrance by other rods. This leads to the consequence that the rods can be more or less completely aligned in a rather small shear field. The rods in such solutions are therefore aligned during flow of the surfactant solution through narrow pipes. All solutions containing rods or discs behave as viscoelastic fluids if the solutions are sheared with a shear rate which is of the same order of magnitude as the reciprocal time constant of the orientation or the rotational diffusion coefficient. In cases of small rods or discs which do not overlap or in systems with overlapping micelles but with weak interaction the shear rate would have to be as high as $10^5\ s^{-1}$ in order for the elastic properties to become effective. For an oscillating deformation the angular frequency would have to be of the order of $10^5\,s^{-1}$. In normal flow velocities in viscometers such shear rates are not encountered, consequently the elastic properties are not observed and the solutions behave as normal Newtonian solutions. But if measurements could be carried out under these conditions the solution should behave as viscoelastic fluids.

S. Hess showed [28] that the shear modulus G^o is given by

$$G^o = \frac{3}{5}\hat{c}\left(1 - \frac{\varepsilon}{kT}\right)\ kT \tag{18}$$

In the network theory G^o is given by $g\nu kT$, where g is a numerical factor in the order of one. These equations show that the shear modulus should be determined by the number ν of the effective elastic chains in the network model or by the number density of rodlike micelles. For systems with these parameters ν or $\hat{c}$ being the same we should expect therefore to observe about the sameG^o-values. Given the same ionic strengths and concentrations the light scattering intensity is a measure for the size and the number of micelles. In figure 2 we have seen that the scattering intensity for surfactants of the same concentration but of different chain length is very similar, we would therefore expect similar G^o-values. Unfortunately the plateau values were experimentally not accessible for all solutions. But it is possible to extrapolate relaxation times and hence to extrapolate G^o-values for some of the solutions. They show that the G^o-values indeed are not very much different for different systems. It is interesting to note that the G^o-values are mainly determined by the number of rods in the solutions. The number of micelles in the overlap region on the other hand is mainly determined by the intermicellar interaction, while this seems to be mainly controlled by the overlap ratio. The micellar density is very similar for different surfactants if the conditions like concentration, ionic strength and temperature are the same. For this reason we can expect to find very similar G^o-values. We see therefore that the specific chemistry disappears in the G^o-values in the overlap region. However the dynamic behaviour of these systems can be quite different because this quantity is controlled by the time constant τ_c which is very sensitive to the interaction energy ε. This energy and consequently τ_c are very much dependent on the headgroup of the detergent, on salt concentration and on temperature. The data which illustrate this behaviour best are the results which were shown in figure 10 and figure 13. The experiments clearly show that while G^o is rather insensitive to chemistry and has very similar values for different systems the time constants τ_c and hence also η^* and η^o vary strongly for different systems.

3. Comparison of the data from different techniques for the system $C_{12}PySal$

For the $C_{12}PySal$ system we used all the available techniques to obtain experimental data in the concentration range which extends at 25° approximately from 10 mM to 100 mM. The data are summarized in table 6. In this concentration range the measurements show the following tendency:

The light scattering intensity decreases with the concentration indicating increasingly strong repulsive interaction between the micelles. Qualitatively this is also shown by the effective diffusion coefficient obtained from the dynamic light-scattering measure-

Table 6. Summary of the results for the system $C_{12}PySal$ at 25 °C in the concentration range 10 – 100 mM.
(c_o: total concentration; R_Θ: Rayleigh factor; r_H: hydrodynamic radius; B: Kerr constant; $f(\omega\tau \approx 1)$: resonance frequency from the AC-Electro-birefringence measurements; $\tau_{rot}(\omega)$: corresponding relaxation time; τ_{n2}: longer orientation time constant; $\tau_{rot}(\chi)$: rotational relaxation time constants from streaming birefringence measurements; $D_{rot}(\chi)$: corresponding rotational diffusion coefficient; τ_{n1}: shorter orientation time constant; $\Delta\nu_{1/2}$: half width of the NMR-resonance signal of the methylene protons; η_o: viscosity of the solutions for zero shear rate)

c_o/mM	R_Θ/cm^{-1}	r_H/Å	$B/V^{-2}\cdot m$	$f(\omega\tau=1)/s^{-1}$	$\tau_{rot}(\omega)/\mu s$	$\tau_{n2}/\mu s$	$\tau_{rot}(\chi)/\mu s$	$D_{rot}(\chi)/s^{-1}$	$\tau_{n1}/\mu s$	$\Delta\nu_{1/2}$/Hz	η_o/cP
10	$7.7 \cdot 10^{-5}$	65	$1 \cdot 10^{-12}$	12000	13	5	–	–	2.5	23	1.05
15	$9.1 \cdot 10^{-5}$	40	$2 \cdot 10^{-12}$	14000	11	20	–	–	1.9	–	1.15
20	$8.7 \cdot 10^{-5}$	38	$3.7 \cdot 10^{-12}$	9000	18	40	24	6580	1.1	31	1.43
30	$7.74 \cdot 10^{-5}$	20	$6.1 \cdot 10^{-12}$	4000	40	75	66	2590	1.1	32	2.45
50	$7.05 \cdot 10^{-5}$	11	$2.2 \cdot 10^{-11}$	440	358	375	256	650	1.1	32	7.38
100	$9.45 \cdot 10^{-5}$	10	$3.3 \cdot 10^{-11}$	400	400	650	150	1100	1.1	30	21.80

ments. The diffusion coefficients increase even though the system becomes more and more crowded. The increasing crowding shows up directly in the rotation diffusion coefficients which were determined from the extinction angle of the flow birefringence. They decrease rapidly with the concentration. The rotation times $\tau_{rot}(\chi)$ from the streaming birefringence measurements are always close to the $\tau_{rot}(\omega)$-values which are determined from the electrical birefringence measurements in the AC-mode.

This is noteworthy and shows that the response of the system is very similar if the rods are aligned by a mechanical shear field or by an electrical field. Considering the large range of these time constants the agreement within a factor two seems good indeed. The differences however might be significant and be an indication that the two time constants do not necessarly have to be exactly the same. For concentrations below 50 mM $\tau_{rot}(\chi)$ seems to be larger than $\tau_{rot}(\omega)$ while it is the other way around for higher concentrations. While the differences could possibly be due to the different alignment mechanism it is also conceivable that they reflect a change of the systems during flow. The size of the micelles is determined by their interaction energy which is bound to change in the shear field. A change of the system can therefore not be ruled out. The time constants $\tau(\omega)$ are the same as those which can be obtained from the long tail of the transient electric birefringence measurements. In order to observe these long tails the field pulse has to be sufficiently long. The time constant for the tail increases with the length of the pulse. The transient measurements indicate further the existence of a short orientation time which varies little with concentration. This time constant is probably due to the free rotation of the rods. The same is true for the line width of the proton signals coming from NMR-measurements. Qualitatively one would expect that both of these parameters should increase somewhat if the dimension of the micelles remain the same and the system is more and more crowded. The increasing crowding should result in increasing hydrodynamic interaction and hence in a slowing down of the rotation and broadening of the lines. It is likely that its effect is roughly compensated for the micelles becoming smaller. The amplitude of the electrical birefringence as given by the Kerr constant increases approximately with the square of the concentration. From the Kerr constant it is difficult to say anything about the size of the aggregates because it depends in a complicated way on the order parameter of the molecules inside the micelles and the orientation degree. Both of these parameters can change with the concentration.

Finally the slowing down of the rotation is expressed in the increasing viscosity of the system. From most parameters it is possible to calculate a length for the micelles. These values are given in figure 15. It is clear from the figure that the agreement is not always good. The probable reason is that the equations which are used for the evaluation are too much simplified and do not take the intermicellar interaction energy into account in the right way.

4. *Approach to the phase boundary*

Most of the data discussed so far were obtained at a concentration which was still more than an order of magnitude away from the phase boundary of the liquid crystal. The lack of data for higher concentrations has to do with the difficulty of handling these solutions at higher concentrations because of their high viscosities and elasticities. It is however interesting to note that if we extrapolate dimensions for the rods from our avail-

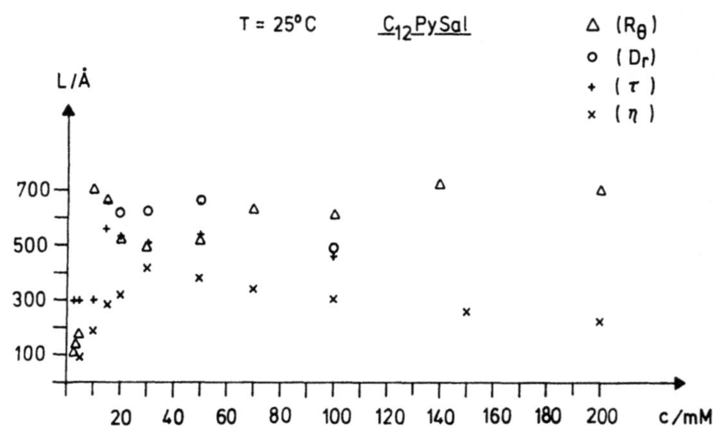

Fig. 15. Concentration dependence of micellar lengths of solutions of C_{12}PySal at 25 °C calculated from light scattering (R_Θ), streaming birefringence (D_r) electric birefringence (τ) and viscosity data (η)

able data to concentrations around the phase boundary we come to rods which have a rather small axial ratio of around 2 – 3. This is interesting because such small rods were also observed in nematic phases from X-ray and small angle neutron scattering measurements by Charvolin et al. [27]. While we could observe for most of the studied systems only a hexagonal phase under the polarization microscope we could observe a texture of a nematic phase for the cetylpyridiniumsalicylate. This makes it likely that the size of the rods does not change when we enter the two phase region from isotropic solutions. However, at present it is not clear why some systems show a nematic phase in a small concentration and temperature range and then a hexagonal phase while other systems form a hexagonal phase directly.

For the dodecylpyridinium systems we have also obtained results for concentrations which are closer to the phase boundary. Some of the data we have so far are given in figure 16. It is remarkable that the concentration dependence of the measured parameters R_Θ, η and D^o_{rot} as a function of the concentration changes again above 100 mM. These might be the sign of an interaction which is not significant at smaller concentrations and which might become pronounced for the higher concentrations. An interpretation of these data cannot be given at present. It will have to wait for further experimental results. The behaviour does not seem specific for the C_{12}-system but is likely to be a general one. First tendencies of this behaviour are also visible in the data of the C_{10} and the C_{14}-system. Furthermore the effect was also seen in the system $C_{14}PyC_7SO_3$ [13] which has been investigated by us. The increase in the light scattering does probably not indicate a change in the growth of the micelles with concentration but rather reflects a change in the interaction between the micelles. It would already be sufficient to explain it by assuming that the S-values increase again with concentration, the absolute values still being on a very low level. The NMR data which should reflect the sizes of the micelles more directly do not indicate a change in the behaviour which started at the cross-over point of the rods. From a theoretical point of view it would also make sense that attractive forces between the micelles should become active when the mean distance between the micelles approaches values of the diameter. In this concentration range the system could no longer be assumed in a semidilute state. It would also make sense from the point of view of the closeness of the phase boundary. At the boundary the intermicellar interaction has to be attractive again, that means $S > 1$ and it is likely therefore that S starts to increase with concentration again considerably before the phase boundary.

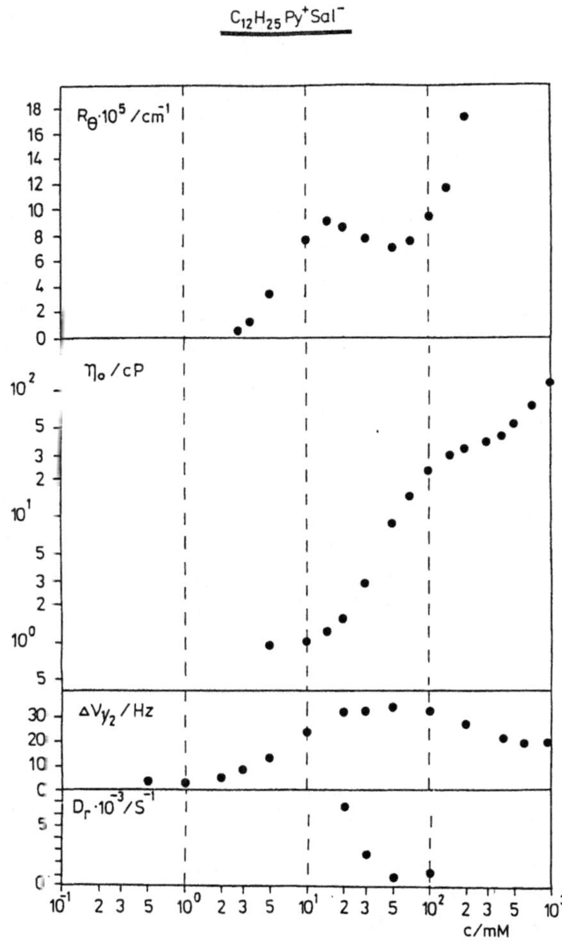

Fig. 16. Dependence of the Rayleigh ratio, the viscosity, the ^{1}H NMR line width and the rotational diffusion coefficient on the concentration

Conclusions

1. Surfactant systems with strongly binding counterions form rodlike micelles from the CMC on.

2. The length of the rods increases at first approximately linearly with the surfactant concentration until the rotational volumes of the micelles overlap. At this concentration the rods can be several thousand Å long when excess salt is present and still be fairly stiff with a persistence length above 2000 Å.

3. For concentrations above the overlap concentration the rods seem to become shorter again with increasing concentration. The rotation of the rods in

these solutions is extremely restricted in comparison to the unhindered rotation of a free rod. The translational motion of the rods is unhindered.

4. The rheological behaviour of these solutions is non-Newtonian. It is mainly determined by the time constant for the rotation τ_c and an elasticity modulus G^o. In the overlap region G^o seems to be relatively independent of the specific chemistry of the system such as the head group, counterion and chain length of surfactant, which can be taken as evidence that the structures and the lengths of the rods are very similar in these solutions when the conditions are the same. The dynamic behaviour in these solutions however can vary many orders of magnitude because τ_c can be slowed down many orders of magnitude from the values of the free rotation by the intermicellar interaction of the rods. For shearing rates $\dot{\gamma}$ in which $\dot{\gamma} \cdot \tau_c > 1$ the rods are aligned in the shear field and the elastic properties of the solutions become strong.

References

1. Tiddy GJT (1980) Physics Reports 57:1
2. Hoffmann H, Platz G, Rehage H, Schorr W, Ulbricht W (1981) Ber Bunsenges Phys Chem 85:255
3. Israelachvili JN, Mitchell DJ, Ninham BW (1976) J C S Faraday Trans II 72:1525
4. Nagarajan R, Ruckenstein E (1979) J Colloid Interf Sci 71:580
5. Hayter JB, Penfold J (1981) J C S Faraday Trans I 77:1851
6. Hoffmann H, Rehage H, Reizlein K, Thurn H (1983) (ed) Shaw DO, Symp Macro- and Microemulsions: Theory and Practice, A C S National Meeting; Washington, in press
7. Mazer NA, Benedek GB, Carey MC (1976) J Phys Chem 80:1075
8. Ikeda S, Hayashi S, Imae T (1981) J Phys Chem 85:106
9. Appel J, Porte G (1981) J Colloid Interface Sci 81:85
10. Nicoli DF, Elias JG, Eden D (1981) J Phys Chem 85:2866
11. des Cloizeaux J (1973) Macromolecules 6:403
12. Hoffmann H, Rehage H, Schorr W, Thurn H (1982) (eds) Mittal KS, Intern Symp Surfactants Solution, Lund, Proceeding in press
13. Hoffmann H, Rehage H, Schorr W, Thurn H, (1984) (ed) Mittal KL, Lindman B, in: Surfactants in Solutions, Plenum Press, New York, London, 425
14. Hoffmann H, Kalus J, Thurn H, Ibel K (1983) Ber Bunsenges Phys Chem 87:1120
15. Mori Y, Ookubo N, Hayakawa R, Wada Y (1982) J Polymer Sci, Polymer Phys Ed 20:2111
16. Peterlin A, Stuart HA (1943) (ed) Eucken A, Wolf KL, Hand- und Jahrbuch der chemischen Physik, Leipzig, Akadem Verlagsgesellschaft, 8:I B
17. Scheraga HA, Backus JK (1951) J Amer Chem Soc 73:5108
18. Ulmius J, Wennerström H (1977) J Magn Resonance 28:309 Stilbs P, Lindman B (1981) J Phys Chem 85:2587
19. Kerker M (1969) The Scattering of Light and Other Electromagnetic Radiation, Academic Press, New York
20. Berne BJ, Pecora R (1976) Dynamic Light Scattering, Wiley u Sons, New York
21. Doi M, Edwards SF (1978) J C S Faraday Trans II, 78:918
22. Hess S (1980) Z Naturforschung 35 a:915
23. Gelbart WM (1983) Intern School Phys, Enrico Fermi, Varenna
24. Porte G, Appel J, Poggi Y (1980) J Phys Chem 84:3105
25. Onsager L (1949) N Y Acad Sci 51:627
26. Flory PJ, Ronca G (1979) Mol Cryst Liq Cryst 54:289
27. Charvolin J, Hendrikx Y, Rawiso M (1984) (eds) Mittal KL, Lindman B, in: Surfactants in Solution, Plenum Press, New York, London, 59
28. Hess S (1983) Physica A, Vol 118 A, 79–108, in Nonlinear Fluid Behavior, Proceedings of the Conference on Nonlinear Fluid Behaviour, held at the University of Colorado, Bolder, USA, June (7–11) 1982

Authors' address:

H. Hoffmann
Lehrstuhl für Physikalische Chemie
der Universität Bayreuth
8580 Bayreuth

Progress in Colloid & Polymer Science Progr Colloid & Polymer Sci 69:29–38 (1984)

Mizellquellung und die Bildung von Solubilisaten, Mikroemulsionen und Emulsionen*

R. Heusch

Bayer AG, Leverkusen, F. R. G.

Zusammenfassung: Solubilisier- und Emulgierversuche von p-Xylol und Styrol mit Hilfe von Polyglykolethern zeigen, daß die Quellung mizellarer Bereiche für diese Prozesse verantwortlich ist.

Löst man den Solubilisator oder Emulgator in der wäßrigen Phase, erhält man bessere Ergebnisse als beim Lösen derselben in der Ölphase, denn nur die in der wäßrigen Phase vorhandenen Mizellstrukturen können solubilisieren und emulgieren. Eine Emulsion wird zerstört durch Überwechseln des Emulgators aus der wäßrigen in die Ölphase.

Durch Zusatz eines artverwandten Alkohols wird in mechanisch wenig beanspruchten Systemen mehr organische Substanz aufgenommen, in mechanisch stark beanspruchten Systemen aber weniger.

Phasendiagramme zwischen Polyglykolethern und Wasser zeigen, daß lamellare Strukturen für die Mizellquellung verantwortlich sind.

Abstract: Solubilizing and emulsifying tests on p-xylene and styrene with the aid of polyglycol ethers show that the swelling of micellar regions is responsible for these processes.

If we dissolve the solubilizer or emulsifier in the aqueous phase, we obtain better results than when dissolving them in the oil phase, since only the micelle structures present in the aqueous phase can solubilize and emulsify. An emulsion is destroyed by the emulsifier changing out of the aqueous into the oil phase.

Through the addition of a related type of alcohol, more organic substance is taken up by systems which are less subjected to mechanical stress, but less organic substance is taken up by systems which are more strongly stressed.

Phase diagrams between polyglycol ethers and water show that lamellar structures are responsible for the swelling of the micelles.

Key words: Mizellen, solubilisieren, Mikroemulsionen, Emulsionen.

1. Einleitung

In der diesjährigen Januarausgabe des Journal of Coll. and Interf. Sci. stellten Mukherjee, Miller und Fort jr. fest, daß es keine allgemein anerkannte Definition für Mikroemulsionen gibt [1] und es ist nicht verwunderlich, daß Friberg und Raymond eine Definition für Mikroemulsionen in ihrem Artikel in der gerade erschienenen „Encyclopedia of Emulsion Technology" vermeiden [2].

In dem vorliegenden Artikel wird über Solubilisier- und Emulgierversuche berichtet, die die Frage anschneiden, ob es einen physikalischen Grund gibt, der die Bezeichnung Mikroemulsionen rechtfertigt. Unsere Versuche führen zu einem Elementarprozeß, der als Mizellquellung bezeichnet wird [3–5]. Wir kommen bei Polyglykolethern zu lamellaren Mizellen, in deren hydrophoben Bereichen organische Substanzen gespeichert werden können. Die Existenz lamellarer Mizellen aber und ihre Quellbarkeit ist von Mc Bain und seiner Schule immer wieder herausgestellt worden [5].

Unsere Versuche greifen auch in die alte Diskussion ein, ob eine Mizellquellung oder die Ausbildung von

* Vortrag, gehalten auf der 31. Hauptversammlung der Kolloid-Gesellschaft, Bayreuth 11. bis 14. Oktober 1983.

KT-W 825

Mischfilmen für die Bildung vom Emulsionen entscheidend ist, eine Gegenüberstellung, die Prince in seinem Buch über Mikroemulsionen exzellent herausgestellt hat [6].

2. Solubilisierung

Wäßrige Lösungen oberflächenaktiver Substanzen besitzen die Fähigkeit, die Löslichkeit von in reinem Wasser schwerlöslichen Substanzen deutlich zu steigern. Die folgenden Abbildungen zeigen dazu drei Beispiele.

In der Abbildung 1 ist die Löslichkeitssteigerung von ®RESOLIN Brilliantorange RS durch Emulgator PV-15, des Umsetzungsprodukts eines kondensierten Phenols mit Ethylenoxid [24], spektroskopisch aufgezeichnet. Der Anstieg der Extinktion drückt die Steigerung der Löslichkeit des Farbstoffes durch den zugesetzten Polyglykolether aus.

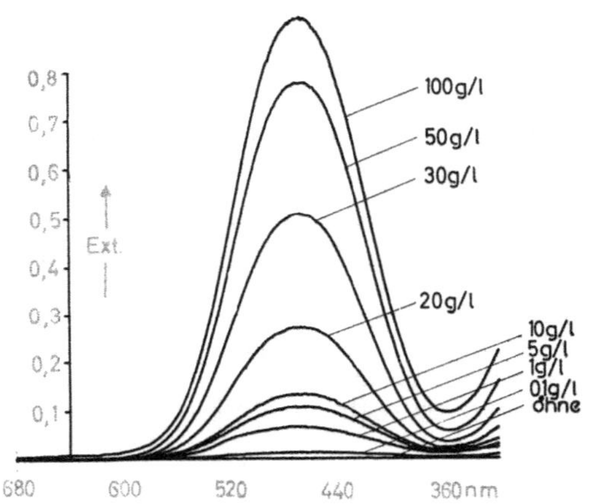

Abb. 1. Solubilisierung von Resolinbrillantorange RS (500 mg/l H_2O) durch Emulgator PV-15

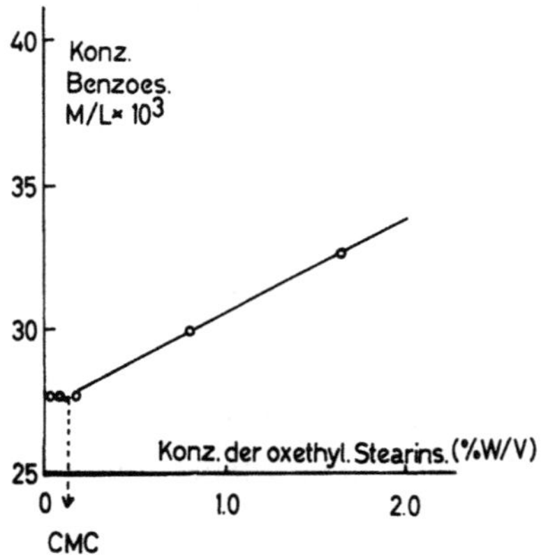

Abb. 2. Solubilisierung von Benzoesäure [7]

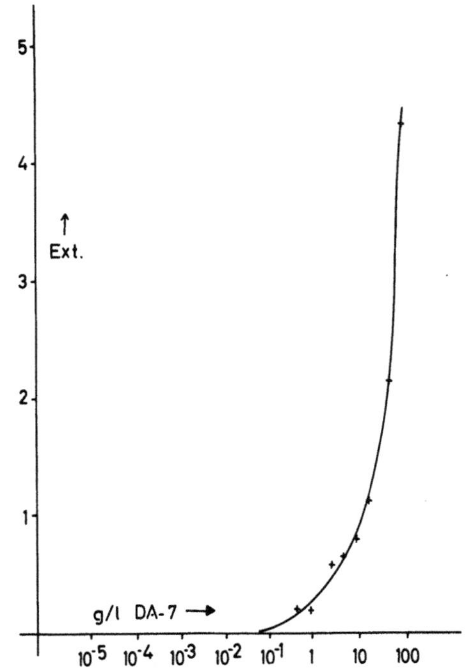

Abb. 3. Solubilisierung eines Pflanzenschutzmittels (250 mg/l H_2O) durch DA-7

Die Abbildung 2 ist der Pharmaliteratur entnommen. Sie zeigt die Solubilisierung von Benzoesäure durch hoch oxethylierte Stearinsäure oberhalb der CMC [7].

In der Abbildung 3 ist ein Pflanzenschutzmittel durch einen oxethylierten Alkohol solubilisiert. Die Löslichkeitssteigerung setzt wieder oberhalb der CMC ein.

Strukturierte Tensidlösungen besitzen also hervorragende anwendungstechnische Eigenschaften, sie können Fremdstoffe wie Farbstoffe, Pharmaka, Pflanzenschutzmittel oder dergleichen mehr aufnehmen, d. h. solubilisieren.

3. Die Solubilisierung und Emulgierung von p-Xylol

Um die Solubilisierung, die Emulgierung und ein eventuelles Auftreten von Mikroemulsionen etwas näher zu untersuchen, wurden Modellversuche durchgeführt, die wir spektroskopisch verfolgen konnten. Dazu haben wir aromatische Substanzen mit einem aliphatischen Emulgator behandelt. Steigende Mengen p-Xylol wurden z. B. in einem 100 ml Mischzylinder mit wäßrigen Lösungen aufgefüllt, die 2, 4 und 6 g n-Dodecylalkoholheptaglykolether (DA-7) im Liter Wasser enthielten.

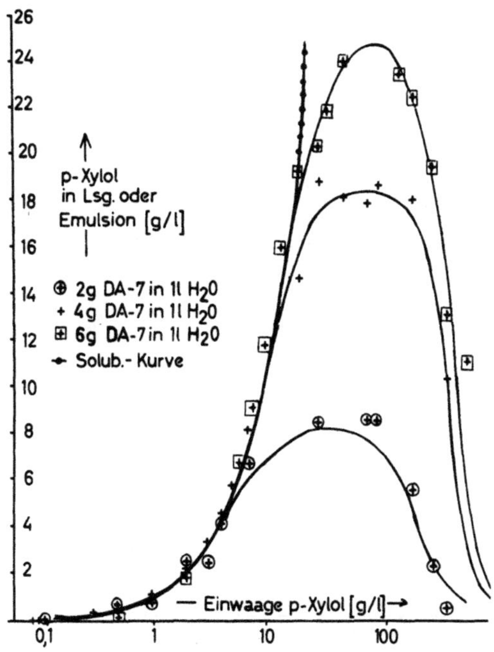

Abb. 4. Solubilisierung und Emulgierung von p-Xylol in DA-7 Lösung bei Raumtemperatur

Nach kräftigem Schütteln und einer Standzeit von 24 Stunden wurde dann das solubilisierte oder emulgierte p-Xylol spektroskopisch bestimmt. Dazu haben wir mit einer Pipette deutlich unterhalb der aufgerahmten Schicht etwa in der Mitte des Kolbens eine Probe gezogen, diese in Methanol aufgenommen und deren p-Xylolmaximum mit einem Eichspektrum verglichen. Wir kommen zu den Ergebnissen der Abbildung 4.

Aufgetragen ist bei der Wiedergabe unserer Modellversuche die spektroskopisch bestimmte Menge des solubilisierten oder emulgierten p-Xylols gegen die Einwaage des p-Xylols. Punktiert gezeichnet ist die Solubilisierungskurve. Auf dieser Kurve ist die eingewogene Menge p-Xylol vollständig in die DA-7-Lösung integriert, vollständig solubilisiert.

Nach einer Standzeit von 24 Stunden sind bei der 0,2 %igen DA-7-Lösung rund 6 g p-Xylol, bei der 0,4 %igen Lösung 13 und bei der 0,6 %igen DA-7-Lösung etwa 19 g p-Xylol ohne Abscheidung oder nennenswerte Trübung solubilisiert. Die Strukturelemente der DA-7-Lösungen, die Mizellen, können also etwas mehr als das 3fache des DA-7-Gewichtes an p-Xylol aufnehmen ohne sichtbar in Erscheinung zu treten.

Erhöhen wir die Zugabe an p-Xylol, werden die Lösungen trübe, und es bilden sich Emulsionen, die zunächst wäßrig beständig sind. Bereits nach 24 Stunden haben sich sahnige Phasen abgeschieden, die aber durch Umschütteln reemulgierbar sind. Bei den eingesetzten Emulgatorkonzentrationen entstehen 2phasige Systeme. Die Ausbildung einer 3. Phase, einer sogenannten „Mittel-" oder „Windsor III. Phase" [8, 9], wurde nicht beobachtet.

Mit steigendem p-Xylolgehalt finden wir Maxima, deren Höhen vom Emulgatorgehalt abhängig sind. In der Lösung, die 2 g DA-7 pro Liter Wasser enthält, sind nach 24 Stunden etwas mehr als 8 g p-Xylol emulgiert. In der 4 g pro Liter Lösung sind 18,5 g und in der 0,6 %igen DA-7-Lösung noch etwa 25 g p-Xylol emulgiert. In verdünnten DA-7-Lösungen können wir also das 4,2fache des Emulgatorgewichtes an p-Xylol emulgieren. Das Abweichen von der Solubilisationskurve zu kleineren Werten hin zeigt, daß die DA-7 Mizellen nur eine begrenzte Menge p-Xylol aufnehmen können.

Oberhalb einer Zugabe von 100 g p-Xylol, bei Konzentrationen also rechts von den Maxima, werden die Emulsionen instabil. Bereits kurz nach der Herstellung findet eine Phasentrennung statt, und durch Umschütteln ist keine Reemulgierung mehr zu erreichen. Das zugegebene p-Xylol scheidet sich über der DA-7-Lösung ab.

Unsere Modellversuche zeigen: Verdünnte DA-7-Lösungen können etwa die 3fache Menge an p-Xylol solubilisieren, die 4,2fache Menge aber emulgieren. Höhere p-Xylol Zusätze zerstören die Emulsionen dadurch, daß der Emulgator aus der wäßrigen Phase in das p-Xylol überwechselt, denn in der unteren, wäßrigen Phase finden wir spektroskopisch nur noch einen Restgehalt an p-Xylol.

Um eine Grenze zwischen Solubilisierung und Emulgierung zu finden, haben wir die Emulsionen eine Woche lang stehen lassen und dann den p-Xylolgehalt wieder in der unteren Phase bestimmt. Die Ergebnisse sind für eine Konzentration von 4 g DA-7 pro Liter Wasser in Abbildung 5 durch Kreise eingetragen.

Schließlich haben wir die Lösungen oder Emulsionen 30 Minuten mit 3500 U/Minute zentrifugiert und den verbleibenden p-Xylolgehalt in der wäßrigen Phase wieder spektroskopiert. Die Ergebnisse sind in Abbildung 5 mit Dreiecken eingezeichnet.

Die Emulsionen werden durch längeres Stehen langsam und durch Zentrifugieren schnell zerstört. Schraffiert ist der Bereich, den wir durch Zentrifugieren nicht zerstören konnten. Hier haben wir es mit solubilisierten Lösungen zu tun, die als thermodynamisch stabil angesehen werden.

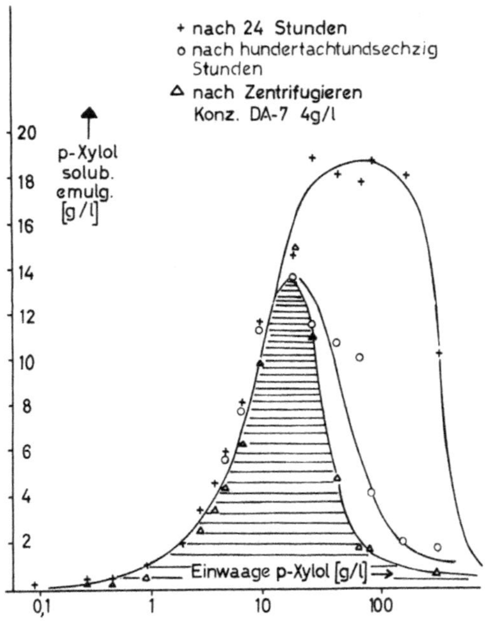

Abb. 5. Solubilisierung und Emulgierung von p-Xylol in DA-7-Lösung bei Raumtemperatur

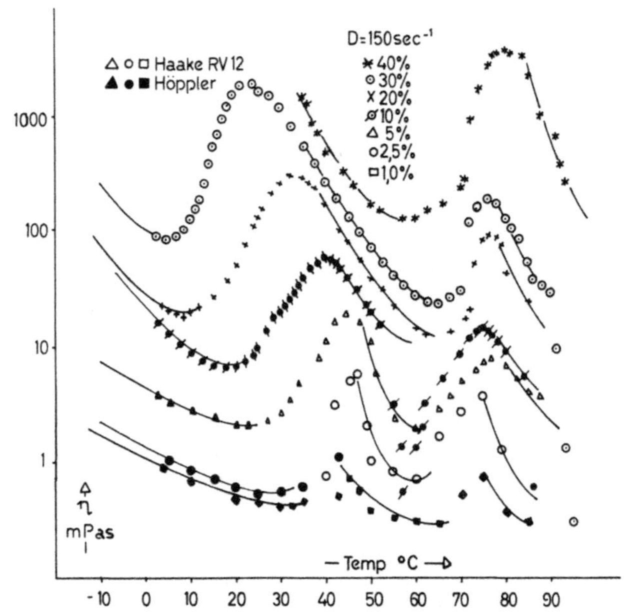

Abb. 7. Temperaturabhängige Viskositätsmessungen an DA-7-Lösungen

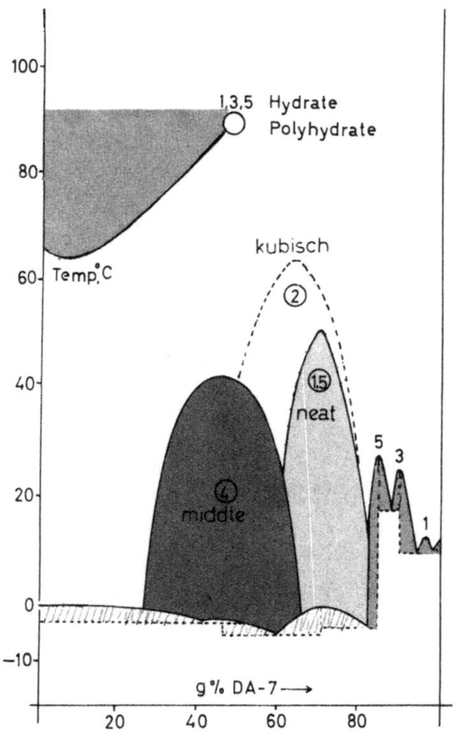

Abb. 6. Flüssig-kristalline Phasen bei DA-7-Wasser-Mischungen

Einen besonderen Bereich für eine Mikroemulsion konnten wir nicht finden, es sei denn, man bezeichnet eine solubilisierte Lösung als Mikroemulsion, nur weil die Teilchengröße unterhalb der Wellenlänge des sichtbaren Lichtes aber oberhalb der Mizellgröße liegt. Die Bezeichnung Mikroemulsion drückt eigentlich nur aus, daß man über die Struktur des vorliegenden Systems wenig oder gar nichts weiß.

In dem vorliegenden Fall liefert das Zustandsdiagramm des DA-7 mit Wasser eine Erklärung für die Speicherung des p-Xylols in den Strukturelementen, die wir als Mizellen bezeichnen. Abbildung 6 zeigt das bekannte Zustandsdiagramm des DA-7 mit Wasser [10].

In diesem Zusammenhang haben wir uns für die strukturierten Lösungen auf der wasserreichen, der linken Seite des Schaubildes interessiert. Dazu wurden in diesem Gebiet temperaturabhängige Viskositätsmessungen durchgeführt, die in Abbildung 7 zusammengestellt sind.

Wir finden zwei Maxima. Das linke Maximum, über das wir bereits berichtet haben [10], kommt durch einen Strukturwechsel zustande, der in der Lösung abläuft. Der exponentielle Abfall links im Bild zeigt das Vorhandensein einer bestimmten Struktur, der Abfall in der Bildmitte zeigt den einer anderen. Zwischen Minima und Maxima findet also ein Strukturwechsel statt.

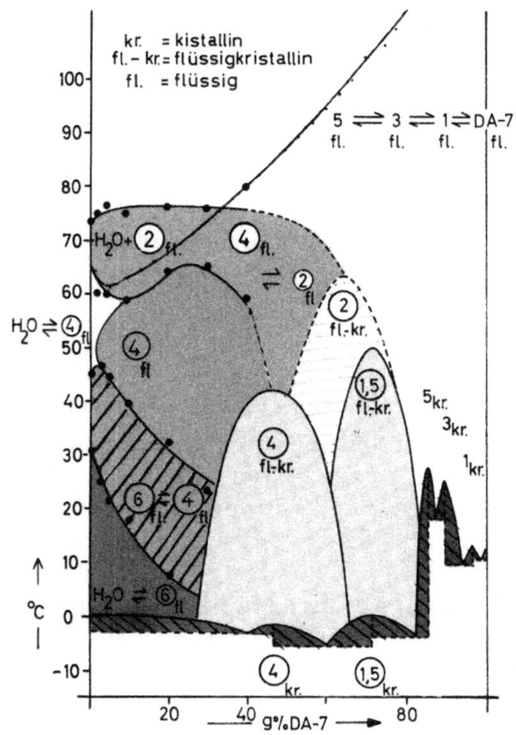

Abb. 8. Zustandsdiagramm von DA-7-Wassermischungen

Verfolgen wir, trotz auftretender Trübe, in unserem Rotationsviskosimeter die Viskositäten zu höheren Temperaturen hin, finden wir erneut Maxima, die wieder einen Strukturwechsel anzeigen, der jetzt im Gebiet der Mischungslücke liegt.

Zeichnen wir nun die Minima- und Maximatemperaturen der Abbildung 7 in das Zustandsdiagramm (Abb. 6) ein, erhalten wir die gesuchten Existenzbereiche der flüssigen Polyhydrate. Abbildung 8 gibt dann das ergänzte Phasendiagramm wieder.

Schraffiert gezeichnet ist das Gebiet, in dem die partielle Dehydratisierung des flüssigen Polyhexa- zum flüssigen Polytetrahydrat stattfindet. Oberhalb des schraffierten Gebietes ist flüssiges Polytetrahydrat existent.

Zwischen 60 und 76 °C verliert dann das flüssige Polytetrahydrat erneut einen Teil seines Hydratwassers und geht in das flüssige Polydihydrat über. In diesem Übergangsgebiet liegt die Mischungslücke, die durch Bestimmung der Trübungspunkte [12] in Abhängigkeit von dem DA-7-Gehalt ermittelt wurde.

Am unteren Rand der Mischungslücke ist im wasserreichen Teil des Phasendiagramms so viel schwerlösliches Polydihydrat entstanden, daß eine Trübung sichtbar wird. In der Mischungslücke selbst finden weitere Dehydratisierungen statt, bis oberhalb von 160 °C wasserfreies DA-7 vorliegt.

Rotationsviskosimetrische Messungen in Abhängigkeit vom Schergefälle belegen unterschiedliche Strukturen bei verschiedenen Temperaturen. In Abbildung 9 ist der Logarithmus der Viskosität (η) gegen den Logarithmus des Schergefälles (D) aufgetragen. Bei der 20 %igen DA-7-Lösung finden wir bei 3 °C eine starke Abhängigkeit der Viskosität vom Schergefälle, ein sogenanntes „pseudoplastisches" Verhalten, bei 50 °C ein „Newtonsches" Verhalten und bei 80 °C tritt eine Abhängigkeit vom Schergefälle erst bei höheren Scherwerten auf.

Die starke Abhängigkeit der Viskosität vom Schergefälle bei 3 °C kommt durch die starke Vernetzung der Polyetherketten mit Wasser zustande. Die vorliegende Hexahydratstruktur, für die die Hydrattheorie ein brauchbares Modell liefert [10, 14], gibt der steigenden mechanischen Belastung nach. Bei 50 °C existiert das Polytetrahydrat, dessen Elemente Newtonsches Verhalten zeigen.

Die Werte bei 80 °C lassen sich der schwach wasservernetzten Lamellenstruktur des Polydihydrats zuordnen, ein Ergebnis, das durch eine elektronenmikroskopische Untersuchung von Kopp erhärtet wird.

Abbildung 10 zeigt eine elektronenmikroskopische Aufnahme einer oberhalb des Trübungspunktes eingefrorenen wässrigen Lösung von ®Triton X-114. (Umsetzungsprodukt von p-Octylphenol mit 7–8 Mol

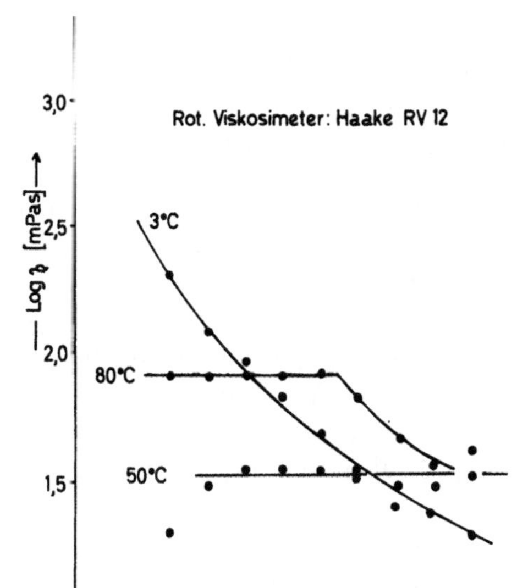

Abb. 9. Rotationsviskosimetrische Messungen an einer 20 %igen DA-7-Lösung

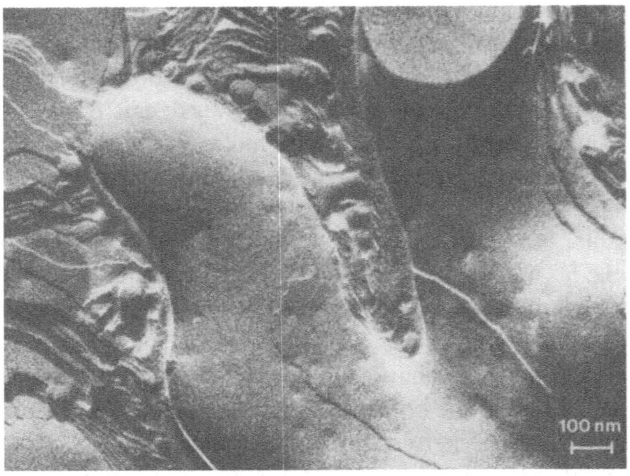

Abb. 10. Elektronenmikroskopische Aufnahme einer lamellaren Triton X 114-Phase, die oberhalb des Trübungspunktes abgeschieden wurden [13]

Ethylenoxid.) Dazu wurde eine einprozentige Lösung von ®Triton X-114 über den Trübungspunkt erwärmt. Die dabei neu entstandene Phase wurde durch kurzes Zentrifugieren angereichert (17 %) und in Portionen von etwa einem Mikroliter rasch in unterkühltem Stickstoff (−210 °C) eingefroren. Im Hochvakuum (besser als 10^{-7} mbar) wurden die gefrorenen Präparate gebrochen. Von der Bruchfläche wird dann durch Aufdampfen von Platin und Kohle ein Abdruck hergestellt. Bedampft wurde von oben. Der Bedampfungswinkel für Platin war 45 Grad. Quergebrochene Lamellenstapel und Flächenschichten von Lamellen sind in Abbildung 10 deutlich zu erkennen [13].

Unsere Ergebnisse lassen den Schluß zu, daß in verdünnten DA-7-Lösungen wie in anderen wäßrigen Polyglykolethersystemen Polyhexahydratlamellen bei Raumtemperatur bis zur CMC hin beständig sind. D. h. aber, Polyhexahydratlamellen sind die Strukturelemente, sind die Mizellen, die organische Substanzen speichern können.

Nach unserer Hydrattheorie [10, 14] aber steht die starke Wasservernetzung der Polyhexahydrate durch Wasserdoppelbrücken einer Quellung der hydrophilen Molekelteile entgegen. Weder organische Substanzen noch Wasser können von den wasserbeladenen Polyetherketten aufgenommen werden.

Die Kohlenwasserstoffketten dagegen sind beweglicher, organische Molekeln wie p-Xylol können in die Kohlenwasserstoffbereiche eindringen. Die Kohlenwasserstoffketten saugen sich mit p-Xylol voll und werden auseinandergeschoben. Die Lamellen werden merklich vergrößert. Sie quellen, bilden Solubilisate und werden schließlich so groß, daß sie als Emulsionströpfchen im Mikroskop sichtbar sind. Wir haben einen Prozeß vor uns, der als Mizellquellung bezeichnet wird.

Unsere Hydrattheorie fordert, daß die Solubilisierung organischer Substanzen in den Kohlenwasserstoffbereichen der Tensidmolekeln stattfindet, ein Befund, der von anderer Seite mehrfach bestätigt worden ist [15–18].

Beim Stehen oder Zentrifugieren der Emulsionen scheidet sich ein Teil des p-Xylols ab. Da die Löslichkeit des DA-7 in p-Xylol aber größer als in Wasser ist, wird der Emulgator aus der wäßrigen Phase herausgelöst. Die Emulsion wird zerstört durch ein Überwechseln des Emulgators aus der wäßrigen in die organische Phase.

Das charakteristische an dem System p-Xylol/wäßrige DA-7-Lösung ist, daß beim Zentrifugieren die Emulsionen zerstört, die Solubilisate aber erhalten bleiben. D. h. aber, ein Teil des p-Xylols ist an die Kohlenwasserstoffketten so fest gebunden, daß er durch Zentrifugieren nicht abgelöst werden kann. Ein anderer Teil ist nur so lose integriert, daß er beim Zentrifugieren herausgepreßt wird.

4. Die Solubilisierung und Emulgierung von Styrol

Die Solubilisierung und Emulgierung von Styrol in 0,2 % DA-7-Lösung führt zu den Ergebnissen der Abbildung 11. Nach 24 Stunden und nach einer Woche finden wir Kurven, die denen der Solubilisierung und Emulgierung von p-Xylol ähnlich sind, nur die von der DA-7-Lösung aufgenommenen Styrolmengen sind wesentlich niedriger als bei der Solubilisierung und Emulgierung von p-Xylol. Nur 2 g Styrol werden solubilisiert (das einfache des DA-7-Gewichtes) und nach 24 Stunden sind maximal 6,2 g Styrol (das 3,1fache des DA-7-Gewichtes) emulgiert.

Das überraschende aber ist, daß beim Zentrifugieren der Styrolphasen nicht nur die Emulsionen, sondern auch die Solubilisate weitgehend zerstört werden. Lediglich der kleine Styrolrest, den die untere

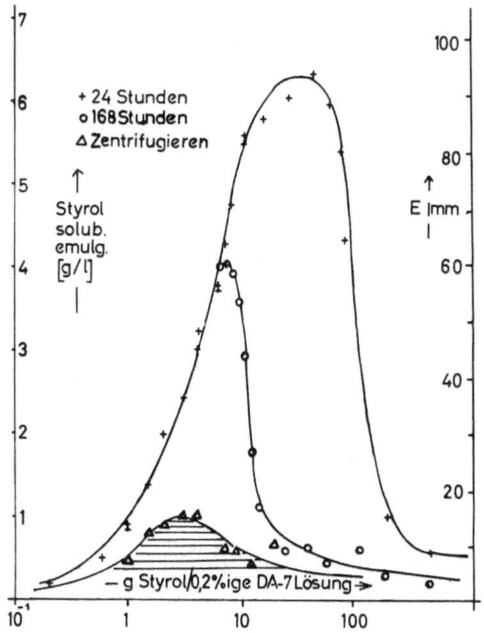

Abb. 11. Solubilisierung und Emulgierung von Styrol in DA-7-Lösung bei Raumtemperatur

Kurve wiedergibt, kann nach dem Zentrifugieren noch in der unteren, wäßrigen Phase nachgewiesen werden.

Gequollene Mizellen zeigen also unterschiedliche Stabilitäten. Es kommen die Wechselwirkungen zwischen den Kohlenwasserstoffketten mit den zu solubilisierenden Substanzen ins Spiel. Unsere Hydrattheorie, die in Abbildung 12 als Blockschema zusammen-

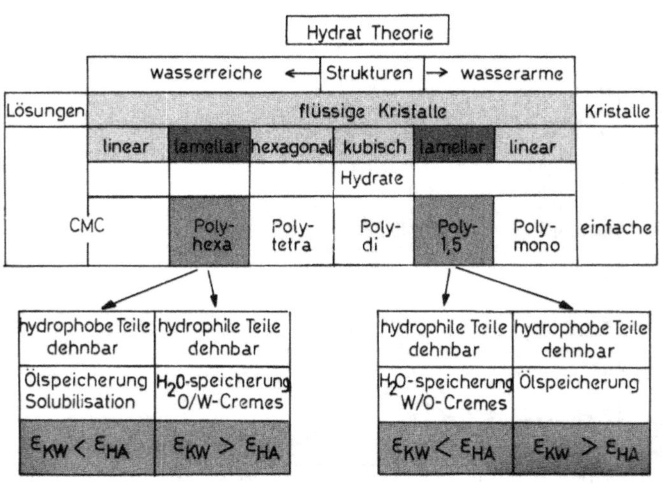

Abb. 12. Die Hydrattheorie im Blockschema

gestellt ist, liefert nicht nur Strukturmodelle für mizellare und mesomorphe Tensidlösungen [10, 14], sondern mit Hilfe der lamellaren Strukturen läßt sich die Speicherung von Wasser oder organischen Substanzen erklären. Die Wasserspeicherung führt zur Gelierung oder Cremebildung und die Aufnahme organischer Stoffe zu der Solubilisierung. In Abbildung 12 ist das Unterschiedsmerkmal zwischen Wasserspeicherung und Solubilisierung in den unteren Kästchen angegeben. Solubilisierung tritt ein, wenn die Bindungsenergien zwischen den Kohlenwasserstoffketten ε_{KW} kleiner sind als die Hydratbindungen ε_{HA}. Organische Substanzen können dann in die Kohlenwasserstoffbereiche eindringen. Die Ketten können sich auseinanderschieben und werden aufgebläht.

Sind umgekehrt die Bindungen der Hydrate ε_{HA} geschwächt, z. B. durch die Ladungen bei ionischen Tensiden, dann können die Wechselwirkungsenergien zwischen den Kohlenwasserstoffketten ε_{KW} überwiegen. Die hydrophilen Molekelteile können dann aufgeweitet werden und bedeutende Mengen Wasser speichern. Wir haben dann Cremes vor uns oder hydrophile Gele, die z. B. beim Lösen von 5 g Natriumstearat in einem Liter Wasser entstehen.

Um die Zentrifugenempfindlichkeit solubilisierter Lösungen zu verstehen, müssen wir zwischen fest, komplexartig an die Kohlenwasserstoffketten gebundenem Solubilisat und Quellsolubilisat unterscheiden und nur das fest an die Kohlenwasserstoffketten gebundene Solubilisat ist zentrifugenbeständig.

Entsprechend unterscheiden wir bei der Wasserspeicherung zwischen festgebundenem Hydratwasser und Quellwasser. Junginger hat mit Hilfe der Thermogravimetrischen Analyse das Verhältnis zwischen freiem und interlamellar gebundenem Wasser bei verschiedenen Systemen untersucht [19] und Noll bezeichnete lose in eine Struktur integriertes Wasser als „Schwarmwasser" bei einer ähnlichen Problematik [20].

Unsere Untersuchungen zeigen, wie gefährlich es ist, von thermodynamisch stabilen Emulsionen zu sprechen. Emulsionen sind immer instabil, auch wenn es gelingt, ihren Zerfall über Wochen und Monate hinaus zu verzögern. Emulsionen sind instabile Systeme, die dem stabilen Zustand, dem thermodynamischen Gleichgewicht zustreben.

Auch Solubilisate – transparente Mikroemulsionen – müssen nicht immer stabil sein. Sie können zentrifugenempfindlich sein oder von der Reihenfolge der Zugabe der Komponenten bei der Herstellung abhängig sein, einen Vorgang, den Larsson bereits 1967 untersucht hat [21].

5. Mizellquellung und Emulgiertechnik

Damit ist ein Problem der Emulgiertechnik angesprochen, das in engem Zusammenhang mit der Mizellquellung steht. In der Regel werden Emulsionen hergestellt, indem man den Emulgator in der inneren Phase löst und diese Lösung in die äußere Phase einarbeitet [22]. Bei unseren Emulgierversuchen aber wurde der Emulgator in der äußeren Phase, in dem Wasser, gelöst und diese Lösung langsam zu dem p-Xylol oder Styrol zugegeben.

Was geschieht aber nun, wenn man nach den Regeln der Emulgiertechnik den Emulgator im p-Xylol löst und die Lösung mit Wasser auffüllt? Die Ergebnisse geben die Abbildungen 13 und 14 wieder.

In Abbildung 13 werden unterschiedlich hergestellte Solubilisate und Emulsionen miteinander verglichen. Die mit Punkten markierte Kurve gibt die Aufnahme von p-Xylol in DA-7-Lösung wieder, die mit Dreiecken gekennzeichnete Kurve zeigt Ergebnisse, die wir erhalten, wenn DA-7 in Xylol gelöst und diese Lösung mit Wasser aufgefüllt wird.

Abbildung 13 zeigt, daß das 3,8fache an p-Xylol solubilisiert wird, wenn p-Xylol in die DA-7-Lösung eingebracht wird. Beim Lösen des DA-7 in p-Xylol wird dagegen nur die 2,5fache Menge solubilisiert. Ähnliche Verhältnisse finden wir beim Emulgieren. Das 4,2fache an p-Xylol wird in DA-7-Lösung emulgiert, das 2,6fache aber nur, wenn DA-7 vor der Emulgierung im p-Xylol gelöst wird. Nach den Regeln der Emulgiertechnik wird also weniger p-Xylol solubilisiert und emulgiert als beim Vorgeben einer wäßrigen DA-7-Lösung.

Abbildung 14 zeigt die Ergebnisse nach dem Zentrifugieren. Die durch Dreiecke markierte Kurve – der Emulgator wurde im p-Xylol gelöst – liegt wieder unterhalb der durch Punkte gekennzeichneten Kurve, die die Ergebnisse von in Wasser gelöstem DA-7 wiedergibt.

Der Grund dieser Unterschiede ist folgender: Wird DA-7 in p-Xylol gelöst, wechselt nur ein Teil des Emulgators aus der organischen in die wäßrige Phase über, wenn wir mit Wasser schütteln, und nur der ins Wasser gelangte Emulgator bildet die Strukturen aus, die für die Solubilisierung und Emulgierung notwendig sind.

Ist das DA-7 aber von vornherein im Wasser gelöst, steht die gesamte Menge der Strukturierung der Lösung zur Verfügung. Es kann entsprechend mehr p-Xylol solubilisiert und emulgiert werden.

Ob sich beim Lösen des Emulgators im p-Xylol und dem anschließenden Schütteln mit Wasser ein Verteilungsgleichgewicht einstellt, kann nicht geprüft werden. Geben wir nämlich den Emulgator nicht ins p-Xylol, sondern ins Wasser, ändern sich die Verhältnisse dadurch, daß für die Mizellquellung mehr Ausgangsmaterial zur Verfügung steht.

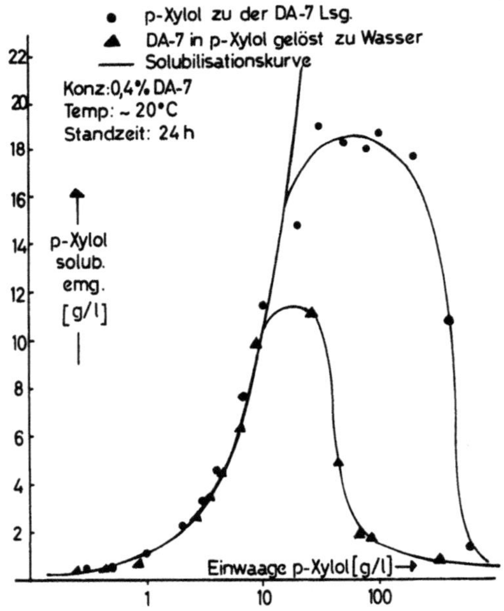

Abb. 13. Solubilisierung und Emulgierung von p-Xylol nach verschiedenen Methoden

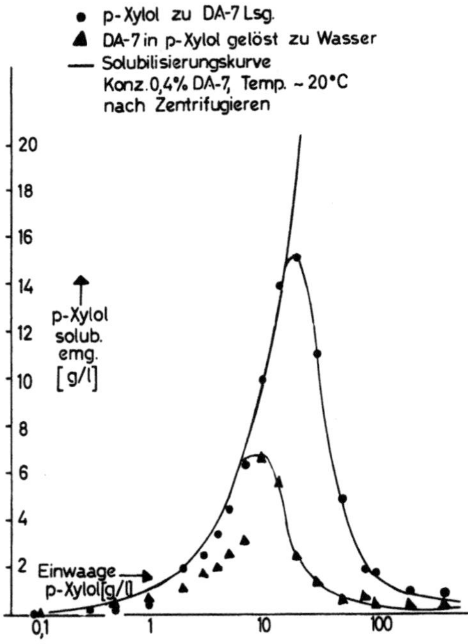

Abb. 14. Solubilisierung und Emulgierung von p-Xylol nach verschiedenen Methoden

6. 4-Komponenten-Systeme

Nach Schulman [23] und vielen anderen Autoren sind Mikroemulsionen 4-Komponenten-Systeme, die aus der zu emulgierenden Substanz, dem Wasser, dem Emulgator und einem Co-surfactant bestehen. Als Co-Tensid wird in der Regel ein Alkohol verwendet.

Abbildung 15 zeigt den Einfluß von Alkoholen bei der Solubilisierung und Emulgierung von p-Xylol.

Die durch Dreiecke markierte Kurve erhalten wir, wenn wir in einer 0,4 % DA-7-Lösung 10 % des DA-7 durch Laurylalkohol ersetzen, die mit Dreiecken umgebenen Punkte finden wir beim Austausch von 10 % DA-7 durch n-Hexanol und die mit einem Kreis umgebenen Punkte geben Ergebnisse einer mit Stearylalkohol gesättigten DA-7-Lösung (Raumtemperatur) wieder. Die durch Punkte gekennzeichnete Kurve zeigt die bereits bekannte Solubilisierung und Emulgierung ohne Zusätze.

Nur der Laurylalkoholzusatz erhöht die solubilisierte und emulgierte p-Xylolmenge. Bei der Solubilisierung steigert sich die p-Xylol-Menge von 13 g auf 21 g um 60 % und bei der Emulgierung von 18,5 g auf 24,5 g um 30 %.

Unsere Modellversuche zeigen, daß durch die Zugabe eines bestimmten artverwandten Alkohols die Lamellenstruktur eines Polyglykolethers so gelockert werden kann, daß mehr Substanz integriert wird. Dann und zwar nur dann sind gemischte Lamellen dehnbarer und quellbarer als Lamellen aus reinen Polyethern.

Abbildung 16 zeigt, wie sich durch Laurylalkohol aufgeweitete DA-7-Lamellen beim Zentrifugieren verhalten.

Die durch Dreiecke markierte Kurve des 4-Komponenten-Systems mit dem Laurylalkoholzusatz liegt jetzt unter der durch Punkte gekennzeichneten Kurve des 3-Komponenten-Systems ohne Alkoholzusatz. Die aufgelockerten Lamellen sind also instabiler und geben leichter solubilisiertes p-Xylol ab als die Lamellen des reinen DA-7.

Es ergibt sich: Durch Alkoholzusatz aufgeweitete Lamellen können in mechanisch wenig beanspruchten Systemen mehr organische Substanz aufnehmen als die alkoholfreien Systeme, in mechanisch stark beanspruchten Systemen dagegen weniger.

Unsere Untersuchungen aber zeigen, daß die Mizellquellung ein Elementarprozeß ist, der für die Bildung von Solubilisaten und Emulsionen verantwortlich ist. Die Ausbildung eines die Emulsionströpfchen umhüllenden Mischfilms dagegen scheint für die Stabilisierung solcher Systeme von Bedeutung zu sein [3]. Unsere Versuche lassen aber auch erkennen, wie man mit einfachen Mitteln Solubilisier- und Emulgierversuche verfolgen kann, um für ein praktisches Problem optimale Bedingungen zu finden.

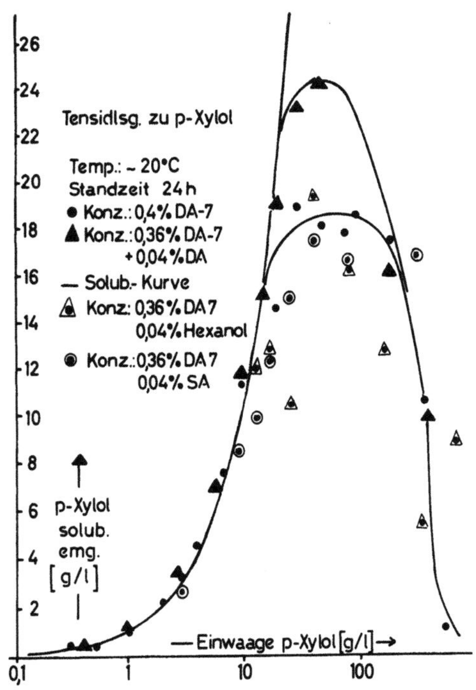

Abb. 15. Solubilisierung und Emulgierung von p-Xylol 4-Komponenten-System

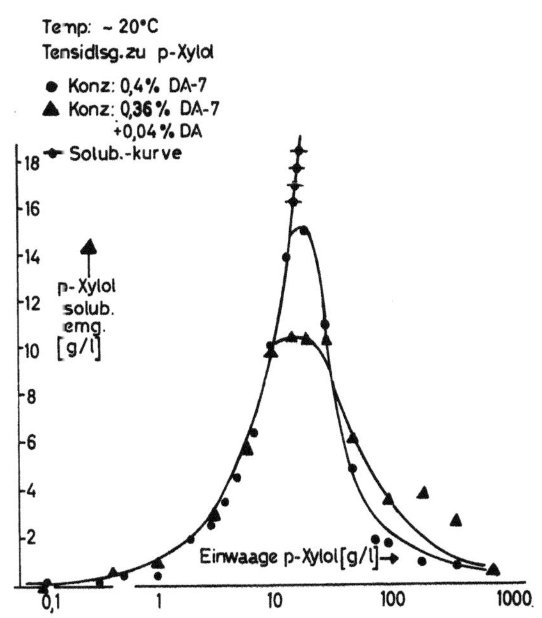

Abb. 16. Solubilisierung und Emulgierung von p-Xylol mit 10 % DA nach dem Zentrifugieren

6. Schlußbemerkung

Kehren wir zur Ausgangsfrage zurück: gibt es einen physikalischen Grund, der die Bezeichnung Mikroemulsion rechtfertigt? Nach unseren Untersuchungen genügt es, zwischen Emulgierung und Solubilisierung zu unterscheiden, Prozesse die ineinander übergehen. Zu Grunde liegt ein Elementarprozeß, der als Mizellquellung bezeichnet wird und für den unsere Zustandsdiagramme, verbunden mit der Hydrattheorie, eine anschauliche Erklärung liefern.

Schließlich konnten wir zeigen, daß solublilisierte Lösungen nicht immer stabil sein müssen und daß es schwierig und vielleicht sogar unmöglich ist, bei strukturierten Lösungen zu prüfen, ob ein thermodynamisches Gleichgewicht vorliegt.

Literatur

1. Mukherjee S, Müller CA, Fort jr T (1983) J Coll Interf Sci 91:223
2. Becher P (1983) Encyclopedia of Emulsion Technology, M Dekker, New York, Kap 4, S 287
3. Heusch R (1983) Chem Ing Techn 55:608
4. Ruckenstein E, Krishman R (1979) J Coll Interf Sci 71:321
5. McBain MEL, Hitchinson E (1955) Solubilisation and Related Phenomena, Academic Press, New York
6. Prince LM (1977) Microemulsions, Academic Press, INC, New York, San Francisco, London
7. Goodhart FW, Martin AN (1962) J Pharm Sci 51:50
8. Obah B, Neumann HJ (1983) Tenside 20:145
9. Windsor AP (1954) Solventproperties of Amphilic Compounds, London
10. Heusch R (1983) Tenside 20:1
11. Heusch R (1981) Makrom Chemie 182:589
12. DIN 53 917 (1981) Jun
13. Kopp F, private Mitteilung (Diabetes Forschungsinstitut Auf'm Hennekamp 65, D-4000 Düsseldorf)
14. Heusch R (1978) Ber Bunsenges Phys Chemie 82:970, (1979) 83:834
15. Aboutaleb AE, Sakor AM, Abdel-Rahman SI (1980) Pharmazie, Berlin 35:99
16. Collet JH, Withington R, Koo L (1975) J Pharmacy & Pharmacol, London 27:46
17. Mukherjee P (1971) J Pharm Sci 60:1528
18. Russell JC, Whitten DC (1982) J Amer chem Soc 104:5937
19. Junginger H, Colloid Polym Sci, in Vorbereitung
20. Noll W, Büchner W, Steinbach, Sucker C (1972) Kolloid Z u Z Polm 250:9
21. Larsson (1967) Brit Patent 1 174 672
22. Stache (1981) Tensidtaschenbuch, Carl Hanser Verlag, München, Wien, Ausgabe 2, S 208
23. Brocott JE, Schulman JH (1955) Z. Elektrochem 59:283
24. DBP 1 094 523 (1960) Bayer AG

Anschrift des Verfassers:

Dr. Rudolf Heusch
c/o Bayer AG
Farbenforschung 4, Geb. H 6
5090 Leverkusen-Bayerwerk

NMR measurements on microemulsions*)

P. Stilbs and B. Lindman[1])

Institute of Physical Chemistry, Uppsala University, Uppsala, Sweden
[1]) Physical Chemistry, Chemical Center, Lund University, Lund, Sweden

Abstract: Fourier transform Nuclear Magnetic Resonance studies, providing information on relative molecular self-diffusion coefficients and nuclear spin relaxation rates in microemulsions, have led to a reassessment of the prevalent „droplet" conception of the molecular organization in microemulsions. An O/W-organized system (e. g. a micellar-like) is necessarily characterized by rapid water diffusion and slow surfactant (and hydrocarbon solubilizate) diffusion. W/O systems must likewise exhibit slow water and surfactant diffusion, while the diffusion processes in the continuous hydrocarbon phase remain rapid. The experimentally observed diffusion behaviour in some microemulsion types (e. g. Aerosol OT-water-hydrocarbon) can, essentially throughout the entire extensive composition stability range, be reconciled with a distinct water-in-oil model while the dynamic properties on the molecular level of others (e. g. medium-chain alcohol cosurfactant types), resemble over wide concentration ranges those of partially aggregated liquids. As clearly shown by our multicomponent self-diffusion data, the degree of molecular organization over large composition regions of cosurfactant microemulsion systems is much lower than that in micellar solutions or normal emulsions, for example. Spin relaxation data indicate, on the other hand, that some supramolecular aggregation prevails during the (organizational) timescale of these experiments (c. 100 ps). In a local frame, this residual organization is probably similar to that found in micelles containing solubilized alcohols or hydrocarbons.

Key words: Microemulsion, micelle diffusion, NMR, surfactant.

1. Microemulsions and their molecular organization

Isotropic, single-phase solutions, containing simultaneously large amounts of water, hydrocarbon and surfactant are usually termed „microemulsions"[2]). The nomenclature dates back to the classic work by Schulman and co-workers, who found that clear solutions could under certain conditions form upon addition of a "cosurfactant" (usually a medium-chain alcohol) to a dispersion of oil, water and surfactant. It was suggested at the time that one obtains a very stable emulsion with small drops. This, in our opinion unfortunate, emulsion-like conception of the molecular organization has since served as a starting point for structural characterizations of microemulsions. The most prevalent picture of microemulsions includes concepts such as oil (or water) droplets, a palisade layer of surfactant (and alcohol, if appropriate) in a continuous water (or oil) phase [2, 3].

Short-chain surfactants remain „nonorganized" up to relatively high concentrations in water and many types of simultaneously water- and hydrocarbon-soluble compounds exist (e. g. alkylamides, substituted aromatics and short-chain alcohols). It is also well known that isotropic solutions can form in systems like isopropanol-water-chloroform over large concentration intervals, even without surfactant. There is certainly no need to invoke extensive molecular organiza-

*) Lecture given at the 31th Conference of the Kolloid-Gesellschaft, Bayreuth October 11–14, 1983.

[2]) A definition of the distinction between emulsions and microemulsions has recently been proposed [1]: A microemulsion is a single phase composed of at least water, oil and surfactant, and this phase is an optically isotropic and thermodynamically stable solution. We will adhere to that definition (which excludes real emulsions, multi-phase systems and surfactantless systems) throughout the present paper.

KT-L 804

tion to explain the phase behaviour of such simple systems. On the other hand, it is certainly not justified to neglect specific intermolecular interactions either, necessarily leading to some molecular organization even in simple solutions.

2. The scope of our NMR investigations

We, and our collaborators, have investigated the structural aspects of several microemulsion types with two principally very different, NMR-based, techniques; multicomponent self-diffusion measurements (here monitoring all total molecular displacements during a period of approximately 200 milliseconds) and ^{13}C spin relaxation measurements (monitoring the dynamic processes on the nano-second to pico-second timescale and the residual molecular organization during typically 100 pico-seconds).

Since the time-scales as well as the quantities measured by spin relaxation and self-diffusion measurements are widely different, the two techniques provide complementary information for the structural and dynamic characterization of microemulsions.

2.1 NMR-based self-diffusion measurements

With the field-gradient spin-echo technique, NMR spectroscopy offers a very general approach towards the study of molecular self-diffusion. In its original form, the principles of this NMR method date back to a classic paper by Hahn in the early fifties [4]. The basic field-gradient spin-echo technique for the determination of molecular self-diffusion coefficients (or more generally, lateral displacement of matter (also flow or geometrically restricted diffusion) has been improved several times; the pulsed field gradient (PFG or PGSE) modification [5] and the Fourier transform (FT) modification of that approach [6] are the most important steps.

Unlike spin-relaxation data (which primarily quantify reorientational processes, ***vide supra***), self-diffusion coefficients characterize the lateral aspects of molecular motion, and do not depend on a model framework for their interpretation; a purely geometric displacement (in the laboratory frame) is monitored. Multicomponent self-diffusion data can be shown to characterize directly many aspects of molecular aggregation or ion binding phenomena and are thus highly desirable experimental quantities.

With regard to experimental aspects, it turns out that the original unresolved PGSE technique is rarely useful in systems with more than one component, since even in favourable cases, considerable difficulties arise with regard to signal assignments and data evaluation procedures. Although straightforward as a concept in modern FT-based pulsed NMR [6], and demonstrated in principle in 1973 [7], the frequency-resolved Fourier transform PGSE technique has only recently been developed into a practical tool [8–11] and extensively applied to physico-chemical problems.

In very convenient and rapid experiments it is now possible to monitor individual self-diffusion coefficients in multicomponent solutions (e. g. microemulsions). Figure 1 illustrates a typical measurement series on a cosurfactant-type microemulsion.

2.2 Spin relaxation measurements

Spin relaxation data contain for many nuclei easily quantifiable information with regard to molecular reorientation processes in solution. With Fourier transform techniques [6], the application field of the methodology was considerably extended and numerous applications of spin relaxation measurements have been presented during the last decades. Extensive developments in spin relaxation theory have also been made, now providing a technique with a very solid theoretical foundation.

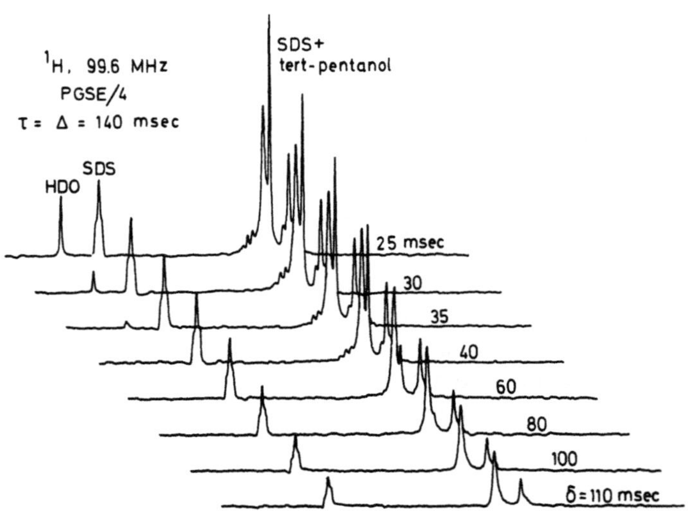

Fig. 1. A sequence of 99.6 MHz pulsed-gradient spin-echo proton NMR spectra on a dilute sodium dodecylsulphate (SDS)/D_2O/tert-pentanol/tetramethylsilane (TMS) sample as a function of the magnetic field gradient duration, δ. Peak heights ($A(i)$) directly reflect self-diffusion coefficients of the individual molecular types according to the relation $A(i) \alpha \exp[-(\gamma G\delta)^2 D(i)(\Delta - \delta/3)]$, where γ, G and Δ are constants of the experiment. See ref. [8–11] for details of the experimental procedures (from ref. [11])

Proper application of spin-relaxation based NMR methods for the study of colloidal and other organized systems is more demanding than for normal molecular solutions, the main reason being that the relatively simple theory and mode of interpretation established for simple systems is not directly applicable to aggregates of dimensions of a nanometer or above. Failure to recognize this has led to many misinterpretations and artifacts in work based on observed NMR parameters in macromolecular and surfactant systems. On the other hand, accounting properly for the influence of a colloidal framework, the potential information obtainable from spin relaxation data is often more complete than for simple systems.

3. Self-diffusion behaviour in surfactant solutions

3.1 Micellar constituents

The characteristic constituent self-diffusion behaviour of an ionic surfactant upon micellization is illustrated in figure 2 a. To a good approximation all individual ion diffusion data obey a (free-micellarly bound) two-site model:

$$D_{obs} = p\, D_{micelle} + (1-p)\, D_{free} \quad (1)$$

in this concentration range and directly provide complete information of the aggregation behaviour of the surfactant. Figure 2b illlustrates the evaluated partitioning of the amphiphile and counterions, as well as the degree of counterion binding.

3.2 Micellar solubilizates

The incorporation of, or binding of, small molecules to supramolecular aggregates in solution can conveniently be monitored by Fourier transform NMR self-diffusion measurements due to the large effect of the binding on the time-averaged self-diffusion coefficient of the substrate, according to the relation given in sub-section 3.1. In a distinct water-micelle situation, a completely solubilized compound assumes a self-diffusion rate equal to that of the micelle itself. Partial solubilization, on the other hand, will manifest itself in time-averaged self-diffusion coefficients of the solubilizate in between those of the micelle and the free molecule (dissolved in water), providing a new and convenient method for the study of solubilization processes and similar phenomena [13, 10, 11]. A two-site model for the time-averaged molecular transport in solution (see preceding subsection) makes possible a direct evaluation of experimental self-diffu-

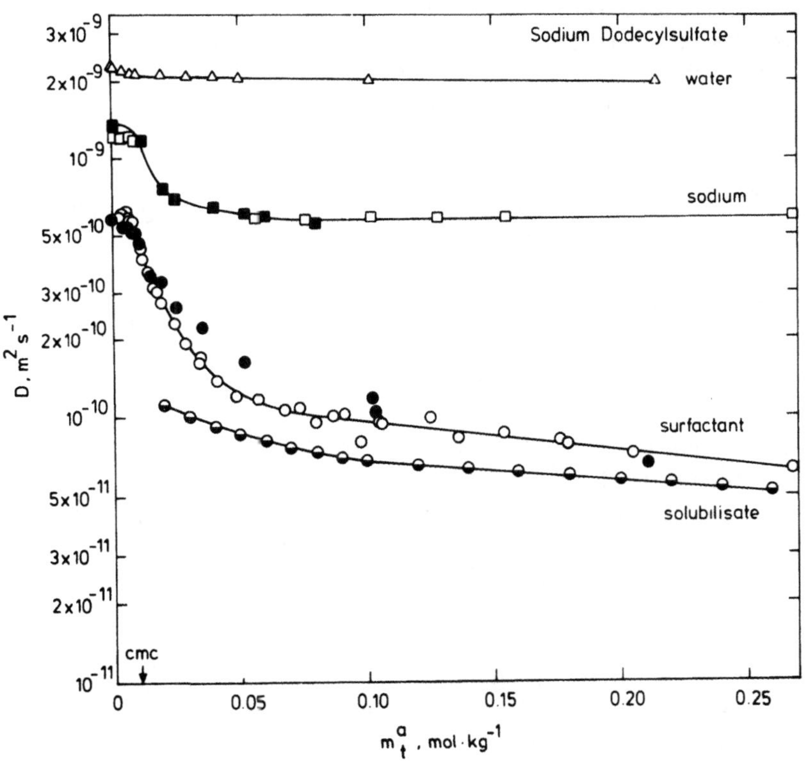

Fig. 2a. Observed ionic constituent self-diffusion data in the sodium dodecylsulphate (SDS) system at 25 °C. (From ref. [12])

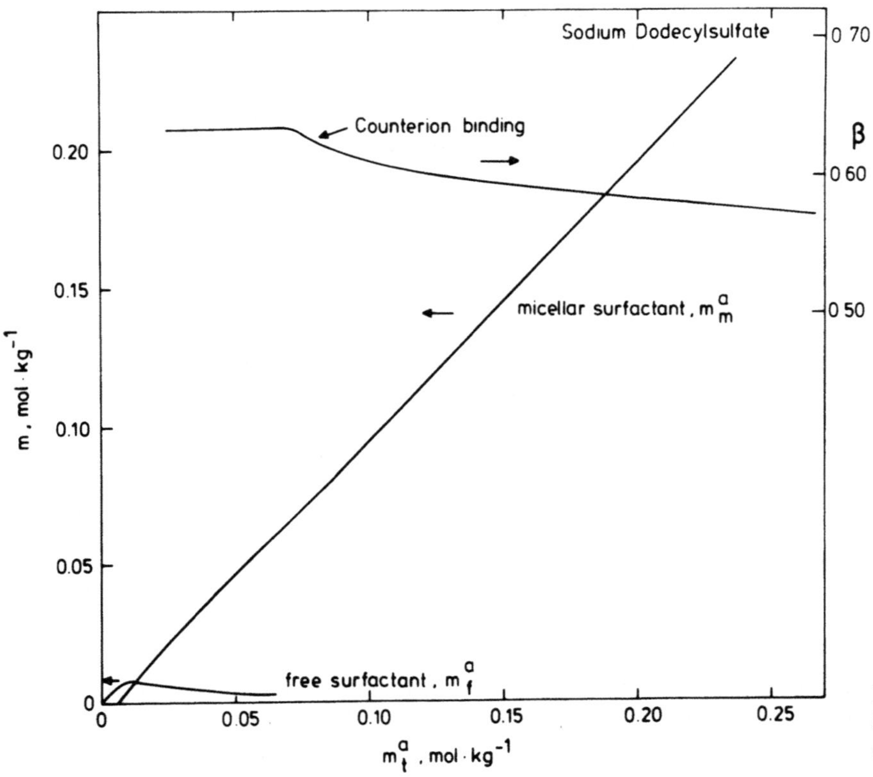

Fig. 2b. Evaluated partitioning of the SDS ionic constituents between the micellar and aqueous phases. (From ref. [12])

sion coefficients in terms of the degree of binding, p. This quantity has a direct physico-chemical meaning as such, but can also be transformed into an equilibrium constant of some desired form. Figure 1 above illustrates a typical experimental set of spectra, recorded within about 10 minutes, leading to direct information on the solubilizate partitioning.

3.3 *Micellar breakdown by solubilizates*

Solubilizates necessarily perturb the basic micellar structure and may, at elevated concentrations, ultimately induce phase transitions. Lamellar mesophases typically form in water-surfactant systems upon addition of medium to long-chain alcohols to a surfactant solution in water.

In a recent paper we demonstrated that a dramatically enhanced self-diffusion rate is observed for trace amounts of solubilized hydrocarbon and for the surfactant itself, upon addition of alcohols to a micellar surfactant-water solution [14]. This indicates that a breakdown of the micellar structure has occurred. Butanol was found to be the most effective "structure-breaker", as illustrated in figure 3. As regards a comparison with the shorter-chain alcohols methanol-propanol, the lower degree of solubilization of these accounts for the relatively lower de-stabilization effect of these compounds.

The actual effect causing the "micellar breakdown" upon the incorporation of alcohols into the solution (most probably into the micelle-water interface) is the

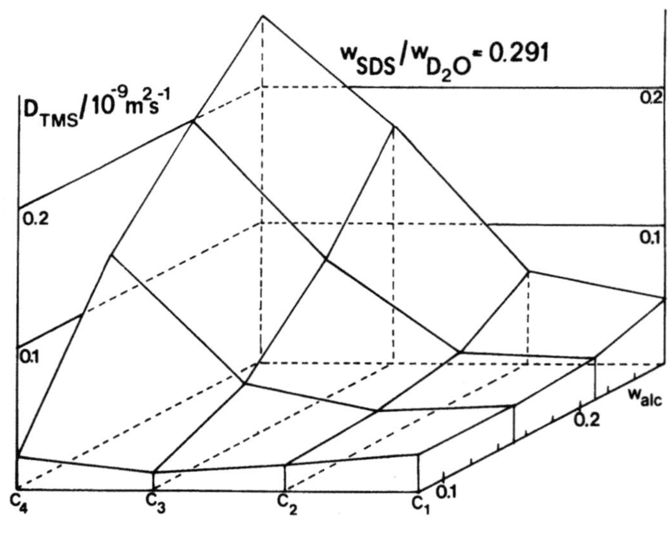

Fig. 3. Self-diffusion behaviour of trace amounts of TMS (tetramethylsilane), as a function of alcohol chain length and alcohol weight fraction in concentrated sodium dodecyl sulphate (SDS)-D_2O solutions. (From ref. [14])

reduction of the barrier for micellar deformation and disruption. Alcohols of this short-to-medium chain type are rather evenly distributed between polar water domains and the micellar pseudophase (as indicated by the solubilizate partitioning studies mentioned in the preceding subsection and independent investigations by other methods), and the effect behind the altered interfacial properties of the micelle-water region is easy to visualize.

As discussed below, a similar effect with a maximum structure-breaking effect for butanol is observed in inverted micellar systems (hydrocarbon-rich cosurfactant microemulsions).

Higher alcohols, on the other hand, cause a mesophase transition to a lamellar phase already at low relative concentrations in micellar systems and are not amenable for a direct comparison. As described below, however, long-chain alcohols do not exhibit such structure-breaking properties, but rather stabilize micelles.

3.4 *Self-diffusion in microemulsions*

According to our investigations, different "microemulsions" exhibit very different self-diffusion characteristics. A wide range of cases are inconsistent with a "closed droplet" model. Figure 4 illustrates the "water-in oil"-type component self-diffusion, regardless of composition, within the large isotropic solution area of an Aerosol OT-water-hydrocarbon system.

The trends in the self-diffusion behaviour in different regions of four-component cosurfactant microemulsions point toward some of the factors governing the phase structure. Referring to figures 5a – d and 6a – d and the displayed self-diffusion-data, one finds a typical aqueous micellar (O/W) behaviour near the "water corner", a rather structureless situation near the butanol-surfactant corner and an inverted micellar near the hydrocarbon-rich corner of the microemulsion region. These figures also strikingly demonstrate the absence of any tendency to a distinct transition at some point from a "normal" (O/W) to an "inverted" (W/O) structural situation. Rather one finds a continuous change from normal micelles, through what would appear to be disrupted micelles to an inverted micellar organization.

The alcohol cosurfactant chain length has a very significant effect on the self-diffusion characteristics of cosurfactant microemulsions. We have systematically investigated a number of systems at constant weight fraction composition, while varying the alcohol cosurfactant chain length. A typical result is illustrated in figure 7. Minimal component segregation manifestations are observed for butanol systems.

The markedly decreasing water self-diffusion with increasing alcohol chain length is also evident and is indicative of more and more distinct aggregates of a discontinuous "water in oil"-type. Microemulsion systems based on higher alcohols can be concluded to be structurally resemblant to inverted micellar systems in this composition range.

4. Spin relaxation in surfactant solutions

4.1 *Micellar solutions*

Wennerström and co-workers have recently [18, 19] demonstrated that spin relaxation in micellar systems may be very significantly influenced by the overall tumbling of the micelles, the rate of which is not far from the Larmor frequency of the NMR experi-

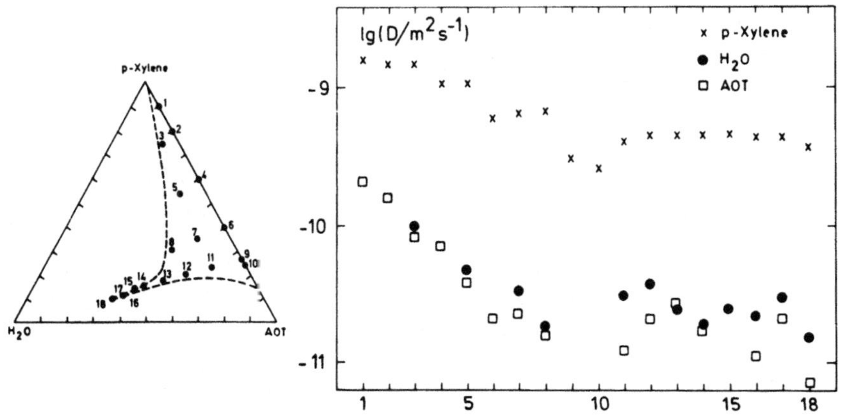

Fig. 4. Observed self-diffusion data in the para-xylene-Aerosol OT-water microemulsion system at 25 °C (weight fractions). (From ref. [15])

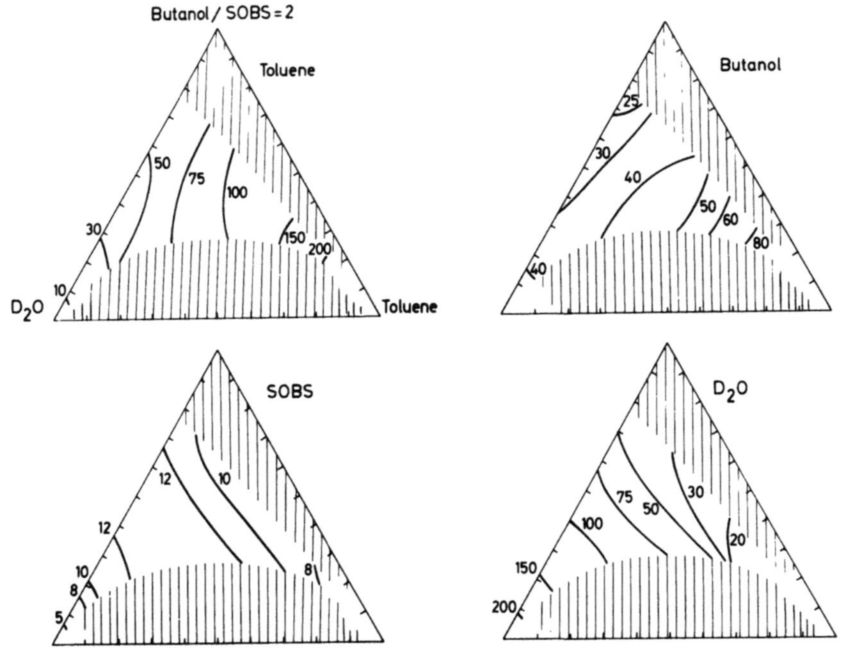

Fig. 5a–d. Observed iso-diffusion lines within the isotropic solution region of the toluene "Lund project" system (self-diffusion coefficients displayed in units of 10^{-11} m^2s^{-1}). The contour lines are based on interpolated values between measured self-diffusion data at approximately 20–30 evenly spaced points within the non-shaded area. Based on data mainly from ref. [17]

ment (ω). The extent of this relaxation effect is related to the degree of ordering of the molecular fragment in question, and can be related to first approximation to the reduced spectral density of motion at the Larmor frequency ($\tilde{J}(\omega)$) by a relation of the form:

$$\tilde{J}(\omega) = S^2\,(2\tau_s/(1 + (\omega\tau_s)^2)) + (1 - S^2)\,(2\tau_f/(1 + (\omega\tau_f)^2)) \qquad (2)$$

where S is the usual Saupe order parameter (here referred to a local director oriented perpendicularly to the micellar surface) and τ_s and τ_f represent the ("slow") correlation time for reorientation of the micelle and the local ("fast") effective correlation time, respectively. (The denominator in the second term usually approximates to 1 under typical conditions in micellar systems). This approach (based on a so-called two-step model of spin relaxation) has found support

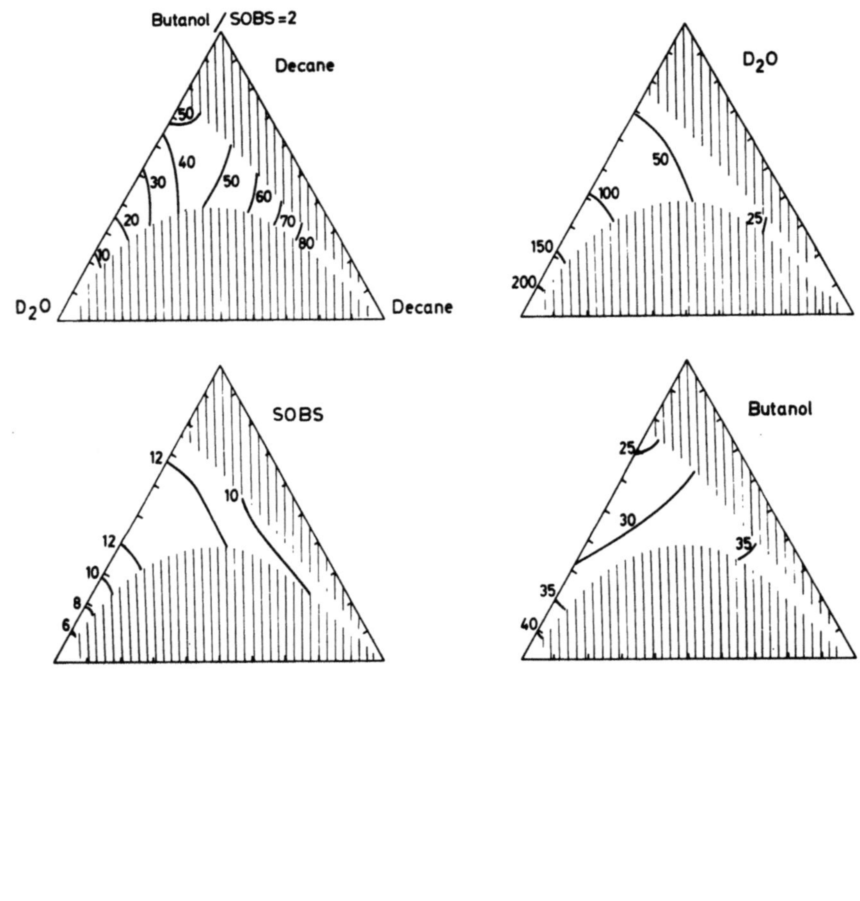

Fig. 6 a–d. The same information as in Fig. 5 a–d, but with decane as hydrocarbon constitutent

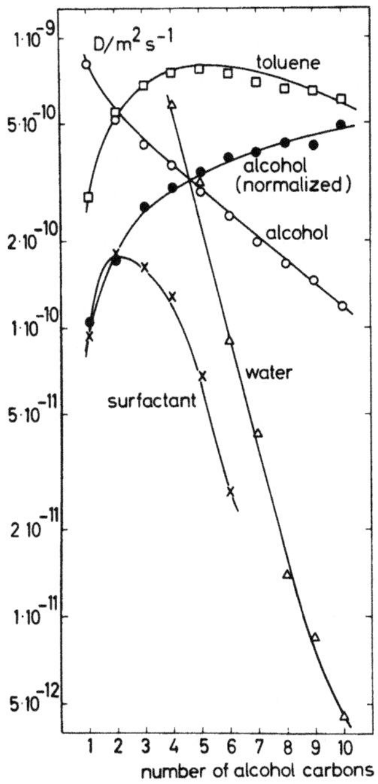

Fig. 7. Self-diffusion data obtained with the FT NMR method on a series of alcohol-sodium dodecyl sulfate-toluene-water systems at a constant (35.0 : 17.5 : 12.5 : 35.0) weight fraction composition. (From ref. [16–17])

in recent experimental studies, where a very significant magnetic field dependence of the carbon-13 relaxation of the alkyl chain carbons of surfactant micelles was demonstrated, despite being in the typical "extreme narrowing" range as judged from the the magnitude of the relaxation rates alone. An analysis of these field-dependent relaxation data makes possible a detailed insight into ordering and alkyl chain dynamics [18, 20] in micelles. Figure 8 illustrates some experimental data from reference [20]. It is seen that a complete picture of the micellar order and dynamics is obtained through an analysis of the spin relaxation data according to equation (2).

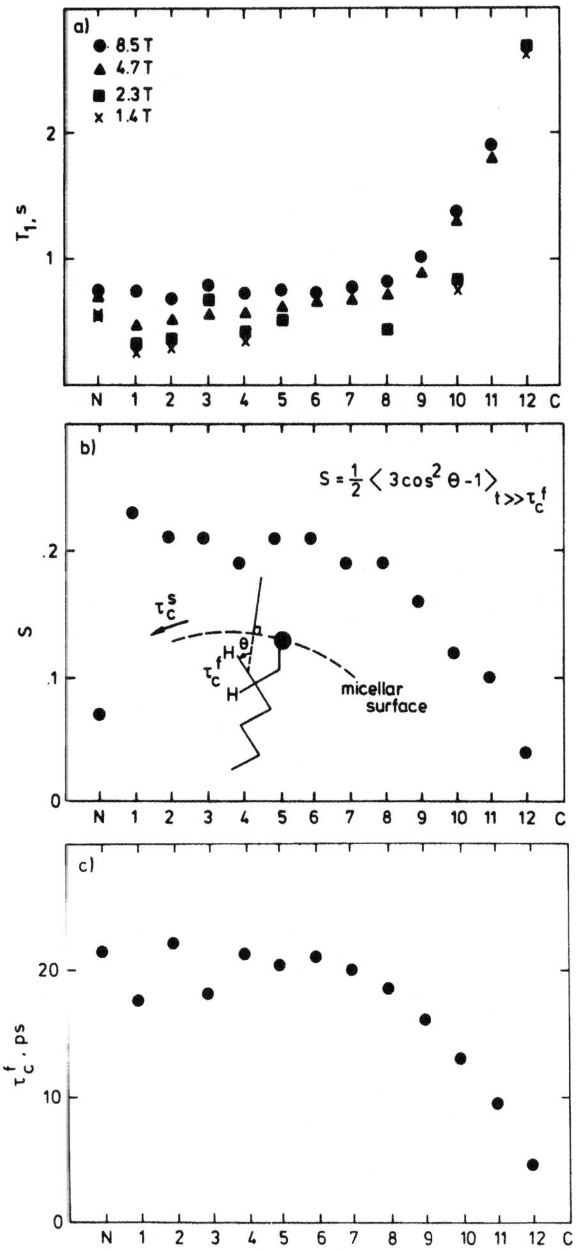

Fig. 8. Carbon-13 nuclear spin relaxation data at different magnetic fields for dodecyl trimethylammonium chloride micelles at 27 °C, together with data from an analysis, in terms of a two-step relaxation model involving the local chain dynamics, chain ordering in the micelle and the (common) overall micellar reorientation rate (data mainly from ref. [20]). The correlation time for the latter process was determined to be 2 nanoseconds

4.2 *Microemulsions*

Bellocq et al. in a ^{1}H-based spin-lattice relaxation study of the butanol-toluene-SDS-water system concluded that all observed spin relaxation data displayed very regular trends with concentration for all components, and suggested on this basis that one most probably has a continuous transition between different modes of aggregation in this system [21]. While this conclusion is most probably correct, the organizational and dynamic information in proton spin relaxation data is difficult to quantify, being affected by intramolecular (reorientational) processes as well as intermolecular processes (related to both reorientation and translational diffusion in the system). In addition, the colloidal framework (if existent) must be accounted for, as seen in the preceding subsection.

For investigations on the structure and dynamics it is necessary to turn to either ^{2}H or ^{13}C spin relaxation, which are overwhelmingly intramolecular in origin. Ahlnäs and co-workers [22, 17] have started studies of the spin relaxation behaviour in microemulsions, utilizing the model framework described in subsection 4.1 as a basis. These investigations demonstrated that, like in micellar systems, spin relaxation in microemulsions is magnetic field dependent, and thus that some form of "slow reorientation" related to supramolecular aggregation in the system must occur. In quantifying these data for cosurfactant microemulsions one finds local chain mobilities and order parameters of alcohol, hydrocarbon and surfactant similar to those observed in micellar systems, containing solubilizates of that kind. The "slow" correlation time from the data analysis (related to the overall aggregate reorientation process, possibly modulated through combination with exchange processes on the nano-second timescale (if existent)) indicated that the aggregation leads to entities smaller than the micelles formed by the surfactant component itself in aqueous solution.

5. Conclusions

One direct conclusion of the multi-component self-diffusion work is that microemulsions, i. e. stable and isotropic solutions of surfactant, water and hydrocarbon, can, dependent on components and composition, (and certainly also temperature) adopt widely different structures. In a number of cases, the microemulsions are distinctly of the droplet type structure (O/W or W/O, dependent on system) over essentially the whole region of existence of the microemulsion phase. For four-component microemulsions with a medium-chain alcohol like butanol or pentanol as cosurfactant the structure approaches the O/W type at high water contents and the W/O type at high hydrocarbon contents. Over most of the stability range of the microemulsion phase, however, the observations are inconsistent with the confinement of either oil or water to closed domains. The bicontinuous behaviour manifested in the molecular translational mobilities can either be explained by special three-dimensional structures with bicontinuous characteristics or by an inapplicability of the conventional assumption for surfactant systems of a distinct separation into hydrophobic and hydrophilic domains. Indeed, from the rather even partitioning of cosurfactant between micelles and the intermicellar solution and the disruption of micelles caused by the cosurfactant, these microemulsions are over wide concentration ranges expected not to be far from a rather structureless situation applicable to simple solutions. The strong dependence of microemulsion structure on alcohol chain length should be noted; a long-chain alcohol giving a distinct droplet-like behaviour (or, alternatively, a liquid crystalline, generally lamellar, phase).

Effectively bicontinuous structures have previously *inter alia* been discussed by Shah et al. [23, 24], by Friberg [25] by Scriven [26–29] and by Auvray [29]. As an example, Shah et al. [23, 24] noted significant structural differences between pentanol and hexanol microemulsions from *inter alia* electrical conductivities, ^{1}H NMR chemical shifts and dielectric relaxation studies. Particularly striking are the water proton chemical shifts; hexanol systems exhibit shifts which are nearly independent of water-to-oil ratio, while with pentanol there is a strong variation. Shah et al. inferred the pentanol system to be similar to a molecular solution, while the hexanol-based system should have a W/O-like structure over a wide composition range.

Scriven [26] examined different periodic minimal surfaces giving rise to bicontinuous structures, investigations which led to a suggestion that submicroscopic bicontinuous structures of a dynamic nature may exist in microemulsions.

As regards the extension and lifetimes of aggregates in microemulsions it is hoped that current multi-field NMR spin relaxation work (mainly ^{13}C and ^{2}H) will be informative [17]. It is very significant that over wide concentration ranges in a butanol cosurfactant system (sodium octylbenzene sulfonate-butanol-toluene-water), ^{13}C T_1 relaxation times lead to a limit of the "slow motion" (see section 4.1) of the order of a few nano-seconds. This would correspond to the reorientational motion of quite small aggregates; i. e. having a radius less than that of a typical extended surfactant molecule. If long-range aggregation beyond that limit is significant, it must then have a very short lifetime or the surfactant residence time must be short.

The combined use of multi-component Fourier transform NMR self-diffusion and spin relaxation is apparently an interesting tool for the characterization of microemulsion structure on different time-scales.

Acknowledgements

It is a pleasure to acknowledge the stimulating collaboration during the past years with our co-authors appearing in the references. This work has been financially supported by the Swedish Natural Sciences Research Council.

References

1. Danielsson I, Lindman B (1981) Colloids and Surfaces, 3:391
2. Microemulsions (1977) (ed) Prince LM, Academic Press, New York
3. Microemulsions (1982) (ed) Robb I, Accademic Press, New York
4. Hahn EL (1950) Phys Rev, 80:580
5. Stejskal EO, Tanner JE (1965) J Chem Phys 42:288
6. Vold RL, Waugh JS, Klein MP, Phelps DE (1968) J Chem Phys, 48:3831
7. James TL, McDonald GG (1973) J Magn Reson 11:58
8. Stilbs P, Moseley ME (1980) Chem Scr, 15:176
9. Stilbs P, Moseley ME (1980) Chem Scr, 15:215
10. Stilbs P (1982) J Colloid Interface Sci, 87:385
11. Stilbs P (1983) J Colloid Interface Sci, 94:643
12. Lindman B, Kamenka N, Puyal MC, Rymdén R, Stilbs P, J Pyhs Chem, in press
13. Stilbs P (1981) J Colloid Interface Sci, 80:608
14. Stilbs P (1982) J Colloid Interface Sci, 89:547
15. Lindman B, Stilbs P, Moseley ME (1981) J Colloid Interface Sci, 83:569
16. Stilbs P, Rapacki K, Lindman B (1983) J Colloid Interface Sci, 95:583
17. Lindman B, Ahlnäs T, Rapacki K, Söderman O, Stilbs P, Walderhaug H (1983) Faraday Discussions (Concentrated Colloidal Dispersions) 76:317
18. Wennerström H, Lindman B, Söderman O, Drakenberg T, Rosenholm JB (1979) J Am Chem Soc, 101:6860
19. Halle B, Wennerström H (1981) J Chem Phys, 75:1928
20. Walderhaug H, Söderman O, Stilbs P (1984) J Phys Chem 88:1655
21. Bellocq AM, Biais J, Clin B, Lalanne P, Lemanceau B (1979) J Colloid Interface Sci, 70:524
22. Ahlnäs T, Söderman O, Hjelm C, Lindman B (1983) J Phys Chem, 87:822
23. Shah DO, Walker Jr RO, Hsieh WC, Shah NJ, Dwivedi S, Nelander J, Papinsky R, Deamer DW (1976) Society of Petroleum Engineers of AIME, paper 5815:243
24. Bansal VK, Chinnaswamy K, Ramachandran C, Shah DO (1979) J Colloid Interface Sci, 72:524
25. Friberg S, Lapczynska I, Gillberg G (1976) J Colloid Interface Sci, 56:19
26. Scriven LE (1977) (ed) Mittal KL, in Micellization, Solubilization and Microemulsions, Plenum, New York, 2:877
27. Kaler EW, Bennet KE, Davis HT, Scriven LE (1983) J Chem Phys 79:5673
28. Kaler EW, Davis HT, Scriven LE (1983) J Chem Phys 79:5685
29. Auvray L, Cotton JP, Ober R, Taupin C, J Physique, in press

Authors' address:

P. Stilbs
Institute of Physical Chemistry
Uppsala University
P.O.B. 532
S-751 21 Uppsala, Sweden

 Progr Colloid & Polymer Sci 69:48–55 (1984)

Zur Kenntnis von 3-Komponenten-Mikroemulsionsgelen*)

3. Mitteilung: Vergleichende Untersuchungen von Mikroemulsionsgelen und verwandten Systemen

E. Nürnberg und W. Pohler

Institut für Pharmazie und Lebensmittelchemie der Friedrich-Alexander-Universität Erlangen-Nürnberg
– Lehrstuhl für Pharmazeutische Technologie –

Zusammenfassung: Die isotrope Gelphase eines Dreikomponentensystems aus Polyoxyethylen(10)oleylether (Brij 96®), flüssigem Paraffin und Wasser – die sog. Mikroemulsionsgele – lassen sich durch verschiedene physiko-chemische Untersuchungsverfahren von anderen Einphasensystemen abgrenzen. Geeignet sind rheologische und differentialthermoanalytische sowie konduktometrische Messungen. Diese Methoden ermöglichen, ebenso wie röntgendiffraktometrische Untersuchungen im Niederwinkelbereich, eine Abgrenzung von lamellaren und hexagonalen Phasen.

Abstract: The isotropic gelphase of a three-component-system composed of polyoxyethylene(10)oleyether (Brij 96®), mineral oil and water – the so called microemulsion gels – can be differentiated from other one-phase-systems by means of different physicochemical investigations e. g. rheological, differential calorimetric and conductometric measurements. These methods as well as small-angle X-ray diffraction studies allow the distinction from hexagonal and lamellar phases.

Schlüsselwörter: Mikroemulsionsgele, Tenside, Isotrope Gele, Differentialthermoanalyse von Tensidgelen, Physikalische Eigenschaften von Oleylalkohol x 10 EO / Wasser-Paraffin-Systeme.

1. Einführung

In früheren systematischen Versuchen wurden die Phasengebiete ternärer Gemische aus polyoxyethylierten-10-oleylether/Wasser/Paraffin ermittelt und in einem Dreikomponentendiagramm graphisch dargestellt (s. Abb. 1) [1, 2].

Man ersieht daraus, daß unter Verwendung von drei Komponenten (Tensid, Wasser und flüssigem Paraffin) Einphasensysteme und Mehrphasensysteme resultieren. Zu ersteren gehören die isotrope Gelphase (G), das Existenzgebiet der sog. Mikroemulsionsgele. Es handelt sich dabei nicht um ein Emulsionssystem, sondern um ein isotropes Einphasengebiet, so daß die Bezeichnung Mikroemulsionsgele nicht zutreffend ist.

*) Vortrag, gehalten auf der 31. Hauptversammlung der Kolloid-Gesellschaft, Bayreuth 11. bis 14. Oktober 1983

KT-W 792

Aufgrund dieser Versuchsergebnisse sollten weitergehende Prüfungen zur Charakterisierung der betreffenden Systeme durchgeführt werden.

2. Untersuchungen der auftretenden Mesophasen

Zur Bestätigung und Ergänzung der bisherigen Untersuchungsergebnisse [2] mußten weitere physikalische Methoden herangezogen werden, die eine Charakterisierung und Differenzierung der einzelnen Mesophasen ermöglichen sollten.

2.1 Rheologische Untersuchungen

Unter Einsatz eines Rotationsviskosimeters mit Platte-Kegel-Meßeinrichtung wurden die Fließkurven jeweils eines typischen Vertreters der isotropen Gelphase (G), der Hexagonalphase (M) und der Lamellarphase (N) bestimmt.

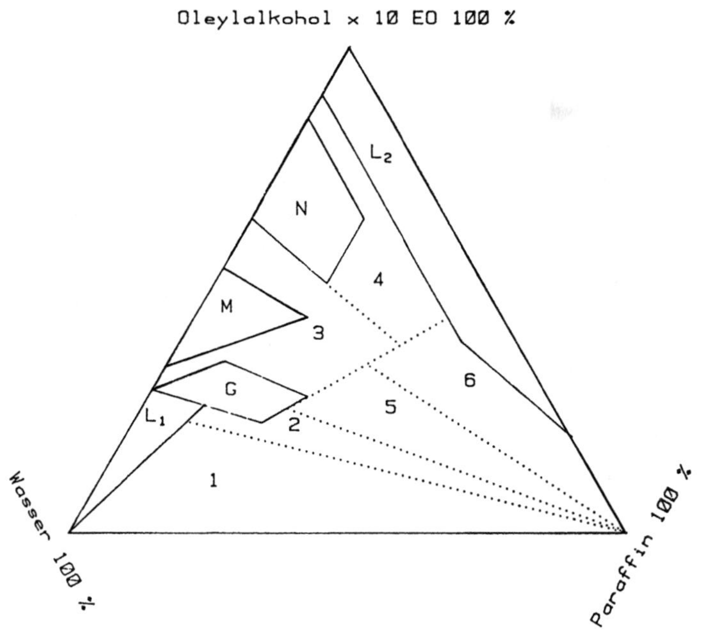

Abb. 1. Phasengebiete beim ternären Gemisch Oleylalkohol x 10 EO/Wasser/Paraffin

Einphasengebiete

L_1: isotrope wässrige Phase
L_2: isotrope ölige Phase
G: isotrope Gelphase
M: hexagonale Phase
N: lamellare Phase

Mehrphasengebiete

1: flüssige Emulsionen
2: Cremes mit ambiphilem Erscheinungsbild
3: zähe anisotrope Gele
4: weiche anisotrope Gele
5: O/W-Cremes
6: zweiphasig-ölige Produkte

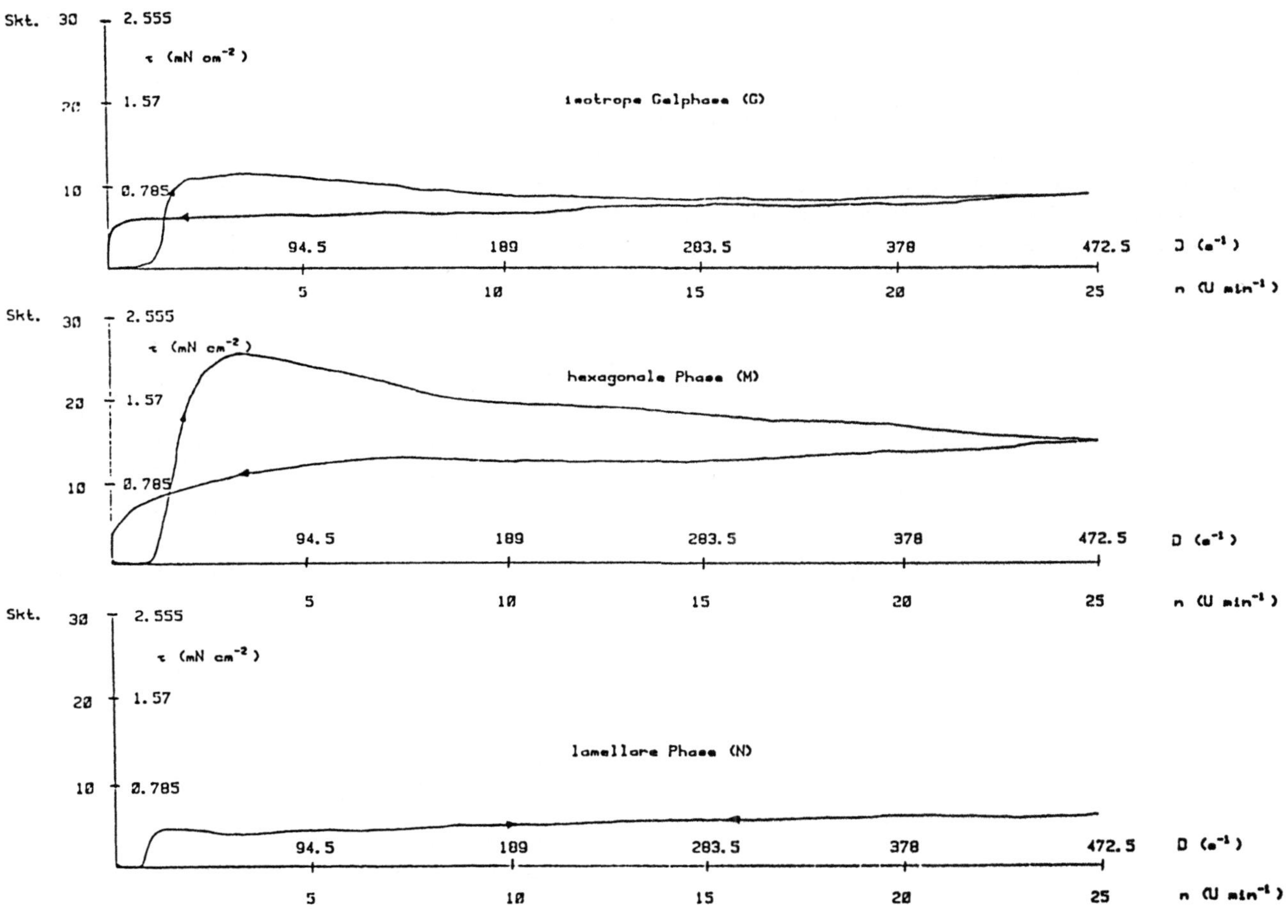

Abb. 2. Fließkurven unterschiedlicher Mesophasen

Die vermessenen Proben hatten folgende Zusammensetzung:

	(N)	(M)	(G)
Oleylalkohol x 10 EO (Brij® 96)	68,6 T	42,8 T	30,0 T
dickfl. Paraffin	11,4 T	7,2 T	5,0 T
Wasser	20,0 T	50,0 T	65,0 T

Wie aus Abbildung 2 zu ersehen ist, können die verschiedenen Mesophasen rheologisch klar differenziert werden.

Die *lamellare Phase N* ist durch ihr *pseudoplastisches* Fließverhalten mit niedriger Fließgrenze und geringer Viskosität gekennzeichnet. Im Gegensatz dazu weisen die *hexagonale M* und die *isotrope Gelphase G thixotropes* Fließverhalten auf und besitzen damit ein wesentliches Kennzeichen von Gelen.

Die hexagonale Phase M zeichnet sich gegenüber der isotropen Gelphase G durch eine wesentlich höhere Fließgrenze und Viskosität sowie stärkere Thixotropie aus.

2.2 *Differentialthermoanalytische Untersuchungen*

Manche thermodynamischen Vorgänge, die nur mit einer geringen Wärmetönung verbunden sind, z. B. Phasen- und Modifikationsumwandlungen, sind nur mit empfindlichen Methoden nachweisbar, wobei eine Differenzmessung zwischen der Probe und einer inerten Referenz unter gleichen Versuchsbedingungen erfolgt. Ein solches Bestimmungsverfahren ist bekanntlich die Differentialthermoanalyse (DTA).

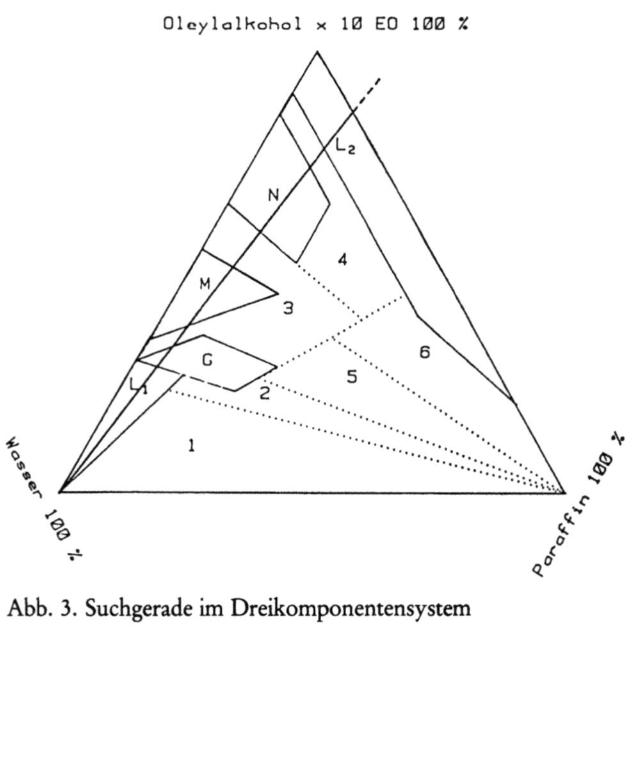

Abb. 3. Suchgerade im Dreikomponentensystem

Ausgangsfrage war, ob die beschriebenen Mesophasen selbst oder möglicherweise auch ihre Mischungen bei der DTA ein unterschiedliches Verhalten zeigen und dadurch gekennzeichnet werden können. Es wurden die reinen Phasen (mizellare Lösung L_1, isotropes Gel G, hexagonale M und lamellare Phase N) sowie ein Mischpräparat im Gebiet 3, in dem die lamellare und hexagonale Phase nebeneinander vorliegen, untersucht.

Die Proben liegen auf einer Suchgeraden, welche alle Einphasengebiete durchquert (s. Abb. 3). Dabei ist das Tensid/Paraffin-Verhältnis konstant 85,7/14,3 und der Wassergehalt variabel.

Diskussion der DTA-Kurven

Abbildung 4 zeigt die DTA-Kurven der untersuchten Proben.

Das *reine Tensid* ist durch einen Schmelzpeak mit Maximum bei 292 K (19 °C) und zwei vorgelagerten Schultern sowie ein kleineres endothermes Signal bei 302 K (29 °C) gekennzeichnet. Unterhalb 292 K (19 °C) liegt ein rein kristalliner Ordnungszustand vor, denn es treten Röntgenreflexe im Weitwinkelbereich auf (s. Abb. 5)

Zwischen 293 und 301 K (20–28 °C) sind im Polarisationsheiztischmikroskop noch anisotrope Nadeln zu erkennen, die beim Signal 302 K (29 °C) schmelzen. In diesem Temperaturintervall sind demnach die Kriterien für flüssigkristalline Zustände erfüllt.

Die Zubereitung aus dem Bereich der *lamellaren Phase* (N) ist durch ein endothermes Signal mit Maximum bei 284 K (11 °C) gekennzeichnet. Hier liegt nicht der Schmelzbereich der lamellaren Struktur, denn im Polarisationsmikroskop sind ihre typischen Texturen auch oberhalb dieser Temperatur zu beobachten. Vielmehr findet ein Übergang vom kristallinen in den flüssigkristallinen Zustand statt. Unterhalb 284 K (11 °C) treten nämlich im Weitwinkelbereich die aus Abbildung 5 bekannten Röntgenreflexe auf. Derartige kristallin-lamellaren Strukturen werden in der Literatur [3] als „crystalline gel phase“ bezeichnet (s. Abb. 6).

Die Kurve aus dem Existenzgebiet der *hexagonalen Phase* (M) zeigt einen scharfen Schmelzpeak bei 335–336 K (62–63 °C).

Die DTA-Kurve des Präparates aus dem *Mischgebiet* [3], in dem sowohl lamellare als auch hexagonale Bereiche – polarisationsmikroskopisch erkennbar –

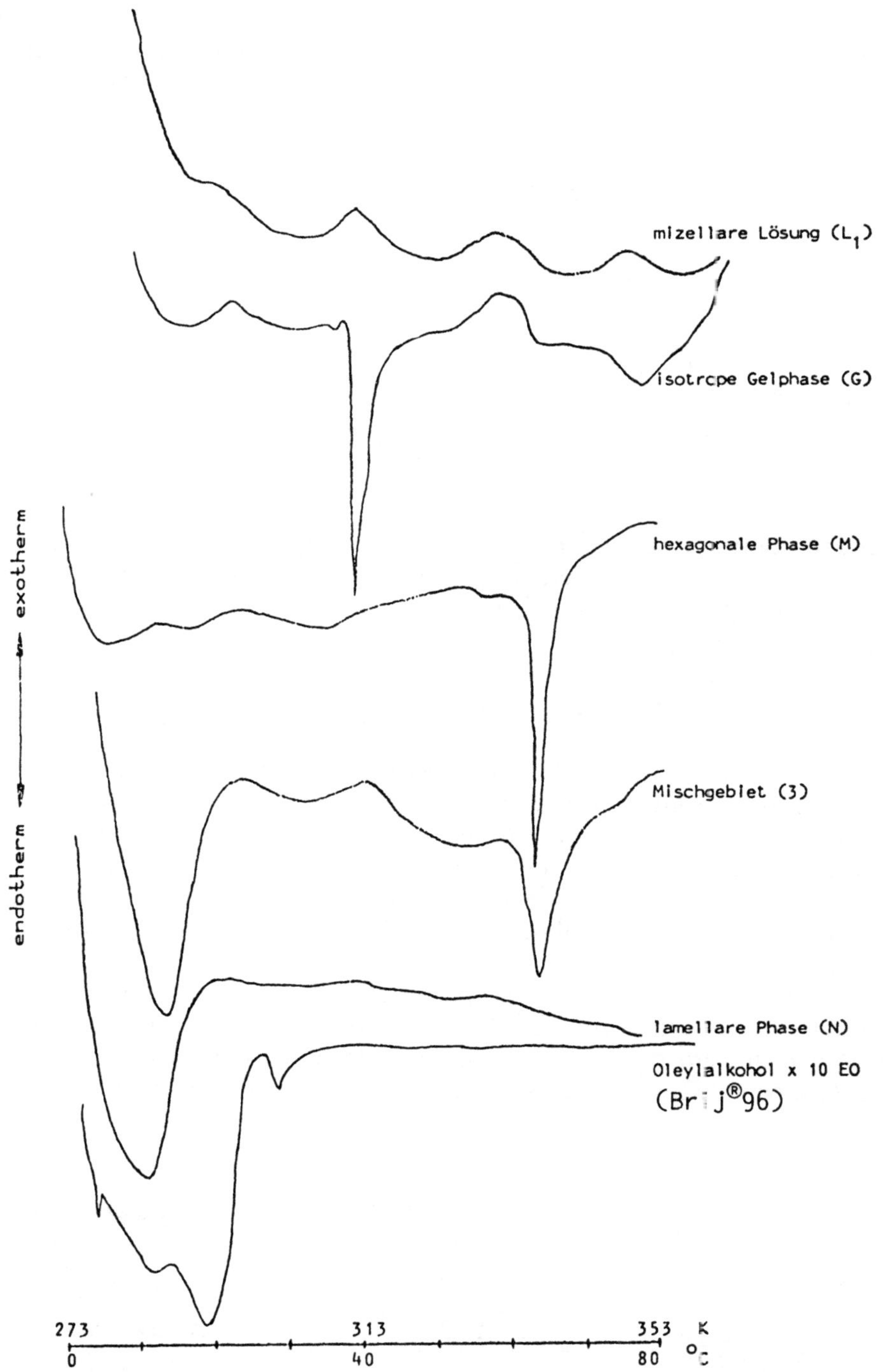

Abb. 4. DTA-Kurven unterschiedlicher Tensidphasen

vorliegen, sind durch die typischen Signale beider Phasen gekennzeichnet. Zusätzlich tritt ein weiterer exothermer Vorgang bei 313 K (40 °C) auf, der polarisationsmikroskopisch als partieller thermotroper Übergang hexagonal/lamellar identifiziert werden konnte.

Im Kurvenverlauf der *isotropen Gelphase* (G) ist ein bei den anderen Systemen nicht beobachteter, scharfer Schmelzpeak bei 310 K (37 °C) erkennbar. Die übrigen, schon bekannten endothermen Signale sind vergleichsweise schwach ausgeprägt. Obwohl auch hier ein exothermer Vorgang bei ca. 329 K(56 °C) auftritt, ist ein Phasenübergang im Polarisationsheiztischmikroskop nur am Rande des Deckglases – also bei Wasserverdunstung – zu beobachten, der sich möglicher-

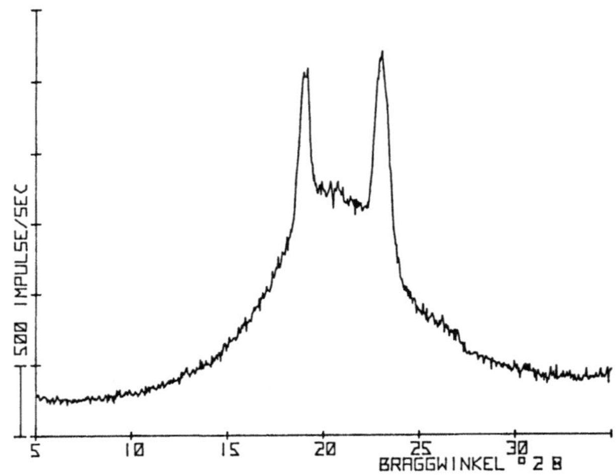

°2 θ	I/Io	d Å
19.15	95.6	4.631
23.05	100.0	3.855

Abb. 5. Röntgendiffraktogramm von Oleylalkohol x 10 EO bei 288 K (15 °C)

weise auch im DTA-Tiegel an der Probenoberfläche abspielt.

Der Kurvenverlauf der *mizellaren Lösung* (L_1) läßt erwartungsgemäß im Gegensatz zu den gelförmigen Produkten keine ausgeprägten Signale erkennen.

Während *Cremesysteme* mit kristallinen Gerüstbildnern noch die DTA-Signale der reinen Emulgatoren erkennen lassen [4, 5], treten bei den hier untersuchten Mesophasen bei Raumtemperatur völlig andere Ordnungszustände auf, die ein unterschiedliches thermisches Verhalten zeigen und differentialthermoanalytisch identifiziert werden können.

Die im Vordergrund des Interesses stehenden Mikroemulsionsgele (G) weisen ein charakteristisches, endothermes Signal bei 308 – 313 K (35–40 °C) auf; dies konnte auch für analoge Systeme mit anderen nichtionischen Komponenten nachgewiesen werden. In allen untersuchten Fällen war bei Mikroemulsionsgelen ein endothermes Signal im Temperaturbereich unterhalb des Schmelzpeaks der hexagonalen Phase erkennbar.

2.3 *Konduktometrische Untersuchungen*

Messungen der spezifischen Leitfähigkeit von Dermatika basieren auf der Beobachtung, daß Wasser als kohärente, äußere Phase den elektrischen Strom leitet und unpolare Phasen isolierend wirken. Auch Unterschiede im Dispersitätsgrad sind u. U. mit dieser Methode feststellbar, wie HOLZNER [6] nachweisen konnte.

Die nachfolgend beschriebenen Untersuchungen sollten klären, ob eine Charakterisierung und Differenzierung der einzelnen Tensidphasen mittels ihrer spezifischen Leitfähigkeit ebenfalls möglich ist.

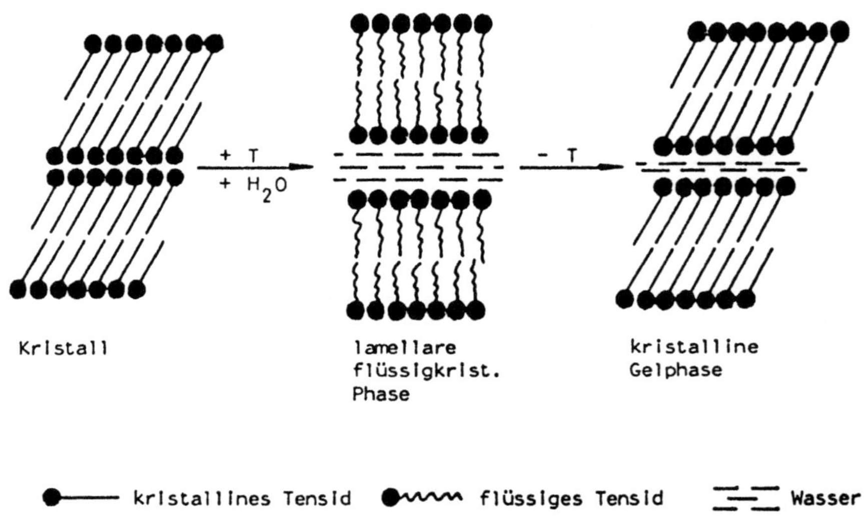

Abb. 6. Schematisches Modell der Bildung lamellarer Mesophasen und kristalliner Gelpahsen

2.3.1 Messung der spezifischen Leitfähigkeit in Abhängigkeit vom Wassergehalt

Zur Untersuchung einer Abhängigkeit der spezifischen Leitfähigkeit vom Wassergehalt wurden Proben hergestellt, die alle auf der in Abbildung 3 dargestellten Suchgeraden liegen. Das Tensid/Paraffin-Verhältnis war konstant, der Wassergehalt wurde von 0 bis 80 % variiert. In Abbildung 7 sind die einen Tag nach Herstellung bei 293 K (20 °C) ermittelten spezifischen Leitfähigkeitswerte gegen den Wassergehalt aufgetragen.

Dabei wies der gelbildende Bereich sprunghafte Übergänge sowohl zu den Zubereitungen mit hohem als auch zu denen mit niedrigem Wassergehalt auf.

Bei Systemen mit *lamellarer Struktur* (N, maximal 35 % Wassergehalt) liegt das Wasser offensichtlich *nicht kohärent*, sondern durch isolierende Lipidschichten getrennt vor. Es ist daher verständlich, daß die gemessenen Leitfähigkeitswerte nicht nennenswert über denen des verwendeten destillierten Wassers (1,7 µS) liegen. Bei 40 % Wassergehalt erfolgt ein abrupter Sprung der Werte, die dann bis 75 % Wassergehalt praktisch linear ansteigen.

In diesem Bereich, der die hexagonale und die isotrope Gelphase (M und G) repräsentiert, verhält sich die Leitfähigkeit offensichtlich proportional zur inkorporierten Wassermenge. Die Systeme (L_1) mit mehr

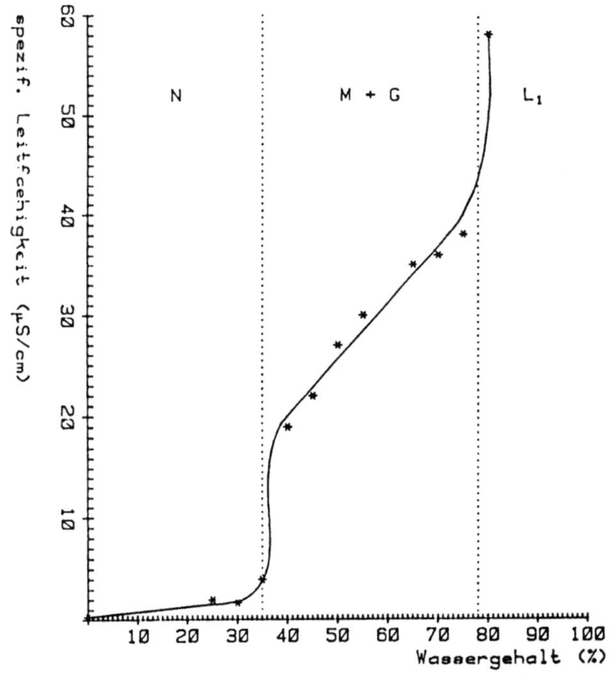

Abb. 7. Spezifische Leitfähigkeit von Tensid-Paraffin-Mischungen in Abhängigkeit vom Wassergehalt

als 75 % Wasser zeigen erneut einen sprunghaften Anstieg der Leitfähigkeitswerte.

Der Kurvenverlauf stellt ein anschauliches Spiegelbild der Einflüsse von Struktur und Viskosität der vermessenen Systeme dar [7]. Während im Bereich der lamellaren Phase (N) die Struktur einer Stromleitung entgegensteht, zeigt sich im Bereich höheren Wassergehalts der *Viskositätsabfall* in einer stetig verminderten Behinderung der Teilchenbeweglichkeit und damit steigender Leitfähigkeit.

Im Grenzbereich der Gelbildung von 65–70 % Wassergehalt macht sich der Struktur- bzw. Viskositätsverlust besonders stark bemerkbar. Hexagonale (M) und isotrope Gelphase (G) können mit dieser Methode jedoch nicht klar voneinander abgegrenzt werden.

2.3.2 Messung der spezifischen Leitfähigkeit in Abhängigkeit von der Temperatur

Wie die differentialthermoanalytischen Untersuchungen gezeigt hatten, weisen die verschiedenen flüssigkristallinen Tensidphasen unterschiedliche Schmelzbereiche auf. Da die strukturbildenden Elemente in diesen Temperaturbereichen kollabieren, erscheint die Frage gerechtfertigt, ob die verschiedenen Phasen auch über den temperaturabhängigen Verlauf der Leitfähigkeit charakterisiert werden können.

Bei den Untersuchungen wurden vier Zubereitungen als typische Vertreter unterschiedlicher Tensidphasen bei Temperaturkompensation vermessen:

- Eine mizellare Lösung L_1
- eine mit hexagonaler Struktur M
- eine mit lamellarer Struktur N
- ein isotropes Gel (Mikroemulsionsgel) G.

Die gelförmigen Proben hatten die gleiche Zusammensetzung, wie die bei den rheologischen Messungen (s. Kap. 2.1) beschriebenen; die mizellare Lösung bestand aus 5 % Tensid in Wasser.

In Abbildung 8 sind die gemessenen Leitfähigkeitswerte über der Temperatur dargestellt.

Der Kurvenverlauf der mizellaren Lösung (L_1) zeigt im Temperaturbereich von 293–318 K (20–45 °C) einen Anstieg der spezifischen Leitfähigkeit mit einem Plateau bei 318–328 K (45–55 °C), dem bei höheren Temperaturen ein starker Abfall folgt. Der erste Teil der Kurve wird durch einen temperaturbedingten Viskositätsabfall geprägt. Im Bereich des Plateaus (318–328 K) liegt die Phaseninversionstemperatur (PIT) des Tensids. Durch die temperaturinduzierte Dehydra-

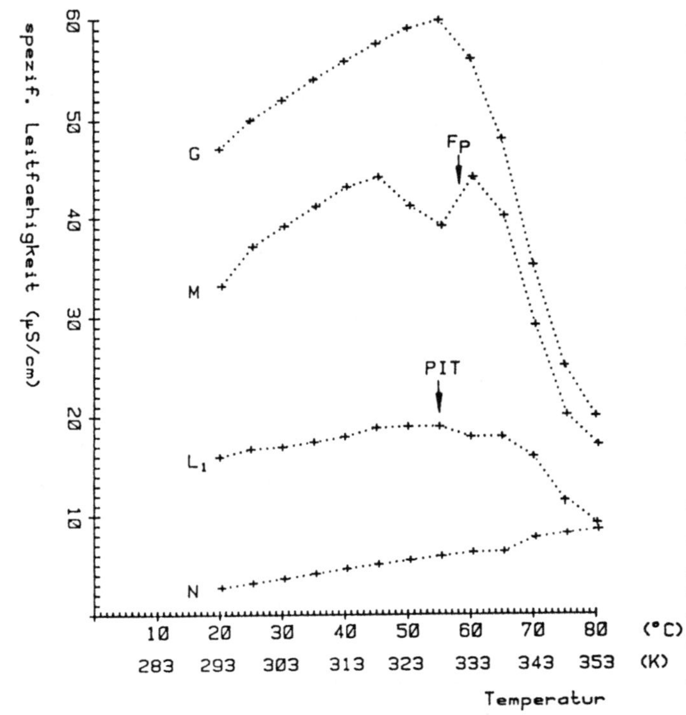

Abb. 8. Spezifische Leitfähigkeit unterschiedlicher Tensidphasen in Abhängigkeit von der Temperatur

tation des hydrophilen Tensidanteils (endothermer Prozeß) nimmt die Löslichkeit des Gesamtmoleküls ab, es fällt – an der eintretenden Trübung erkennbar – aus; die spezifische Leitfähigkeit der Lösung nimmt infolgedessen ab.

Ähnlich verhält sich das *isotrope Gel* (G). Bis zur PIT erfolgt ein stärker ausgeprägter, kontinuierlicher Anstieg der Kurve mit anschießendem rascherem Abfall. Erstaunlicherweise beeinflußt die Gelstruktur des isotropen Systems die Leitfähigkeitswerte nicht, denn ihr Schmelzbereich bei 308–313 K (35–40 °C) macht sich im Kurvenverlauf nicht bemerkbar.

Auch bei der Zubereitung mit *hexagonaler Struktur* (M) ist ein Anstieg und späterer Abfall im Kurvenverlauf erkennbar. Im Unterschied zu den anderen Systemen zeigt sie jedoch bei Temperaturen zwischen 318 und 333 K (45–60 °C) ein abweichendes Verhalten. Wie aus DTA-Untersuchungen und Beobachtungen im Polarisationsheiztischmikroskop hervorgeht, liegt hier der Schmelzbereich des anisotropen Systems.

Der Strukturverlust macht sich hier offensichtlich durch einen Abfall der spezifischen Leitfähigkeit bemerkbar. Nach Abschluß dieses Vorgangs erhält die Kurve wieder den bekannten Verlauf.

Die Zubereitung mit *lamellarer Struktur* (N) zeigt bei Temperaturerhöhung einen kontinuierlichen Anstieg der spezifischen Leitfähigkeit. Hier übt nur der Viskositätsabfall einen Einfluß aus. Da das Tensid in diesem Fall keinen positiven Beitrag zur Leifähigkeit des Systems leistet, macht sich die PIT im Kurvenverlauf nicht bemerkbar.

Wie diese Untersuchungsergebnisse zeigen, ist die Methode zur Differenzierung der verschiedenen Tensidphasen gut geeignet. Dafür zeigen die Kurvenverläufe der mizellaren Lösung (L_1) und des isotropen Gels (G) eine gewissen Analogie, was auf eine Verwandtschaft beider Systeme hinweist.

2.4 *Röntgendiffraktometrische Untersuchungen*

Mit einer Kiessig-Kamera wurden Röntgenkleinwinkelaufnahmen jeweils eines exemplarischen Vertreters der Mikroemulsionsgele (G), der Hexagonalphase (M) und der Lamellarphase (N) durchgeführt. Der Präparat-Filmabstand betrug 420 mm, die Belichtungszeit 48 h. Weiter Einzelheiten können dem Anhang (Kap. 3) entnommen werden.

Die Aufnahmen erbrachten folgendes Resultat:

Die *Lamellarphase (N)* zeigt einen starken Reflex mit einem Durchmesser D = 21,0 mm sowie zwei schwache mit D = 19,5 und 10,5 mm.

Die *Hexagonale Phase M* ist durch vier Interferenzringe gekennzeichnet, von denen der mit 14,1 mm Durchmesser deutlich und die mit 11,7, 23,5 sowie 37,0 mm nur schwach erkennbar sind.

Das *isotrope Gel G* zeigt ebenfalls vier Reflexe mit D = 10,0 14,1 16,0 und 18,6 mm, wobei die beiden äußeren nur geringe Intensitäten aufweisen.

Die Ergebnisse sind in folgender Tabelle zusammengefaßt:

Tabelle 1: Braggsche Abstände der Proben verschiedener Phasenzugehörigkeit

Zusammensetzung (%) Tensid – Wasser – Paraffin			Phase	Braggsche Abstände (Å)
68,6	20	11,4	N	61,7; 67,5; 123,3;
42,8	50	7,2	M	35,1; 55,2; 91,8; 110,78
30	65	5	G	68,9; 80,9; 91,9; 129,5

Im Weitwinkelbereich wiesen die Präparate nur einen breiten, diffusen Reflex (Halo) auf, dessen Maximum ein Braggscher Abstand von 4,5 Å zugeordnet werden kann und der auch beim flüssigen Paraffin auftritt.

Alle hier untersuchten Proben zeigen die charakteristischen Merkmale flüssigkristalliner Zustände, nämlich das Fehlen eines hohen Ordnungsgrades bei größeren Braggwinkeln und das Auftreten im Niederwinkelbereich, denen größere Abstände zuzuordnen sind.

Die Intensität und Anzahl der erhaltenen Reflexe ist jedoch zum Teil relativ gering. Zudem wurden, je nach Positionierung der Kapillaren im Probenhalter (vertikal oder horizontal), Textureffekte beobachtet. Dies könnte auf eine Erwärmung der Proben durch Absorption während der Aufnahme hindeuten. Daher ist die Erstellung eines Strukturmodells an Hand der erhaltenen Daten durch Vergleich mit literaturbekannten Systemen problematisch.

Wenn auch für konkrete Aussagen über die Struktur von Mikroemulsionsgelen zusätzliche Untersuchungen notwendig waren, so konnte doch mit dieser Methode der Nachweis für das Vorliegen flüssigkristalliner Ordnungszustände erbracht werden.

3. Verwendete Geräte

3.1 Rheologie

Rotationsviskosimeter RV3, Fa. Haake, Berlin
Messkopf 500
Platte-Kegel-Messeinrichtung PKI:

Faktoren: $A = 78{,}5 \cdot 10^{-4} \dfrac{N}{cm^2 \cdot Skt}$

$M = 18{,}9 \text{ min} \cdot s^{-1}$

$G = 4150 \dfrac{mPa \cdot s}{Skt \cdot \text{min}}$

Temperiereinrichtung: Umlaufthermostat Typ FE (Haake)

Schreiber: xy-Schreiber BW 133, Rikadenki, Tokyo
xt-Schreiber Typ 1100, W + W Elektkronik, Basel

Zur Berechnung verwendete Formeln:

Schubspannung $\tau = A \cdot S \; (N \cdot cm^{-2})$

Schergeschwindigkeit $D = M \cdot n \; (s^{-1})$

Viskosität $\eta = \dfrac{G \cdot S}{n} \; (mPa \cdot s)$

(S = abgelesene Skalenteile, n = Umdrehungsgeschwindigkeit U/min; A, M und G = Faktoren der Messeinrichtung).

3.2 DTA

Mettler TA 2000 (B) als Tieftemperaturausführung (– 175 bis + 550 °C).

Heiz-/Kühlrate	5 K/min
Meßbereich	100 µV
Probenbehälter	Alutiegel zu, 40 µl
Referenz	Alutiegel zu, 40 µl
Spülgas	Luft
Einwaage	10–15 mg

3.3 Konduktometrie

Konduktometer LF 42, Wiss.-Techn. Werkstätten, Weilheim.
Schaufelelektrode LTA/S mit Zellkonstante 0,878
Temperaturkompensation mit Fühler N 500/410
Messbereich $0{,}2 - 11 \cdot 10^4$ µS
Messfrequenz 40 Hz und 4 kHz
Messspannung max. 2 V

3.4 Röntgenniederwinkelmessungen

Kiessig-Kamera, Fa. Seifert, Hamburg
Abstand Probe/Film: 420 mm
Röntgengenerator Kristalloflex IV, Fa. Siemens, Karlsruhe
Röntgenröhre mit Cu-Anode, CuK_α = = 1,54178 Å
Nickelfilter im Primärstrahl.
Glaskapillaren: Länge 80 mm, Außendurchmesser 2,0 mm, Wanddicke 0,01 mm
Artikel Nr. 1420, Hildenberg-Glas, Malsfeld
Betriebsbedingungen: Anodenspannung 35 kV
Heizstrom 38 mA
Belichtungszeit 48 h

Literatur

1. Nürnberg E, Pohler W (1983) Pharmazeut Ztg 128:2601
2. Nürnberg E, Pohler W (1983) Dtsch Apoth Ztg 123:1993
3. Krog N, Lauridsen JB (1976) In: Food Emulsions, Marcel Dekker Inc., New York, Basel, Chap 3, p 84 ff
4. Kohl P (1979) Dissertation Universität Erlangen-Nürnberg
5. Nürnberg E, Muckenschnabel R (1982) Dtsch Apoth Ztg 122:2093
6. Holzner G (1966) Seifen-Öle-Fette-Wachse 92:399
7. Pohler W (1983) Dissertation Universität Erlangen-Nürnberg

Für die Verfasser:

Prof. Dr. E. Nürnberg
Institut für Pharmazie und Lebensmittelchemie
der Friedrich-Alexander-Universität Erlangen-Nürnberg
Lehrstuhl für Pharmazeutische Technologie
Schuhstraße 19
8520 Erlangen

Struktureller Aufbau von Cholesterol-Polyoxyäthylenfettalkoholäther-Wasser-Mischungen*)

B. Usselmann und C. C. Müller-Goymann

Institut für Pharmazeutische Technologie der TU Braunschweig

Zusammenfassung: Die Wechselwirkungen von Cholesterol mit nichtionischen Tensid-Wasser-Mischungen wurden untersucht, wobei als nichtionische Tenside Polyoxyäthylenfettalkoholäther mit unterschiedlichem Äthoxylierungsgrad (2,4 und 23 PÄG-Einheiten) und verschiedenen Kettenlängen des Alkohols (12 C- und 16 C-Atome) eingesetzt wurden. Die untersuchten Fettalkoholpolyglykoläther bilden mit Wasser flüssigkristalline Lamellarphasen bzw. isotrope Gele. Die Bildung dieser Strukturen hängt vom Wassergehalt und vom eingesetzten Tensid ab.

In binären Mischungen aus Cholesterol und Fettalkohol-PÄG-äthern wurde eine Mesophase nur bei Verwendung des Hexadecanol-PÄG2-äthers detektiert. In den anderen untersuchten Systemen liegen die kristallinen Ausgangsmaterialien nebeneinander als Kristallmischung vor, bzw. Cholesterol kann in flüssigem Fettalkohol-PÄG-äther suspendiert sein.

Am Beispiel ternärer Mischungen mit Wasser wird gezeigt, daß bei geringen Wassergehalten kristallines Cholesterolmonohydrat neben hydratisiertem Fettalkohol-PÄG-äther vorliegt. Letzterer ist bei Raumtemperatur je nach Kettenlänge kristallin oder flüssig. Bei Erhöhung des Wasseranteils bildet sich eine aus allen drei Komponenten aufgebaute flüssigkristalline Lamellarphase. Bei weiterer Wasserzugabe werden zunehmend multilamellare Vesikel detektiert. Die flüssigkristallinen Strukturen wurden durch Röntgentechnik und Transmissionselektronenmikroskopie nachgewiesen.

Abstract: The interaction between cholesterol, water and different fatty alcohol-PEG-ethers was investigated by varying both the chain length of the fatty alcohol (12 and 16 carbon atoms) and the degree of ethoxylization (2,4 and 23 PEG-units). The fatty alcohol-PEG-ethers form lyotropic liquid crystals in water. Depending on the concentration of water and the type of the surfactant the mixtures are anisotropic liquid crystals of the lamellar type or isotropic gels which are liquid crystals as well.

Binary mixtures of cholesterol and fatty alcohol-PEG-ether will only form thermotropic liquid crystals if hexadecanol-PEG2-ether is used. Binary mixtures of cholesterol and the other ethoxylated alcohols are mixtures of crystals of the two surfactants, or cholesterol crystals are dispersed in the outer phase of liquid alcohol-PEG-ether.

Ternary mixtures of all components show different behaviour on changing the water content. With a low concentration of water crystals of cholesterolmonohydrate are detected in addition to hydrated fatty alcohol-PEG-ether. The hydrated fatty alcohol-PEG-ether will be liquid, however, if dodecanol-PEG4-ether is used. By increasing the concentration of water, lamellar liquid crystals occur consisting of all the components. These liquid crystals were detected as multilamellar vesicles by transmission electron microscopy. The swelling of the lamellar liquid crystals with water was proved by X-ray technique.

Key words: Fettalkoholpolyglykoläther, Cholesterol, isotrope Gele, multilamellare Vesikel, Transmissionselektronenmikroskopische Aufnahmen.

*) Vortrag, gehalten auf der 31. Hauptversammlung der Kolloid-Gesellschaft, Bayreuth 11. bis 14. Oktober 1983

KT-W 793

Einführung

In pharmazeutischen Zubereitungen wie Emulsionen, Salben und Cremes sind in der Regel Tenside eingearbeitet. Sie erhöhen einerseits die Stabilität der Systeme, andererseits nehmen sie auch Einfluß auf den strukturellen Aufbau der Mischungen. Je nach gewünschtem Typ der Zubereitung Wasser/Öl oder Öl/Wasser finden stärker lipophile bzw. hydrophile Tenside Anwendung. Auch die Kombination von lipophilem und hydrophilem Tensid ist vielgeübte Praxis.

Vorausgegangene Untersuchungen [1] an einem betont lipophilen System, das in Anlehnung an die Monographie für Hydrophilic Petrolatum USP XX [2] die Komponenten Cholesterol, Fettalkohol, Vaseline und Wasser enthielt, haben gezeigt, daß bei Einhaltung definierter Konzentrationsbereiche flüssigkristalline Lamellarphasen der eingesetzten Tenside nachzuweisen sind. In Abhängigkeit von der Kettenlänge des Fettalkohols existiert die Mesophase entweder bei Raumtemperatur oder ist nur bei höheren Temperaturen stabil. Die Anwesenheit von Wasser führt zu einer begrenzten Quellung der Lamellarphase.

Ziel der vorliegenden Arbeit ist die Untersuchung von stärker hydrophilen Systemen hinsichtlich ihres strukturellen Aufbaus im Vergleich zum bereits untersuchten ausgeprägt lipophilen Modellsystem. Die Variation des hydrophilen Charakters wurde dadurch erreicht, daß unter Beibehaltung des Cholesterol der Fettalkohol durch einen mehr oder weniger stark äthoxylierten Fettalkohol ersetzt wurde. Zusätzlich zum Äthoxylierungsgrad wurde die Fettalkoholkettenlänge variiert. Es wurden die Wechselwirkungen der Tenside in Kombination miteinander und mit Wasser untersucht.

Material und Methoden

Substanzen

Cholesterol USP XX, Merck, D-Darmstadt
Polyoxyäthylen-4-lauryläther (Brij 30)
Polyoxyäthylen-23-laurylather (Brij 35)
Polyoxyäthylen-2-cetyläther (Brij 52), Atlas Chemie, D-Essen

Folgende Abkürzungen werden im weiteren Text verwendet:

Für Brij 30 = C 12E04
Brij 35 = C 12E023
Brij 52 = C 16E02

Aqua bidestillata

Röntgenuntersuchungen

Röntgenkleinwinkeluntersuchungen nach Kiessig [3]
Aufnahmeabstand: 300 mm
Strahlung: Cu Kα, λ = 0.1541 nm, Ni-Filter
Generator: Müller Mikro 111, Philips
Beschleunigungsspannung: 40 kV
Anodenstrom: 20 mA
Filmmaterial: X-ray Film, NS-2T, Kodak
Expositionszeiten: 2–4 Tage, je nach Intensität der Interferenzen

Röntgenweitwinkeluntersuchungen
Debye-Scherrer Kammer mit 360 mm Umfang
Generator: Philips PW 1730
Strahlung: Cu Kα, λ = 0.1541 nm, Ni-Filter
Beschleunigungsspannung: 60 kV
Anodenstrom: 40 mA
Filmmaterial: X-ray Film NS-2T, Kodak
Expositionszeiten: 1 Stunde

Polarisationsmikroskopie

Zeiss Photomikroskop III mit Temperiereinrichtung Mettler FP5/FP52

Transmissionselektronenmikroskopie

Die Proben wurden nach einer von Moor et al. entwickelten Technik präpariert [4]. Eine ausführliche Beschreibung des Präparationsvorgangs ist bei Junginger und Heering [5] zu finden. Gefrierbruchätzanlage: BAF 400, Balzers AG, CH-Balzers Transmissonselektronenmikroskop: EM 300 T, Philips Beschleunigungsspannung: 80 kV

Ergebnisse und Diskussion

Mischungen mit C12 E04

C12 E04 mit einem HLB-Wert von 9.7 besitzt nahezu ausgewogene lipophile und hydrophile Eigenschaften. Die Reinsubstanz ist bei Raumtemperatur flüssig und zeigt in der Röntgenkleinwinkeluntersuchung einen breiten Reflex, der etwa 26–34 Å entspricht. Diese Größenordnung ist der mittleren Kettenlänge zuzuordnen. Es muß also bereits eine wenn auch stark gestörte Schichtung der Moleküle in der Flüssigkeit vorliegen.

Binäre Mischungen: C12 E04 – Cholesterol

In binärer Mischung mit Cholesterol gibt es keine Wechselwirkung zwischen den Tensiden im Sinne einer Bildung von Flüssigkristallen. Bei Raumtemperatur sind bis zu 10 % Cholesterol in C12 E04 löslich. Höhere Cholesterolkonzentrationen stellen sich als mehr oder weniger konzentrierte Suspensionen dar, die bei Temperaturerhöhung in homogene Schmelzen übergehen.

Binäre Mischungen: C12 E04 — Wasser

Mit Wasser bildet C12 E04 in einem Konzentrationsbereich von 20–80 % bei Raumtemperatur stabile flüssigkristalline Lamellarstrukturen aus, deren Existenz auch schon von Bostock et al. [6] beschrieben wurde. In Abbildung 1 sind die Ergebnisse der Röntgenkleinwinkeluntersuchung nach Kiessig zu sehen. Es sind die Netzebenenabstände in Å gegen den Wassergehalt der Mischungen abgetragen.

Bis zu 15 % Wasser wird in C12 E04 eingearbeitet, ohne das flüssigkristalline Strukturen nachgewiesen werden können. Es liegt eine homogene isotrope Lösung vor. Die Röntgenuntersuchung zeigt ebenso wie bei der Reinsubstanz C12 E04 nur einen breiten diffusen Reflex, der aber zu etwas größeren d-Werten verschoben ist. Das eingearbeitete Wasser wird zur Hydratisierung der hydrophilen Kopfgruppe des Äthers benötigt. Erst bei weiterer Wasserzugabe kann es zu einer Wechselwirkung zwischen verschiedenen Polyäthylenglykol-Enden über Wasserstoffbrücken kommen. Es resultiert die bis über 10 nm quellfähige Lamellarphase. Mischungen mit mehr als 70 % Wasser sind zweiphasig, isotrop und röntgenamorph.

Ternäre Mischungen: C12 E04 — Cholesterol — Wasser

In den untersuchten ternären Mischungen wird ein konstantes Gewichtsverhältnis der Tenside vorgegeben und der Wassergehalt variiert. Das eingesetzte Tensidverhältnis beträgt 2 Teile C12 E04 zu 1 Teil Cholesterol, da diese Mischung eine nicht zu hoch konzentrierte Suspension darstellt, und mögliche Wassereinflüsse bereits polarisationsmikroskopisch gut erkannt werden können.

Bei niedrigem Wassergehalt liegt eine Suspension von Cholesterolmonohydrat-Kristallen dispergiert in schwach hydratisiertem C12 E04 vor. Die zugegebene Wassermenge reicht nicht zur Solubilisierung des Cholesterols aus.

Im Konzentrationsbereich von 20–90 % Wasser bildet sich eine aus allen drei Komponenten aufgebaute flüssigkristalline mit Wasser quellbare Lamellarphase, wobei dann Cholesterol solubilisiert vorliegt. Kristallines Cholesterol ist in diesen Mischungen nicht mehr vorhanden.

Abbildung 2 zeigt die graphische Darstellung der Netzebenenabstände der ternären Mischungen in Abhängigkeit vom Wassergehalt. Binäre Mischungen aus C12 E04 und Wasser sind — wie oben schon erwähnt — bei einem Wassergehalt von mehr als 70 % polarisationsoptisch isotrop. In den untersuchten ternären Mischungen werden dagegen auch noch bis 90 % Wasseranteil flüssigkristalline Strukturen festgestellt. Es handelt sich hierbei um Systeme, in denen eine äußere wäßrige Phase eine dispergierte gequollene Lamellarphase einschließt. Die nachweisbaren Strukturen werden transmissionselektronenmikroskopisch als multilamellare Vesikel detektiert, die bereits von

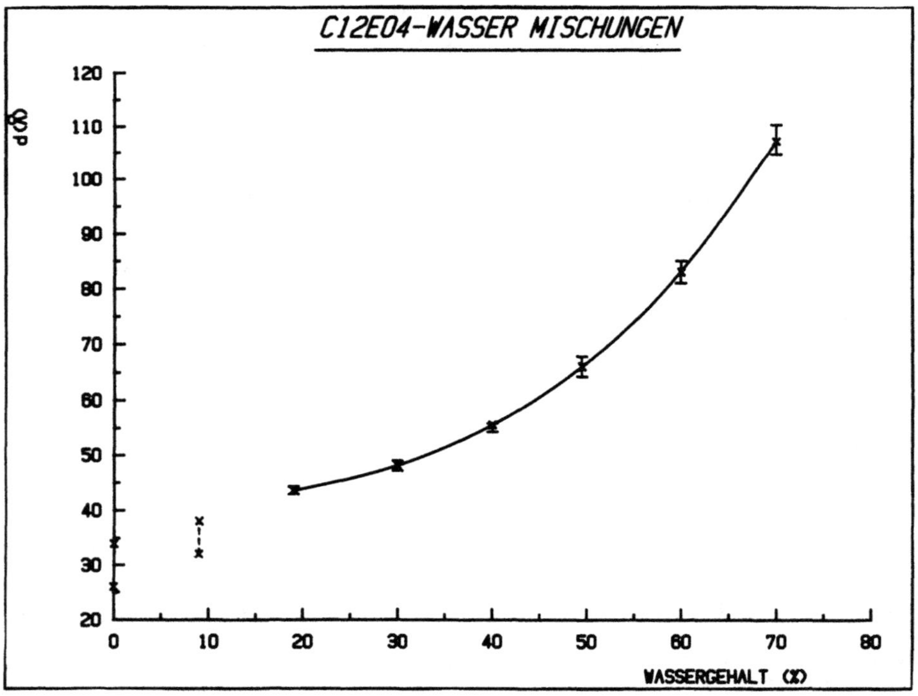

Abb. 1. Netzebenenabstände der binären C12 E04 — Wasser Mischungen bei 25 °C

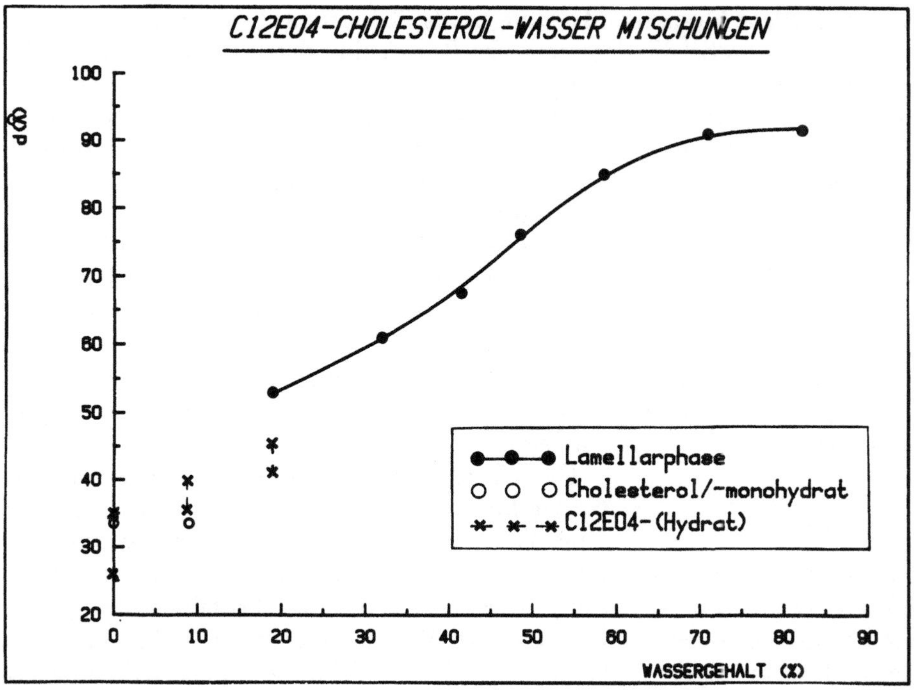

Abb. 2. Netzebenenabstände der C12 E04 – Cholesterol – Wasser Mischungen bei 25 °C. Das eingesetzte Tensidverhältnis beträgt 2 Teile C12 E04 zu 1 Teil Cholesterol

Müller-Goymann in ausgeprägt lipophilen cholesterolhaltigen Cremes beschrieben wurden [7].

Abbildung 3 zeigt ein angebrochenes etwa 600 nm großes Vesikel in einer Mischung mit hohem Wassergehalt. Es sind mehrere Schichten erkennbar. Bei geringen Wassergehalten ist die lamellare Mesophase transmissionselektronenmikroskopisch in planarer und nicht in vesikulärer Anordnung erkennbar, wie aus Abbildung 4 zu entnehmen ist. Man sieht die quergebrochenen lamellaren Schichten. Die Stufenhöhe korreliert gut mit den Röntgendaten.

Mischungen mit C12 E023

Als weiteres Tensid wurde C12 E023 (HLB-Wert 16,9) untersucht. C12 E023 selbst liegt bei Raumtemperatur kristallin vor. Im Weitwinkelbereich erhält man das für den kristallinen Zustand typische Interferenzmuster. Im Kleinwinkelbereich werden zwei Langperioden von 184 ± 8 Å und 105 ± 3 Å gefunden. Kehren und Rösch [8], die ebenfalls bei äthoxylierten Fettalkoholen zwei Reflexfolgen fanden, erklären dieses damit, daß nebeneinander Zickzackstruktur und Mäanderstruktur vorliegen. Nach unserem Verständ-

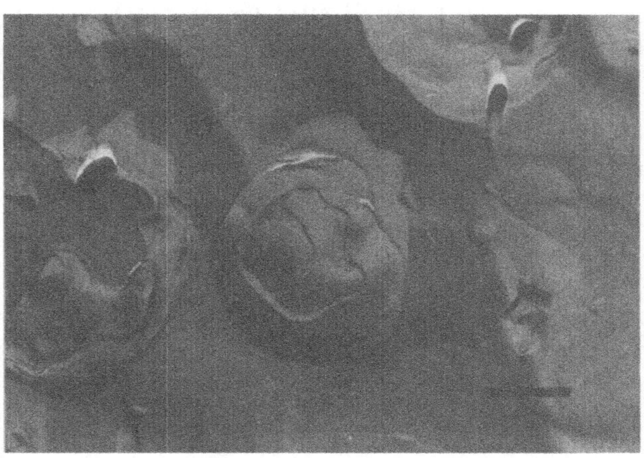

Abb. 3. TEM-Aufnahme eines quergebrochenen Vesikels bei 85000-facher Vergrößerung, Balken 250 nm, Beschattungsrichtung ↗. Ternäre Mischung aus 6.4 % C12 E04, 3.2 % Cholesterol und 90.4 % Wasser

Abb. 4. Lamellarphase aus 54 % C12 E04, 27 % Cholesterol und 19 % Wasser. Vergrößerung 117000-fach, Balken 250 nm, Beschattungsrichtung ↗

nis handelt es sich dabei um zwei verschiedene Modifikationen, eine gestreckte und eine gefaltete, wobei die Faltung im Bereich der Äthylenoxid-Helix lokalisiert sein müßte. Diese Interpretation steht in Übereinstimmung mit Untersuchungen von Gilg et al. [9] sowie Spegt [10] an reinen Polyäthylenglykolen.

Binäre Mischungen: C12 E023 – Cholesterol

In den binären Mischungen mit Cholesterol liegen beide Tenside bei Raumtemperatur in kristalliner Form vor. Beim Aufheizen gehen die kristallinen Mischungen ohne Bildung von Flüssigkristallen direkt in isotrope homogene Schmelzen über.

Binäre Mischungen: C12 E023 – Wasser

Wird dagegen C12 E023 mit Wasser anrezeptiert, so kann man in Abhängigkeit vom Wassergehalt Strukturen beobachten, deren Existenz bereits von Kehren und Rösch bei der Untersuchung anderer äthoxylierter Fettalkohole beschrieben wurde [8]. In Abbildung 5 sind die gemessenen Netzebenenabstände der binären Mischungen in Abhängigkeit vom Wassergehalt dargestellt.

Wird wenig Wasser eingearbeitet und zwar bis zu 18 %, dann liegt eine Suspension von kristallinem C12 E023 in einer isotrop viskosen Lösung vor, in der ein Teil des Tensids gelöst ist (Gebiet A). Man kann diese Mischungen als übersättigte Lösungen ansehen. Die gemessenen Röntgeninterferenzen konnten dem kristallinen C12 E023 zugeordnet werden. Dieser kristalline Anteil wird mit steigendem Wassergehalt geringer, so daß im Bereich B ab 18 % Wasser eine isotrope viskose Lösung vorliegt, die nur noch einen Reflex von 60 Å aufweist.

Bei einem Wasseranteil von 40–60 % bildet sich ein isotropes Gel aus (Gebiet C). Die in der Abbildung abgetragenen d-Werte sind diejenigen mit der höchsten Intensität.

Mehr als 70 % Wasser ergeben isotrope Flüssigkeiten (Gebiet D), die einen breiten Reflex von 65–70 Å aufweisen.

Der Übergang von der viskosen Lösung über das Gel zur klaren isotropen Lösung kann mit einer unterschiedlichen Hydratisierung erklärt werden. Im Bereich von etwa 20–35 % Wasser ist die Hydratisierung ausreichend, um das kristalline Gefüge des C12 E023 bei Raumtemperatur aufzubrechen. Man kann dieses als eine Schmelzpunkterniedrigung verstehen, verursacht durch die zweite Komponente Wasser. Es resultiert eine isotrope viskose und homogene Schmelze bzw. Lösung. Aufgrund ihres Anfärbeverhaltens mit lipophilen Farbstoffen wird deutlich, daß das gesamte Wasser für die Hydratation verbraucht wird. Gebundenes Wasser verfügt nämlich über zu wenig freie Valenzen, wodurch die Löslichkeit hydrophiler Farbstoffe gegenüber lipophilen Farbstof-

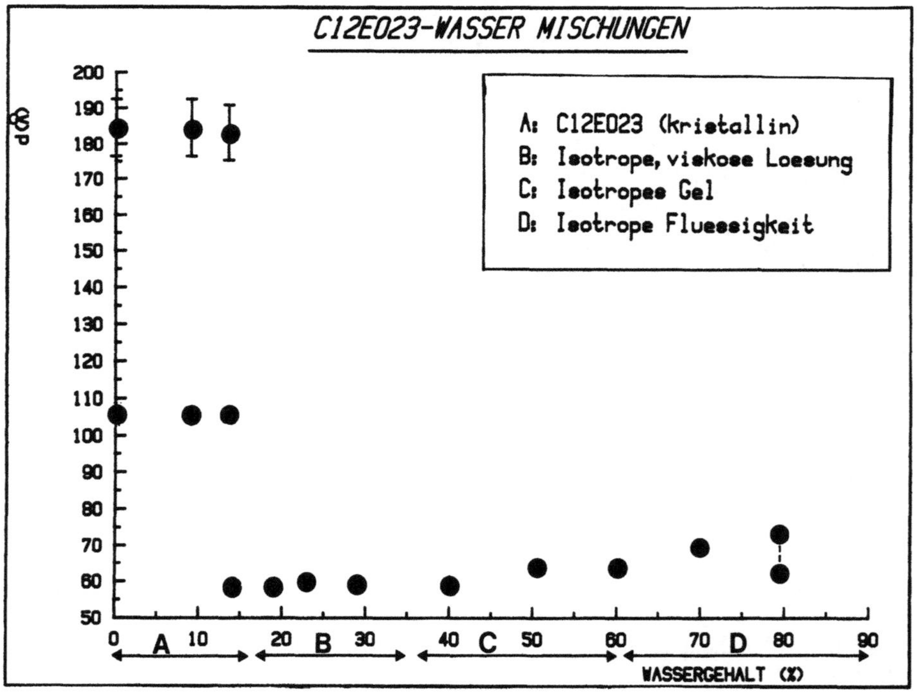

Abb. 5. Netzebenenabstände der binären C12 E023 – Wasser Mischungen bei 25 °C

Abb. 6. TEM-Aufnahme eines isotropen Gels bei 209000-facher Vergrößerung, Balken 100nm, Beschattungsrichtung ↗. Binäre Mischung aus 53 % C12 E023 und 47 % Wasser

fen behindert wird. Es ist aber auch eine Abschirmung der polaren Bereiche durch die lipophilen Molekülteile denkbar.

Bei weiterer Wasserzugabe entsteht ein isotropes Gel. Hierbei muß die Hydratisierung so weit gehen, daß über Wasserbrücken ein zusammenhängendes dreidimensionales System resultiert. Transmissionselektronenmikroskopisch (Abb. 6) kann man ein hochdisperses System aus dicht gepackten Partikeln erkennen, deren Durchmesser mit 5–6 nm gut mit den Röntgendaten korreliert. Wir interpretieren diese Partikel als Mizellen, wobei die hydrophilen, stark hydratisierten Molekülteile nach außen weisen. Über Wasserstoffbrücken wird der feste Zusammenhalt dieser Mizellen bewirkt, so daß sich diese Mischungen als hoch elastische Gele – sogenannte swinging gels – darstellen. Aufgrund der geringen Partikelgröße im Nanometerbereich sind sie optisch transparent und isotrop.

Ab 70 % Wassergehalt liegt eine mizellare L1-Phase vor. Der Zusammenhalt zwischen den Mizellen ist infolge des großen Wasserüberschusses, der für keine weitere Hydratisierung genutzt werden kann, verloren gegangen. Es resultieren daher niedrigviskose Systeme.

Ternäre Mischungen: C12 E023 – Cholesterol – Wasser

In Abbildung 7 sind die Netzebenenabstände der ternären Mischungen in Abhängigkeit vom Wassergehalt abgetragen. Bis zu einem Wassergehalt von 40 % gibt es keine Unterschiede zu den binären C12 E023 – Wasser – Mischungen (siehe Gebiete A–C in Abbildung 5 und Abbildung 7). Der zusätzliche Reflex in den ternären Mischungen bei 33.5 Å ist Cholesterolmonohydrat zuzuordnen.

Erst wenn der Wassergehalt mindestens 40 % beträgt (Gebiete D und E), treten alle drei Komponenten unter Bildung einer gequollenen Lamellarphase in Wechselwirkung. Ein Überschuß an kristallinem Cho-

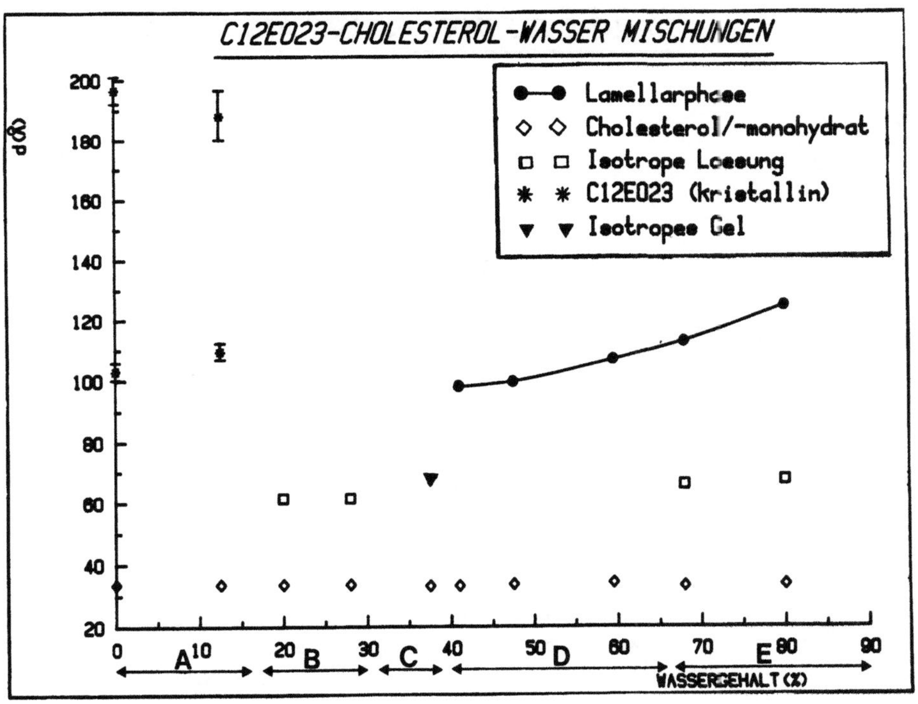

Abb. 7. Netzebenenabstände der ternären C12 E023 – Cholesterol – Wasser Mischungen bei 25 °C. Das eingesetzte Tensidverhältnis beträgt 2 Teile C12 E023 zu 1 Teil Cholesterol

lesterolmonohydrat liegt dispergiert in der Lamellarphase vor. Ab 70 % Wasser (Gebiet E) tritt als zusätzliche Phase eine mizellare Lösung auf, in der Cholesterolmonohydratkristalle und die Mesophase in Form multilamellarer Vesikel dispergiert sind.

Beim Vergleich dieser ternären Mischungen mit den Systemen, die das lipophilere C12 E04 enthalten, ergeben sich folgende Unterschiede:

1) C12 E023 solubilisiert wenig Cholesterol. In allen Mischungen sogar bis zu 90 % Wassergehalt konnte kristallines Cholesterolmonohydrat nachgewiesen werden. Wird dagegen C12 E04 als Tensid eingesetzt, dann liegt schon bei einem Wassergehalt von nur 20 % kein kristallines Cholesterol mehr vor. Je niedriger der HLB-Wert des Äthers ist, umso leichter wird das relativ lipophile Cholesterol mit einem HLB-Wert von 3 solubilisiert.

2) Die flüssigkristalline Lamellarphase unter Beteiligung des hydrophilen C12 E023 konnte ausschließlich in Form von multilamellaren Vesikeln nachgewiesen werden. Wird C12 E04 verwendet, treten die Vesikel erst bei einem Wassergehalt von mindestens 75–80 % Wasser auf. Bei niedrigen Wassergehalten stellt sich die Lamellarphase in planarer Anordnung dar.

Mischungen mit C16 E02

Als weiterer Fettalkoholäther wurde C16 E02 untersucht, der bei Raumtemperatur flüssigkristallin ist. Es handelt sich um eine thermotrope smektische Mesophase, die bei Zugabe von Wasser eine Aufweitung ihrer lamellaren Abstände erfährt. Bei 50 % Wassergehalt ist die maximale Quellung erreicht.

Binäre Mischungen: C16 E02 – Cholesterol

In binären Mischungen aus C16 E02 und Cholesterol ist die Ausbildung einer gemeinsamen thermotropen Mesophase zu beobachten. Diese Flüssigkristalle sind nur beim Abkühlen aus der Schmelze um 50 °C stabil. Bei Raumtemperatur liegt kristallines Cholesterol neben flüssigkristallinem C16 E02 vor.

Ternäre Mischungen: C16 E02 – Cholesterol – Wasser

Wird in die binäre Mischung – das Tensidverhältnis beträgt wieder 2 Teile Äther zu 1 Teil Cholesterol – Wasser eingearbeitet, dann erhält man eine bei Raumtemperatur mehr oder weniger stabile aus allen drei Komponenten aufgebaute Mesophase vom lamellaren Typ.

In Abbildung 8 sind die Netzebenenabstände der ternären Mischungen in Abhängigkeit vom Wassergehalt dargestellt. Bereits ab 5 % Wassergehalt sind die lamellaren Flüssigkristalle nachzuweisen. Bei niedrigen Wassergehalten von etwa 5–20 % rekristallisieren allerdings nach 1–3 Tagen Lagerung Cholesterolmonohydrat und C16 E02-Hydrat. Wird der Wasseran-

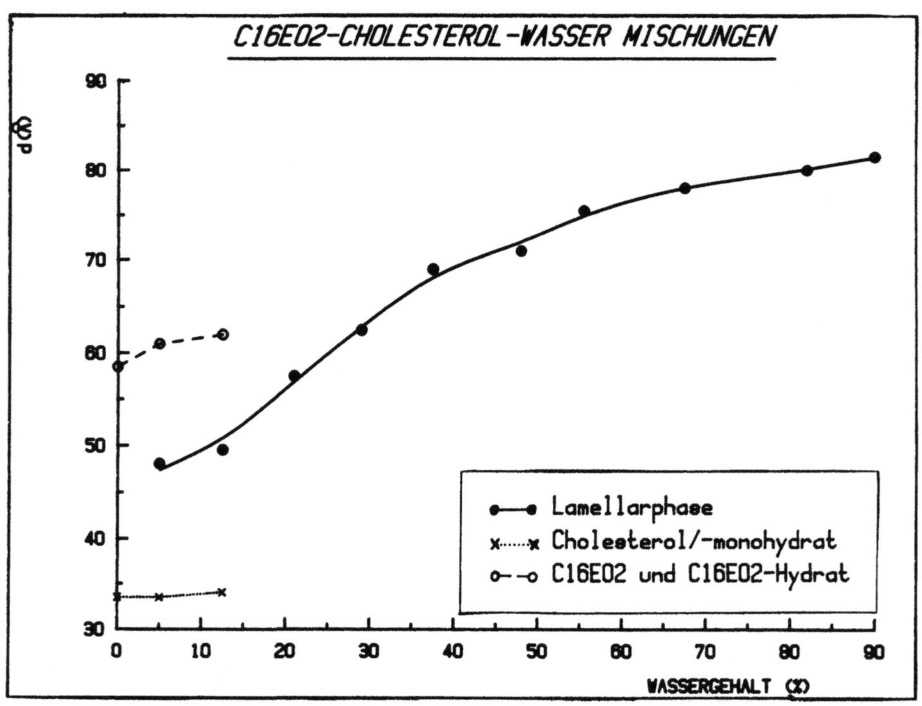

Abb. 8. Netzebenenabstände der ternären C16 E02 – Cholesterol – Wasser Mischungen bei 25 °C. Das eingesetzte Tensidverhältnis beträgt 2 Teile C16 E02 zu 1 Teil Cholesterol

teil erhöht, stabilisiert sich die Lamellarphase, indem weiter Wasser interlamellar eingebaut wird. Ab ca. 60 % Wasseranteil konnte auch in diesem System die Mesophase in Form multilamellarer Vesikel festgestellt werden. Von allen untersuchten Fettalkoholäthern zeigt C16 E02 die größte Solubilisierungsfähigkeit für Cholesterol.

Vergleich der Mischungen mit äthoxyliertem bzw. nicht äthoxyliertem Fettalkohol

In den binären Fettalkoholäther-Wasser Mischungen können lamellare und isotrop-gelige Mesophasen detektiert werden. Fettalkohole bilden dagegen mit Wasser kristalline Semihydrate sowie eine entsprechend hydratisierte α-Phase als Hochtemperaturmodifikation [11]. Erst wenn die Alkohole äthoxyliert vorliegen, das heißt die Lipophilie des Tensids abgeschwächt ist, kommt es bereits bei Raumtemperatur zu einer erhöhten Wechselwirkung zwischen Äther und Wasser im Sinne einer Hydratation. Es bilden sich abhängig vom Wassergehalt flüssige Kristalle.

In den binären Tensidmischungen liegt der umgekehrte Fall vor. Nur bei hinreichender Lipophilie des Fettalkohol-PÄG-äthers – z. B. C16 E02 – gibt es eine Wechselwirkung mit Cholesterol, und es bilden sich Flüssigkristalle. Dieses Ergebnis steht in Einklang mit Literaturdaten, die die Ausbildung von Flüssigkristallen aus den beiden relativ lipophilen Komponenten Cholesterol und Fettalkohol in definierten Konzentrations- und Temperaturbereichen bestätigen [12, 13].

In allen ternären Systemen mit äthoxyliertem Fettalkohol wird eine mit Wasser stark quellfähige lamellare Mesophase unter Beteiligung beider Tenside gebildet. Ein bestimmter Mindestwassergehalt, der von der Art des eingesetzten Fettalkohol-PÄG-äthers abhängt, ist jedoch für die Solubilisation des Cholesterols in der gemeinsamen Mesophase notwendig. Mit zunehmendem Wassergehalt können multilamellare Vesikel in den Mischungen detektiert werden. Die Systeme sind als mehrphasig mit dispergierter Lamellarphase in Vesikelform anzusehen, wobei das Dispersionsmedium eine wässrige Lösung darstellt. Die vesikuläre Anordnung der flüssigkristallinen Lamellarphase wird auch festgestellt, wenn anstelle der Fettalkohol-PÄG-äther die lipophilen Fettalkohole eingesetzt werden. Allerdings ist die Mesophase dann nur sehr begrenzt um maximal 3 Å quellbar. Wasserkonzentrationen, die über die wenigen in die multilamellaren Vesikel eingelagerten Prozent hinausgehen, werden in den stärker lipophilen Systemen in Tropfenform dispergiert und beteiligen sich nicht mehr an der Mesophase.

Obwohl die Hydrophilie bzw. Lipophilie der eingesetzten Tensidmischungen stark variiert, lassen sich in allen Systemen bei Gegenwart von Cholesterol multilamellare Vesikel nachweisen. Sie bilden damit ein gemeinsames Strukturmerkmal.

Schlußbemerkung

Alle ternären Mischungen sind vom makroskopischen Aussehen her in der Regel weiße bis transparente mehr oder weniger weiche Zubereitungen, die pharmazeutisch technologischen Gesichtspunkten einer Creme gerecht werden. Da bei der Anwendung und in der Wirkung aber Creme nicht gleich Creme ist, liegt in der Kenntnis der Feinstruktur möglicherweise der Ansatzpunkt für eine Beurteilung von Unterschieden.

Danksagung

Herrn Prof. Dr. C. Führer wird herzlich für die stete Unterstützung der Arbeitsgruppe gedankt.

Literatur

1. Müller-Goymann C (1981) Strukturuntersuchungen an 4 – Komponenten – Mischungen als Beitrag zur Aufklärung des W/O Creme – Zustandes, Dissertation, TU Braunschweig
2. The United States Pharmacopeia Twentieth Revision (1980)
3. Kiessig H (1957) Kolloid Z 152:62
4. Moor H, Mühlenthaler K, Waldner H, Frey-Wyßling A (1961) J Biophys Biochem Cytol 10:1
5. Junginger H, Heering W (1983) Acta Pharm Technol 28:85
6. Bostock RA, Donald MP, Tiddy GJT, Waring L (1979) Surface Active Agents, Soc of Chem Ind, Symposium Nottingham, 26.9–28.9
7. Müller-Goymann C, Führer C (1982) Acta Pharm Technol 28:243
8. Kehren M, Rösch M (1957) Fette, Seifen, Anstrichm 59:80
9. Gilg B, Spegt P, Skoulios A (1965) C R Acad Sci Paris 261:5482
10. Spegt P (1970) Makromol Chemie 140:167
11. Murase N, Gonda K, Kagami I, Koga S (1974) Chemistry Letters, 4:333
12. Mlodziejowsky A (1923) Z Physik 20:317
13. Lawrence ASC (1969) Symposium ordered fluids liquid crystals 2nd, Proceedings American Chem Soc Symposium, 289

Anschrift der Verfasser:

B. Usselmann
Institut für Pharmazeutische Technologie
der TU Braunschweig
Mendelssohnstraße 1
D-3300 Braunschweig

Zur Kenntnis von Transparenten 3-Komponenten-Tensidgelen

4. Mitteilung: Der Gelcharakter optisch isotroper Tensid-H_2O-Paraffinsysteme*)

E. Nürnberg und W. Pohler

Institut für Pharmazie und Lebensmittelchemie, Lehrstuhl für Pharmazeutische Technologie der Friedrich-Alexander Universität Erlangen-Nürnberg

Zusammenfassung: Rheologische Messungen lassen erkennen, daß die optisch isotropen Tensidgele (Mikroemulsionsgele) ein kohärentes Gerüst besitzen: Sie zeigen eine Fließgrenze und Thixotropie. Die Viskosität des Systems ist temperaturabhängig und die Struktur erweist sich als thermoreversibel. Es treten Eigenschaften auf, die für Nebenvalenzgele charakteristisch sind. Der Gel-Sol-Übergang kann differential-thermoanalytisch bestimmt und rheologisch bestätigt werden. Weniger deutlich fällt dieser Nachweis refraktormetrisch aus.

Elektronenmikroskopische Untersuchungen lassen einen Gelaufbau aus sphärischen Assoziaten mit einem mittleren Durchmesser von ca. 38 nm erkennn. Die Teilchen liegen in einer unregelmäßig verzweigten, perlschnurartigen Anordnung vor und die Gerüststruktur entsteht über Wassermoleküle durch Wasserstoffbrücken. Für diese Annahme spricht die aus der Schmelzenthalpie errechnete Bindungsenergie von 30 kJ/Mol.

Abstract: Rheological measurements are showing the optical isotropic surfactants based gels (Microemulsiongels) to have coherent framwork: They show yield point and thixotropy. The system's viscosity depends on temperature, its structure turns out to be thermoreversible, properties, which are characteristic for secondary bond gels. The gel-sol-transition was analysed by differential thermal analysis (DTA), confirmed by rheological investigations, less by refractometer.

Investigations by Electron microscope showed the gel-structure as spherical associations with a diameter of about 38 nm. Particles are irregulary linked, its structure results on hydrogen bonding via watermolecules, demonstrated by calculated linkage energy of 30 kJ/Mol.

Key words: Microemulsions, gels, surfactants, electron microscopy.

1. Untersuchungen zum Gelcharakter

Gele sind definiert als bikohärente, disperse Systeme fest/flüssig (Lyogele) oder fest/gasförmig (Xerogele), wobei der Feststoff ein kohärentes, dreidimensionales Gerüst, das als Matrix oder Gelgerüst bezeichnet wird, ausbildet, in dem die zweite Phase, ebenfalls kohärent, fixiert ist. Zum Aufbau dieser Struktur müssen die gerüstbildenden Teilchen untereinander Bindungen ausbilden können, wobei je nach Bindungsart zwischen Hauptvalenz- und Nebenvalenzgelen unterschieden wird. In Nebenvalenzgelen, zu denen die meisten streichfähigen Arzneistoffsysteme gehören, werden die Bauelemente der Gerüstphase vor allem durch van der Waalssche Kräfte und Wasserstoffbrückenbindungen zusammengehalten. Solche Produkte gehen beim Erhitzen thermoreversibel in den Solzustand über.

Aber auch durch mechanische Beanspruchung wird das Gelgerüst abgebaut. Die flüssige Phase ist in der Regel in Form von Hydrat- bzw. Solvathüllen relativ fest an die Gerüststruktur fixiert. In Fortführung der Untersuchungen von transparenten 3-Komponenten-Tensidgelen [1–3] aus POE-10-oleylether, flüssigem Paraffin und Wasser mußten rheologische, refraktrometrische und elektronenmikroskopische Prüfungen

*) Vortrag, gehalten auf der 31. Hauptversammlung der Kolloid-Gesellschaft, Bayreuth 11. bis 14. Oktober 1983.

KT-W 827

durchgeführt werden, um Strukturvorschläge aufzeigen zu können.

1.1 Rheologische Untersuchungen

Für die im folgenden beschriebenen Prüfungen wurde ein transparentes Gel aus folgenden Komponenten verwendet:

– Oleylalkohol × 10 EO in Form von Brij 96® HLB-Wert: 12,4	30 T
– dickflüssiges Paraffin	5 T
– Wasser	65 T

Wie aus vergleichenden Untersuchungen des Fließverhaltens verschiedener Tensid-Mesophasen hervorging [4], zeigen die isotropen Gele des Modellsystems, die sog. Mikroemulsionsgele, zwei charakteristische Eigenschaften: Fließgrenze und Thixotropie.

Zur Auslösung des Fließvorgangs muß eine gewisse Mindestschubspannung (τ_0) einwirken, um die Haftkräfte der Matrix zu überwinden. Die Fließgrenze wird durch die elastischen Eigenschaften des Systems bzw. seines Gelgerüsts hervorgerufen. Auch thixotropes Fließverhalten, das der Fließkurve die Form einer Hystereseschleife verleiht, ist an die Existenz einer inneren Gerüststruktur gebunden.

Den Matrixaufbau erkennt man besonders deutlich bei Messung der Viskosität des Systems als Funktion der Scherzeit. Daher wurde eine Probe der isotropen Gelphase (Rezeptur 1) im Rotationsviskosimeter bei konstanter Umdrehungsgeschwindigkeit (n = 50/min) geschert. Abbildung 1 zeigt die Schubspannung (τ) des Präparates in Abhängigkeit von der Zeit.

Deutlich erkennbar ist ein für Gele charakteristischer Kurvenverlauf: In der Anfangsphase ein starker Abfall der Schubspannungswerte, die schließlich einem konstanten Endwert zustreben.

Bei Nebenvalenzgelen ist ein solcher Strukturabbau nicht nur auf mechanischem Wege möglich. Die Haftkräfte zwischen den einzelnen Bauelementen des Gelgerüsts können auch durch Zufuhr thermischer Energie überwunden werden. Differentialthermoanalytische Untersuchungen hatten für das gleiche isotrope Gel einen scharfen Schmelzpeak bei etwa 308 K (35 °C) ergeben [4].

Nun war rheologisch zu überprüfen, ob dieses Signal tatsächlich den Schmelzbereich des Gelgerüsts

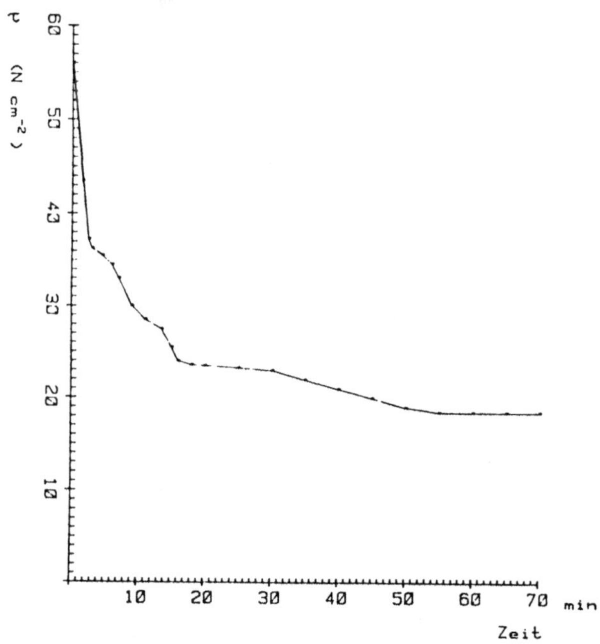

Abb. 1. Abnahme der Schubspannung in Abhängigkeit von der Zeit (Rezeptur 1)

anzeigt. Der alleinige Nachweis einer Viskositätsabnahme mit steigender Temperatur ist dazu jedoch nicht ausreichend. Die Viskosität ist bekanntlich eine stark temperaturabhängige Größe. Mit zunehmender kinetischer Energie der Teilchen werden diese beweglicher, wobei die Viskosität sinkt. Tritt jedoch zusätzlich ein temperaturabhängiger Übergang erster oder zweiter Ordnung auf, so weist dieser auf eine Änderung der inneren Struktur, z. B. die Zerstörung eines Gelgerüstes, hin [5].

Da in der Literatur keine entsprechende rheologische Untersuchungen zu finden waren, mußten zur Klärung der Frage, ob ein solcher Übergang bei sog. Mikroemulsionsgelen rheologisch nachweisbar ist, die Fließkurven der Modellrezeptur (Rezeptur 1) im Temperaturbereich von 293 bis 313 K (20 – 40 °C) aufgezeichnet werden (siehe Abb. 2).

Das System zeigt im unteren Bereich von 293 bis 303 K (20 – 30 °C) ein thixotropes Fließverhalten.

Mit steigender Temperatur nimmt das Ausmaß der Thixotropie deutlich ab und ist bei 308 K (35 °C) und darüber nicht mehr nachweisbar. In diesem Bereich weisen die Proben nur noch pseudoplastisches Fließverhalten auf.

Der sprunghafte Übergang zwischen 303 und 308 K (30 – 35 °C) wird noch deutlicher, wenn man aus den Fließkurven die *Fließgrenze* (τ_0) und die *Endviskosität* (η_E) am Umkehrpunkt zwischen Auf- und Abwärts-

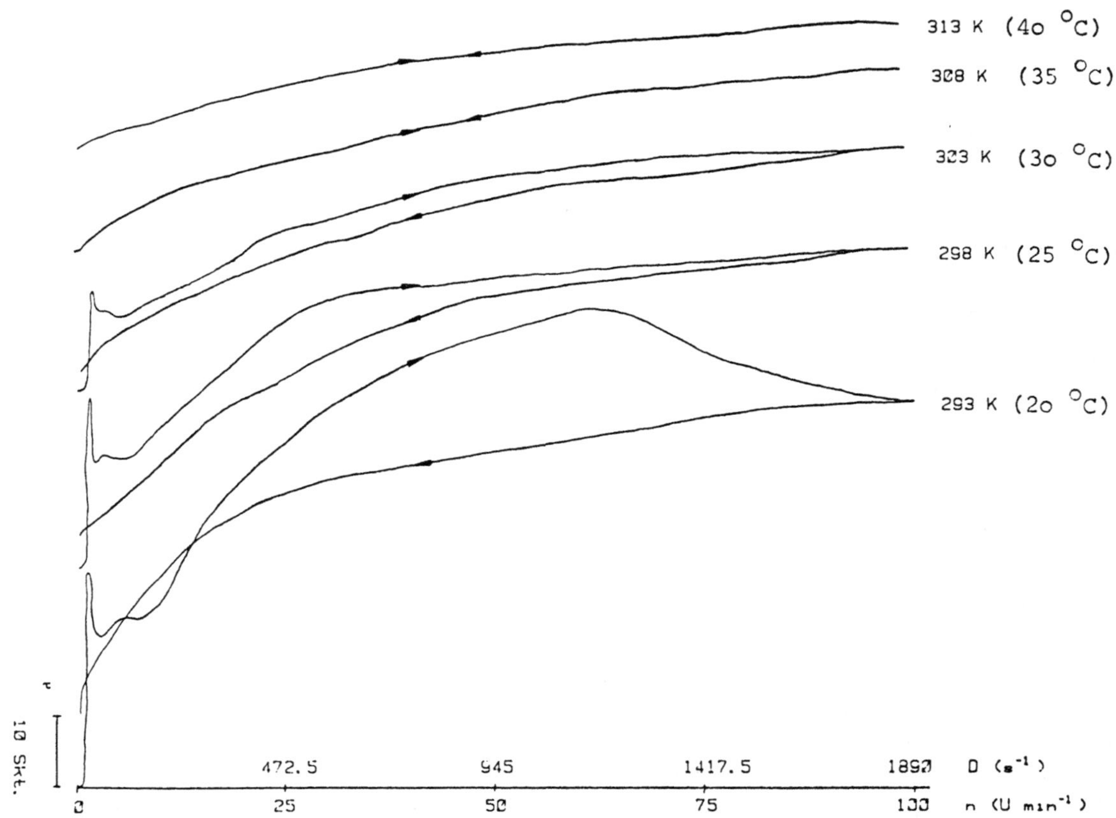

Abb. 2. Fließkurven der Modellrezeptur bei unterschiedlichen Temperaturen

kurve (Schergeschwindigkeit $D = 1890\ s^{-1}$) errechnet. Die so erhaltenen Werte sind in Abbildung 3 graphisch gegen die Temperatur aufgetragen.

Erwartungsgemäß weisen Fließgrenze und Endviskosität eine starke Temperaturabhängigkeit auf. Beide Meßgrößen nehmen mit steigender Temperatur praktisch linear ab. Bei den Werten der Fließgrenze ist jedoch zusätzlich im Bereich zwischen 303 und 308 K (30 - 35 °C) ein deutlicher Übergang erkennbar, der einen temperaturabhängigen Strukturzusammenbruch anzeigt.

Verständlicherweise tritt dieses Phänomen im Kurvenverlauf der Endiviskosität (η_E) nicht auf, weil hier der Gerüstabbau schon mechanisch erfolgt ist.

1.2 Refraktometrische Untersuchungen

Zur Identifizierung und Reinheitsprüfung flüssiger oder streichfähiger Substanzen kann bekanntlich die Bestimmung ihres Brechungsindexes herangezogen werden.

Im Rahmen der Untersuchungen über Strukturänderungen von transparenten Tensidgelen bei Temperaturerhöhung sollte geprüft werden, ob auch mit dieser Methode der Nachweis eines veränderten Ordnungszustandes erbracht werden kann. Dabei ist zu berücksichtigen, daß die gemessenen Werte stark temperaturabhängig sind und bei organischen Flüssigkeiten in der Regel eine Abweichung von ca. $5 \cdot 10^{-4}$/Grad beobachtet wird [6].

Mit Hilfe eines temperierbaren Abbé-Refraktometers wurden die Berechnungsindices der Modellrezeptur im Temperaturbereich zwischen 293 und 323 K (20 - 50 °C) bestimmt. Die Ergebnisse sind in Abbildung 4 graphisch dargestellt.

Eine lineare Abhängigkeit der Berechnungsindices von der Temperatur ist gut ersichtlich. Weniger ausgeprägt – aber eindeutig – ist ein im Bereich der thermischen Destruktion erkennbarer Übergang. Während der „Schmelzvorgang“ offensichtlich nur eine geringe Änderung des Brechungsindexes zur Folge hat, zeigt die DTA-Kurve einen bei verwandten Systemen nicht auftretenden scharfen Peak bei 310 K (37 °C) [4].

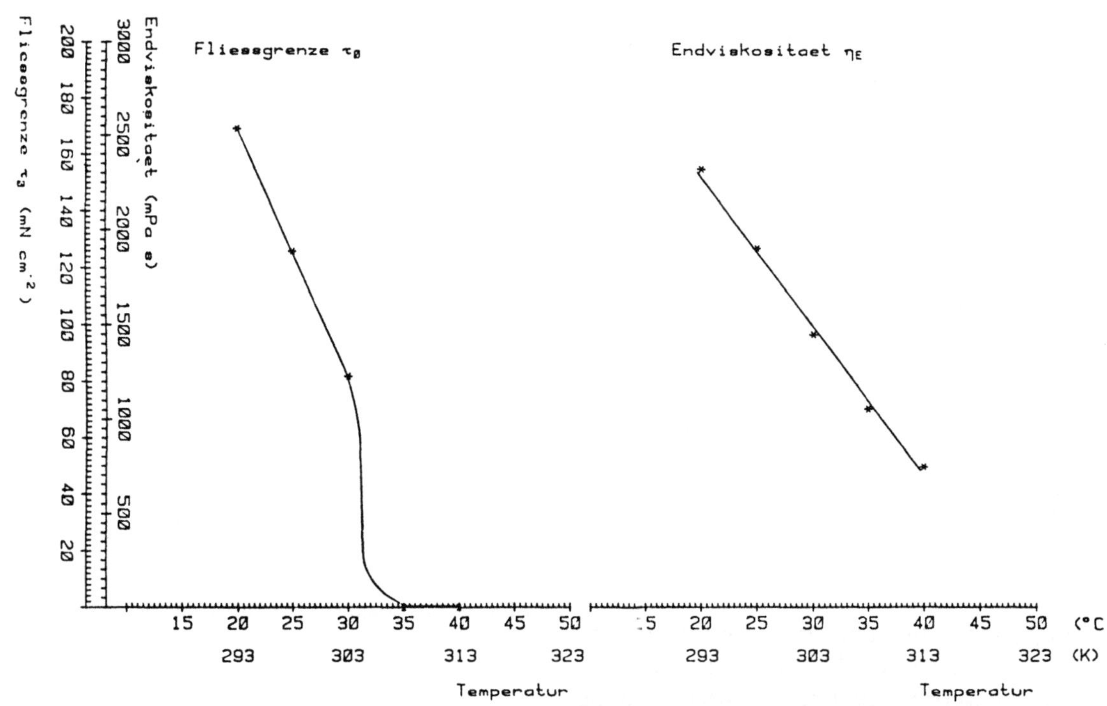

Abb. 3. Fließgrenze und Endviskosität der Modellrezeptur (Rp. 1) in Abhängigkeit von der Temperatur

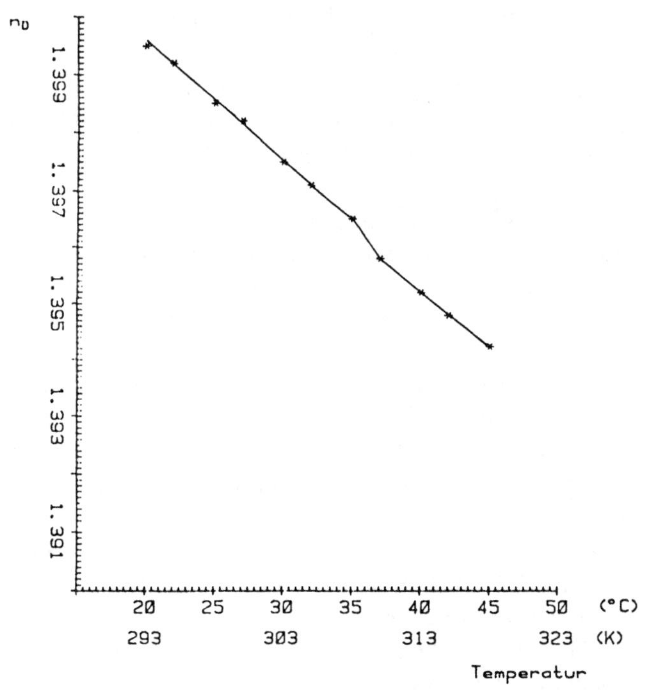

Abb. 4. Brechungsindex der Modellrezeptur in Abhängigkeit von der Temperatur

1.3 Elektronenmikroskopische Untersuchungen

Aufgrund der bisher gewonnen Daten war davon auszugehen, daß dem untersuchten Modellsystem bei Raumtemperatur eine Gelstruktur zukommt, die flüssigkristallinen Charakter aufweist. Damit war jedoch für präzise Strukturvorstellungen noch zu wenig über den strukturellen Feinbau bekannt.

Weiterreichende Aussagen über den kolloidchemischen Aufbau des Systems sollten elektronenmikroskopische Aufnahmen ermöglichen. Über derartige Untersuchungen ist bisher in der pharmazeutischen Literatur nur wenig publiziert worden. Offenbar sind nur unter Verwendung einer relativ aufwendigen Präparationstechnik, der Gefrierbruch-Ätztechnik, die auch im vorliegenden Fall eingesetzt wurde, zuverlässige Ergebnisse zu erwarten.

Abbildung 5 zeigt die Replica einer Bruchoberfläche der Modellrezeptur.

Deutlich sind Tensidaggregate als sphärische Teilchen, die einen Durchmesser von ca. 38 nm aufweisen, zu erkennen. Bei genauerer Betrachtung kann man eine perlschnurartige Anordnung der Partikel postulieren. Die Vorstellung eines dreidimensionalen Aufbaus dieser Aggregate ergibt das Bild eines Gelgerüstes.

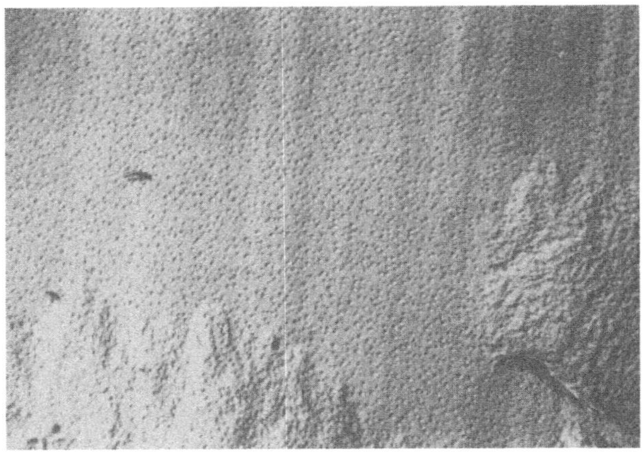

Abb. 5. Elektronenmikroskopische Aufnahme der Modellrezeptur (Vergrößerung 1 : 15500)

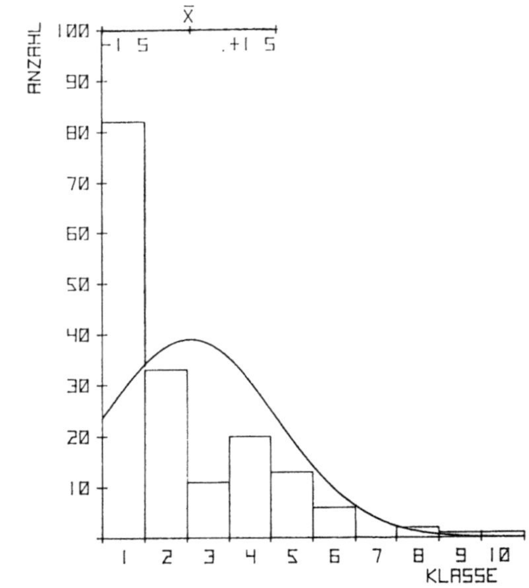

Abb. 6. Statistische Auswertung der Teilchenabstände (Histogramm)

Der möglicherweise subjektive Eindruck einer derartigen Sekundarstruktur konnte durch eine statistische Datenauswertung bestätigt werden. Mit der Meßlupe wurden auf einer photographischen Abbildung (Vergrößerung 1 : 32815) die Abstände jeweils benachbarter Teilchen bestimmt und in einem Statistik-Rechnerprogramm ausgewertet.

Aus dem Histogramm ist zu erkennen, daß die Teilchenabstände keine *Gaußverteilung* aufweisen.

Vielmehr beträgt für fast 50 % aller Partikel (Klasse 1) der Abstand zum Nachbarteilchen auf der Fotografie 0,1 mm und weniger, d. h. 3 nm in der Realität und für 68 % (Klasse 1 + 2) 7,1 nm.

Damit konnte der Nachweis erbracht werden, daß die sphärischen Teilchen im untersuchten System nicht in einer statistischen Verteilung vorliegen, sondern miteinander zu einer Sekundärstruktur verknüpft sind.

Tabelle 1. Statistische Datenauswertung der Teilchenabstände

A	B	C	D
1	0,1	82	3,0
2	0,2	33	6,1
3	0,3	11	9,1
4	0,4	20	12,2
5	0,5	13	15,2
6	0,6	6	18,3
7	0,7	0	21,3
8	0,8	2	24,4
9	0,9	1	27,4
10	1,0	1	30,5

A: Klasse; B: Gemessene max. Abstände in mm (Foto); C: Klasseninhalt; D: Berechnete Abstände nm (Werte aus B/Vergrößerungsfaktor)
$n = 169$, $\bar{x} = 0{,}3071$ mm, Min. 0,1 mm; Max. 1,0 mm, Variationsbreite: 325,6 %, $s = 0{,}2057$; $s_{rel} = 66{,}9954$ %

2. Strukturvorschlag für das Transparente 3-Komponenten-Gel

Aufgrund der durch Auswertung der elektronenmikroskopischen Aufnahmen gewonnenen Daten sind über die Struktur des untersuchten Mikroemulsionsgels folgende Aussagen möglich:

1. Das Gel besteht aus sphärischen Assoziaten mit einem mittleren Durchmesser von 37,5 nm.

2. Diese kugelförmigen Partikel liegen in einer unregelmäßig verzweigten, perlschnurartigen Anordnung vor, so daß eine Sekundärstruktur entsteht, die übereinstimmt mit üblichen Modellvorstellungen von dreidimensionalen Gelgerüsten.

3. Der häufigste Abstand zwischen benachbarten Teilchen in einer solchen „Perlenschnur" beträgt ca. 3 nm.

4. Die Kettenabstände sind größer und bewegen sich in der Größenordnung des 3- bis 5fachen der Assoziatdurchmesser.

Abbildung 7 zeigt das schematische Modell eines solchen Gelgerüstes.

Der Abstand zwischen den sphärischen Matrixbausteinen ist mit ca. 3 nm kleiner als die doppelte Poly-

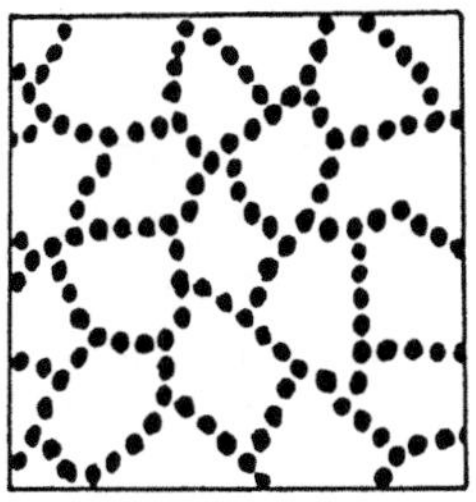

Abb. 7. Schematisches Modell der Gerüststruktur des untersuchten Mikoemulsionsgels

oxyethylenkettenlänge des verwendeten Tensids, die bei 10 Ethylenoxydeinheiten ca. 2 nm beträgt. Von der räumlichen Entfernung her wäre eine partielle Überlappung der Polyoxyethylenketten von Tensidmolekülen benachbarter Assoziate möglich und ihre Verknüpfung über Wassermoleküle durch Wasserstoffbrückenbindungen denkbar.

Die Abbildung 8 zeigt das Kalottenmodell einer solchen von Rösch [7] vorgeschlagenen Vernetzung.

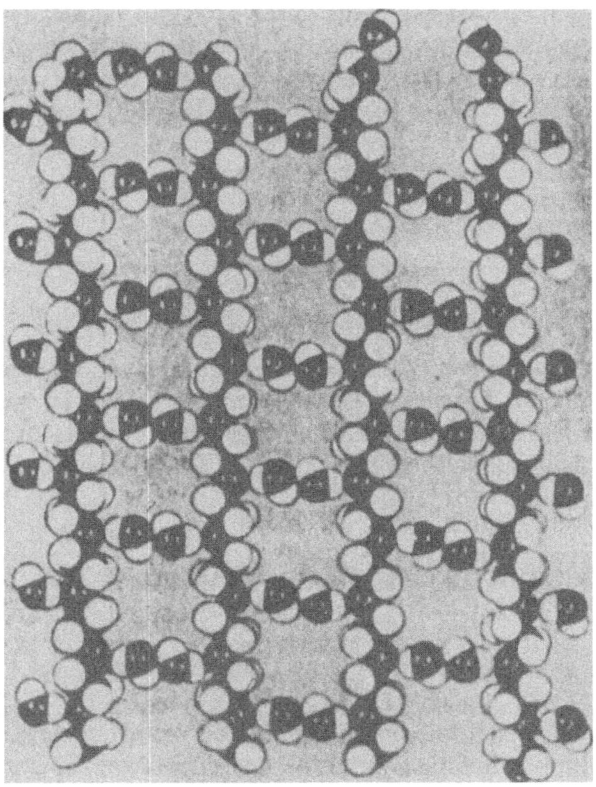

Abb. 8. Vernetzung von PEG-Ketten über Wassermoleküle durch Wasserstoffbrückenbindungen (nach Rösch)

Die kohärente wässrige Phase hätte man sich in den von der Gerüststruktur gebildeten Hohlräumen vorzustellen.

3. Untersuchungen über die Bindungskräfte

3.1 Literaturhinweise

Bekanntlich müssen zum Aufbau eines Gelgerüstes die dispergierten Teilchen untereinander Bindungen ausbilden, deren Stärke maßgeblich die Eigenschaften des Systems bestimmt. Zur weiteren Charakterisierung von Mikroemulsionsgelen war es daher von Interesse, Untersuchungen vorzunehmen, welche Rückschlüsse auf die Natur der auftretenden Haftkräfte zulassen.

Die Temperatur- und Scherempfindlichkeit der Systeme gibt zu erkennen, daß die Bindungskräfte zwischen ihren Gerüstbausteinen nur gering sind und daher vermutlich Nebenvalenzgele vorliegen. Dies steht in Übereinstimmung mit Literaturangaben [8, 9], wo als wahrscheinliche Ursache für die hier auftretende Gelbildung van der Waalssche Kräfte oder Wasserstoffbrückenbindungen genannt werden.

Über solche allgemeinen Angaben hinaus sind auch speziellere Betrachtungen und Untersuchungen über die Gelbildung polyoxyethylierter Tenside mit Wasser bekannt geworden, denen gemeinsam ist, daß für dieses Phänomen Wasserstoffbrückenbindungen verantwortlich gemacht werden: Schulman et al. [10] berichten über eine Strukturbildung in solchen Systemen durch Verknüpfung des Polyoxyethylenanteils der Tenside über Wassermoleküle.

Boehmke und Heusch [11] gelang bei polyoxyethylierten Tensiden der Nachweis, daß sich eine bestimmte Anzahl von Wassermolekülen an die Sauerstoffatome der Polyethylenglykolketten zu definierten Hydratkomplexen anlagert und dabei bevorzugte Hydratstufen auftreten.

Heusch et al. [12] schreiben das Auftreten von Netzstrukturen in flüssigkristallinen Tensidsystemen einer Verknüpfung von Strukturelementen über Wasserdipole zu. Nach Feger [13] spielen Wasserstoffbrükkenbindungen auch eine entscheidende Rolle bei der Ausbildung von flüssigkristallinen Phasen polyoxyethylierter Tenside.

3.2 Differentialthermoanalytische Bestimmung der Schmelzenthalpie

Aufgrund der vorliegenden Erkenntnisse gab es also eine Reihe von Hinweisen auf eine Verknüpfung der

Gerüstpartikel in Mikroemulsionsgelen über Wasserstoffbrücken, ohne daß jedoch an diesen Systemen exakte Messungen durchgeführt worden wären. Eine Bestätigung dieser Hypothese gelang mit Hilfe quantitativer differentialthermoanalytischer Untersuchungen.

Als Maß für die Stärke von Bindungen können deren Energien herangezogen werden [14]. Mit einer Bindungsenergie von 8–40 kJ/Mol sind Wasserstoffbrückenbindungen relativ schwach. Im Vergleich erweisen sich Atombindungen mit 200–600 kJ/Mol und Ionenbindungen mit weit über 400 kJ/Mol als wesentlich stärker. Über die Höhe der Bindungsenergie, die im vorliegenden Fall differentialthermoanalytisch bestimmt wurde, sind also u. U. Rückschlüsse auf die Art der Bindung möglich.

Das DTA-Signal ΔU ist eine elektrische Spannung, die vom Schreiber als Funktion der Zeit aufgetragen wird.

Über die Beziehung

$$dH/dt = \Delta U/E$$

dH/dt = Wärmefluß zur Probe
ΔU = Differenzthermospannung
E = kalorische Empfindlichkeit

erhält man als neue Ordinate den momentanen Wärmestrom zur Probe dH/dt.

Eine Fläche in diesem Koordinatensystem hat die Dimension Wärmestrom × Zeit und stellt die Enthalpieänderung der Probe dar. Ihre SI-Einheit ist 1 mW × 1 s = 1 mJ.

Man errechnet die *Schmelzenthalpie* ΔH einer Probe nach der Gleichung:

$$\Delta H = \frac{F}{E}$$

F = Peakfläche
E = kalorische Empfindlichkeit

Für die Modellrezeptur wurde eine Schmelzenthalpie von ca. 13 J/g ermittelt. Das entspricht einer Bindungsenergie von 30 kJ/Mol Tensid bei einem Tensidgehalt von 30% und einem Molekulargewicht von 694. Somit liegt der gefundene Wert im Bereich der Wasserstoffbrückenbindungen.

3.3 Gezielte Erzeugung von Störungen

Auch gezielte Störungen der Gelstruktur erlauben Aussagen über die Natur der Bindungskräfte. So ist bekannt, daß Harnstoff oder Ionen auf Wasserstoffbrücken strukturbrechend wirken, indem sie die Orientierung der Lösungsmittelmoleküle in unmittelbarer Nachbarschaft beeinflussen [15]. Harnstoff besitzt in wässriger Lösung eine starke Neigung zur Bildung von Wasserstoffbrücken mit Wassermolekülen und verändert die Polarität des Wassers nur geringfügig. Eine Beeinflussung der Grenzflächeneffekte findet dabei nicht statt.

Daher wurde in der Modellrezeptur eine Substitution des Wasseranteils durch eine 4 M Harnstofflösung bzw. eine 4 M Kochsalzlösung vorgenommen.

Infolge des behinderten Gerüstaufbaus wiesen die sonst gelförmigen Produkte nun erwartungsgemäß gießfähige Konsistenz auf.

Von Glycerol oder anderen mehrwertigen Alkoholen wird berichtet, daß sie, aufgrund ihres gegenüber Wasser höheren Dipolmoments, Wassermoleküle aus der Hydrathülle von polyoxyethylierten Tensiden verdrängen und selbst mit letzteren assoziieren [15]. Obwohl solche Stoffe die Viskosität der „Wasserphase" erhöhen, kann erwartet werden, daß ihr Zusatz zur Modellrezeptur eine Konsistenzminderung der Produkte zur Folge hat, wenn das Gelgerüst durch den o. a. Effekt geschwächt wird. Eine Überprüfung ergab, daß dies tatsächlich der Fall ist. Bei partieller Substitution des Wasseranteils durch Glycerol oder 1,2-Propandiol nahm die Viskosität der Gele mit zunehmender Konzentration dieser Verbindungen ab. Bei Zusatz von 20% Glycerol bzw. 30% 1,2-Propandiol, entsprechend 45 bzw. 35% verbleibendem Wasser, resultierten nur noch flüssige Zubereitungen.

3.4 Bedeutunng für die pharmazeutische Praxis

Der Einsatz solcher Verbindungen, wie Harnstoff und mehrwertige Alkohole, in Dermatika ist allgemein üblich. Bei den sog. Mikroemulsionsgelen, muß daher ihre viskositätsmindernde Wirkung berücksichtigt werden.

Dieser Effekt kann auch gezielt zur Konsistenzeinstellung der Präparate herangezogen werden. Z. B. ergaben rheologische Messungen in Abhängigkeit vom 1,2-Propandiolgehalt folgende Fließgrenzen bzw. Quasiviskositäten bei einer Schergeschwindigkeit $D = 945\ s^{-1}$ (s. Tab. 2).

Danach wird ersichtlich, daß bei praxisüblichen 1,2-Propandiolkonzentrationen von 5–15% Produkte resultieren, die immer noch eine deutlich ausgeprägte Fließgrenze aufweisen und gut verstreichbar sind. Die Konsistenz ist über den Glykolgehalt steuerbar.

Tabelle 2. Einfluß von 1,2-Propandiol auf die Konsistenz von Mikroemulsionsgelen

Rezeptur Nr.	1,2-Propandiol-gehalt %	Fließgrenze ($mN \cdot cm^{-2}$)	Quasiviskosität ($mPa \cdot s$)
1	0	1254	3127
3	5	1023	2821
5	15	709	956

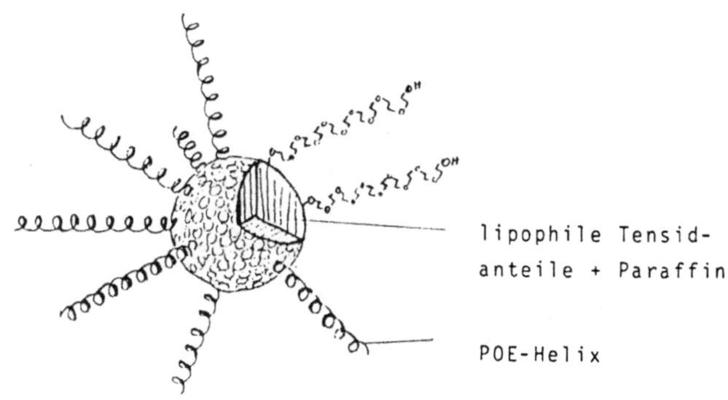

Abb. 10. Schematische Darstelllung eines Primär-Assoziates aus POE-Oleylether und Kohlenwasserstoff (in Anlehnung an A. N. Martin)

4. Diskussion des Strukturmodells

Aus den gewonnenen Daten können folgende Vorstellungen abgeleitet werden:

– Tensid (Abb. 9) und dickflüssiges Paraffin bilden unter den genanntenn Konzentrationsbedingungen in Wasser Assoziate unter Einschluß von dickflüssigem Paraffin. Sie weisen einen Durchmesser von ca. 38 nm auf und können gemäß Abbildung 10 schematisch dargestellt werden.

– Diese Primärassoziate sind durch Wasserstoffbrücken über Wassermoleküle, z. B. gemäß Abbildung 11, miteinander verknüpft und bilden so ein dreidimensionales Gelgerüst.

– Es entsteht ein bikohärentes System (siehe Abbildung 12) aus einer unregelmäßig verzweigten, perlschnurförmigen Matrix, in deren Hohlräumen sich das inkororierte Wasser befindet.

5. Verwendete Geräte

Rheologie und Konsistenzmessungen

Rotationsviskosimeter RV3, Fa. Haake, Berlin
Messkopf 500
Platte-Kegel-Messeinrichtung PKI:

Faktoren:

$$A = 78{,}5 \cdot 10^{-4} \frac{N}{cm^2 \cdot Skt}$$

$$M = 18{,}9 \text{ min} \cdot s^{-1}$$

$$G = 4150 \frac{mPa \cdot s}{Skt \cdot min}$$

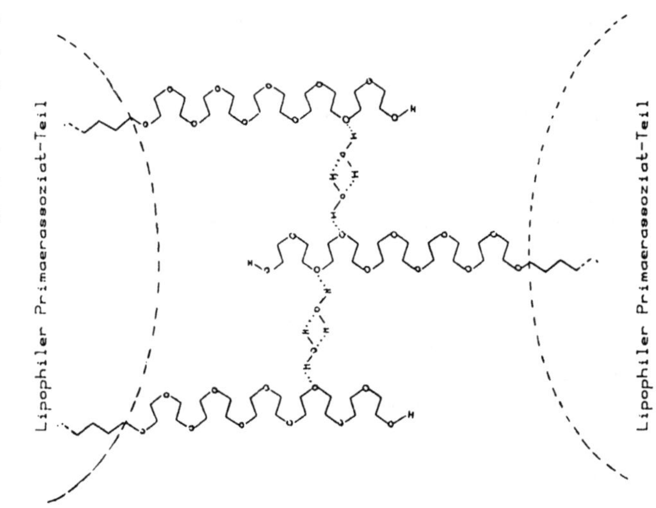

Abb. 11. Schematische Darstellung einer Verknüpfung der Primärassoziate durch Wasserstoffbrückenbindungen über Wassermoleküle

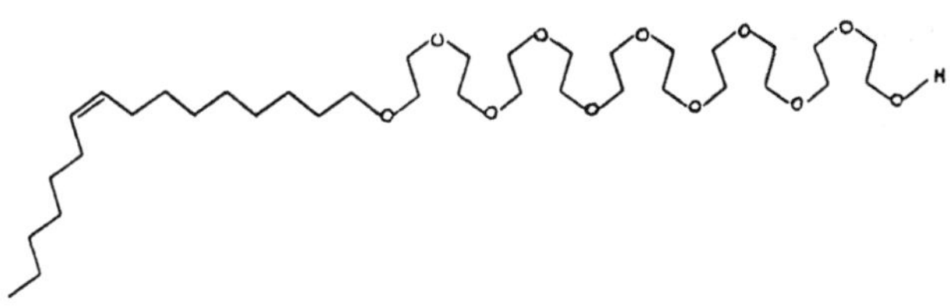

Abb. 9. Polyoxyethylen-10-oleylether (Brij 96®)

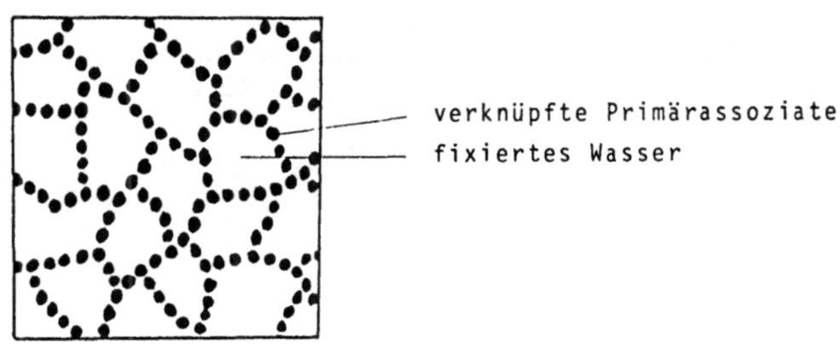

Abb. 12. Schematische Darstellung des räumlichen Aufbaus der Sekundärstruktur (Gelgerüst)

Koaxiale Zylindermesseinrichtung profiliert SVIIP

Faktoren:

$$A = 34{,}8 \cdot 10^{-4} \frac{N}{\mathrm{cm}^2 \cdot \mathrm{Skt}}$$

$$M = 0{,}78 \,\mathrm{min} \cdot s^{-1}$$

$$G = 4461 \frac{mPa \cdot s}{\mathrm{Skt} \cdot \mathrm{min}}$$

Externer Drehfrequenz-Programmgeber PG 128 mit linearem Drehfrequenzprogramm

Temperiereinrichtung: Umlaufthermostat Typ FE (Haake)

Schreiber: xy-Schreiber BW 133, Rikadenki, Tokyo, xt-Schreiber Typ 1100, W + W Elektronik, CH-Basel

Zur Berechnung verwendete Formeln:

Schubspannung $\tau = A \cdot S \; (N \cdot \mathrm{cm}^{-2})$

Schergeschwindigkeit $D = M \cdot n \; (s^{-1})$

Viskosität $\eta = \frac{G \cdot S}{n} \; (mPa \cdot s)$

(S = abgelesene Skalenteile, n = Umdrehungsgeschwindigkeit U/min; A, M und G = Faktoren der Messeinrichtung).

Mikro-Penetrometer

A. H. Thomas Co., Philadelphia/USA

Plexiglas-Spezialkonus 15 g, Best.-Nr. 18-036.1, Sommer u. Runge KG, Berlin.

Eindringzeit: 5 Sekunden

Meßtemperatur: 294 K ± 1 (21 °C)

Die Ergebnisse stellen den Mittelwert aus $n = 5$ dar.

DTA

Mettler TA 2000 (B) als Tieftemperaturausführung (−175 bis + 550 °C).

Heiz-/Kühlrate	5 K/min
Meßbereich	100 µV
Probenbehälter	Alutiegel zu, 40 µl
Referenz	Alutiegel zu, 40 µl
Spülgas	Luft
Einwaage	10 - 15 mg

Elektronenmikroskopie

Durchstrahlungs-Elektronenmikroskop EM 400, Philips, Kassel.

Gefrierätz-Präparation:
BALZERS Gefrierätz-Präparationsanalge BA 360 M.

Schockgefrierung in schmelzendem Stickstoff (ca. −210 °C, Abkühlrate 10^4 bis 10^5 $K \cdot s^{-1}$).

Ätzung: Eine Minute.

Beschattung: 1. Pt-C-Gemisch
2. Kohlebedampfung

Refraktometrie

Abbé-Refraktometer Zomess G 110, Zeiss Opton, Oberkochen.
Umlaufthermostat Typ FE, Haake, Berlin.

Literatur

1. Nürnberg E, Pohler W (1983) Pharmazeut Ztg 128:2601
2. Nürnber E, Pohler W (1983) Dtsch Apoth Ztg 123:1993
3. Nürnberg E, Pohler W (1984) Progr Colloid and Polymer Sci 69
4. Pohler W (1983) Dissertation Unversität Erlangen-Nürnberg
5. Bremecker KD (1980) Acta Pharm Technol 26:231
6. Schwetlick K et al (1974) Organikum, 13 Aufl, S 92ff, VEB Deutscher Verlag der Wissenschaften, Berlin
7. Rösch M (1963) Fette-Seifen-Anstrichmittel 65:223
8. Courtney DL (1972) Am Perfumer and Cosmetics 87:31
9. Becher P (1967) Am Perfumer and Cosmetics 82:41
10. Schulman JH, Matalon R, Cohen M (1951) Discuss Faraday Soc 11:117
11. Boehmke G, Heusch R (1960) Fette-Seifen-Anstrichmittel 62:87
12. Heusch R, Stessel A, Schwendke H (1976) Fette-Seifen-Anstrichmittel 78:359
13. Feger M (1978) Dissertation TU Braunschweig
14. Martin AN, Swarbrick J, Cammarata A (1980) Physikalische Pharmazie, 2 Aufl, S 65ff, Wissenschaftl Verlagsges, Stuttgart
15. Schönfeld N (1976) Grenzflächenaktive Äthylenoxid-Addukte, 1 Aufl, S 117 u 155ff, Wissenschaftl Verlagsges, Stuttgart

Anschrift der Verfasser:

Prof. Dr. E. Nürnberg
Institut für Pharmazie und Lebensmittelchemie
Lehrstuhl für Pharmazeutische Technologie der
Friedrich-Alexander-Universität Erlangen-Nürnberg
Schuhstraße 19
D-8520 Erlangen

Über das Phasen- und Struktursystem thermotroper Flüssiger Kristalle*)

H. Sackmann

Sektion Chemie der Martin-Luther-Universität Halle-Wittenberg, Halle (Saale), DDR

Zusammenfassung: In einer Übersicht wird das Phasen- und Struktursystem thermotroper Flüssiger Kristalle behandelt. Die Phasen werden auf Grund ihrer Mischbarkeit in Phasentypen zusammengefaßt und diese mit Buchstabensymbolen gekennzeichnet. Die Phasen erscheinen in einer bestimmten Folge auf der Temperaturskala. Dem Phasensystem entspricht ein Struktursystem, welches beschrieben wird. In den Umwandlungswärmen wird die Änderung der Strukturordnung der beteiligten Phasen sichtbar. Besonderheiten im Phasenverhalten und der Strukturen bestimmter polarer Substanzen werden erörtert (reentrant Phasen und A-Phasenpolymorphie).

Abstract: A review of phases and structures in thermotropic liquid crystals is given. Studies on the miscibility of phases have shown a system of l. c. phases with a characteristic phase sequence on the temperature scale. The phase types are coded by letter symbols. A system of structures corresponds to the system of phase types. The changes of order in phase transitions are reflected in the transition heats. Special features in the phase behaviour and the structures of more polar substances are described (reentrant phase and A-phase polymorphism).

Key words: Thermotropic Liquid Crystals, Classification of l. c. Phases, System of l. c. Structures, Phase Transitions.

1. Einleitung

Flüssige Kristalle sind kondensierte Stoffzustände, die in Ein- und Mehrstoffsystemen als Phasen in bestimmten individuellen Temperatur-, Druck- und Konzentrationsbereichen existieren. Sie erscheinen auf der Temperaturskala aus den gewöhnlichen isotropen Schmelzen beim Abkühlen und gehen bei tiefen Temperaturen in kristallisierte Festkörper über.

Die Strukturelemente dieser Phasen sind molekulare Einheiten mit starker Formanisotropie. Das erlaubt den Aufbau von zwei prinzipiellen Phasenstrukturen, die man als nematische und smektische Flüssige Kirstalle zu bezeichnen pflegt (Abb. 1).

In den nematischen Phasen besteht eine durch die Orientierung der Längsachsen der Moleküle gegebene Vorzugsrichtung (Direktororientierung). In den smektischen Phasen ist die Schichtanordnung der Moleküle gemeinsames, wichtiges Strukturmerkmal.

Die Moleküle mit starker Formanisotropie sind Strukturen, die aus Kohlenwasserstoffgruppen vom aliphatischen und aromatischen Typ in einer weitgehend gestreckten Anordnung zusammengesetzt sind.

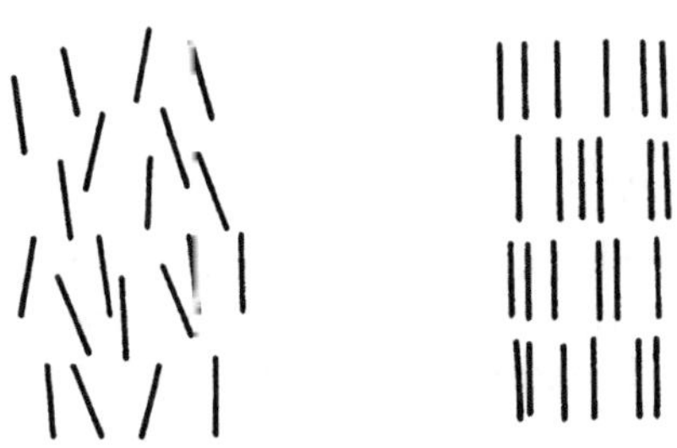

Abb. 1. Schema der nematischen Struktur und der smektischen Schichtstrukturen Flüssiger Kristalle

*) Vortrag, gehalten auf der 31. Hauptversammlung der Kolloid-Gesellschaft, Bayreuth 11. bis 14. Oktober 1983

KT-W 794

Der Einbau von Heteroatomen in diese Anordnungen ist verbreitet, wenn das Grundprinzip der gestreckten Anordnung nicht entscheidend verletzt wird. Einige wenige Verbindungen mit ihren Molekülstrukturen sind in den folgenden Abschnitten genannt. Jedoch geht es bei einer Übersicht über Phasen und Strukturen in erster Linie um die Organisation der als Ganzes stäbchenförmig vereinfachten molekularen Strukturelemente. Eine umfassende Darstellung der molekularen Spezies kann man [1] entnehmen.

Die stoffliche Seite der kristallin-flüssigen (kr. fl.) Systeme spielt auch eine Rolle bei der Unterscheidung zwischen thermotropen und lyotropen Flüssigen Kristallen. In lyotropen Systemen sind Stoffe beteiligt, welche selbst keine kr. fl. Phasen aufweisen, aber für den Aufbau der kr. fl. Phasen in Mehrstoffsystemen und ihre Eigenschaften von Bedeutung sind. Eine große Klasse dieser Art sind Stoffsysteme mit amphiphilen Substanzen und Wasser als Komponenten.

Eine scharfe Unterscheidung zwischen thermotropen und lyotropen Systemen ist jedoch nicht möglich. Die übernommene Bezeichnungsweise ist für eine physikalisch präzisierte Definition nicht ausreichend, was eine in der Wissenschaft mit zunehmender Erkenntis durchaus häufige Entwicklung ist. Der Gebrauch der Bezeichnungen gliedert aber derzeit in genügender und nützlicher Weise die stofflichen Bereiche, die zum Teil eigenständige Problemstellungen haben, eine Bevorzugung bestimmter methodischer Hilfsmittel aufweisen und auch in Anwendungen verschiedener Art und in verschiedene Disziplinen einmünden. Gerade auf dem Strukturgebiet der Phasen sind aber die Zusammenhänge besonders eng.

Im folgenden wird eine elementare Übersicht über Phasen und Strukturen der thermotropen Systeme gegeben. Die smektischen Phasen mit ihren Schichtstrukturen stehen dabei im Vordergrund. Eine ausführlichere Darstellung ist an anderer Stelle erschienen [2]. Hier findet sich auch ein weitergehender Zugang zur Originalliteratur.

2. Das Phasensystem

2.1. Texturen, Polymorphie

Die Erkennung der kr. fl. Phasen erfolgt im allgemeinen durch mikroskopische Betrachtung einer Substanzprobe auf einem Heiztisch im polarisierten Licht. Diese Bilder von kr. fl. Phasen (Texturen) erlauben oft eine erste Abschätzung über das Vorliegen eines bestimmten Phasentyps. Insbesondere an dem Auftreten von charakteristischen optischen Diskontinuitäten kann eine Unterscheidung zwischen nematischen und smektischen Phasen oder auch eine Unterscheidung zwischen verschiedenen smektischen Phasen erfolgen. Das Verschwinden oder Erscheinen bestimmter Texturen bei reproduzierbaren Temperaturen ergibt die Schmelz- oder Erstarrungstemperatur (Ep), die Temperatur des Überganges in den isotropen Zustand (Klärtemperatur, Klp) oder die Umwandlungstemperaturen (U) von verschiednen kr. fl. Phasen. Eine zusammenfassende Darstellung kr. fl. Texturen findet sich in [3].

Das Auftreten mehrerer kr. fl. Phasen an einem Stoff wird, wie bei den Kristallphasen, als Polymorphie bezeichnet. Die Registrierung des Trends dieser charakteristischen Temperaturen in homologen Reihen von Stoffklassen erlaubt oft eine erste Zuordnung individueller Phasen zu Phasentypen (Abb. 2) [2, 4].

2.2 Die auswählende Mischbarkeit (Mischbarkeitsauswahlregel)

Es hat sich gezeigt, daß die Existenz lückenloser Mischbarkeit zwischen kr. fl. Phasen der Komponenten in einem binären System als ein Kriterium für die Verwandtschaft individueller Phasen benutzt werden kann: Sind zwei kr. fl. Phasen durch ein Gebiet *lücken-*

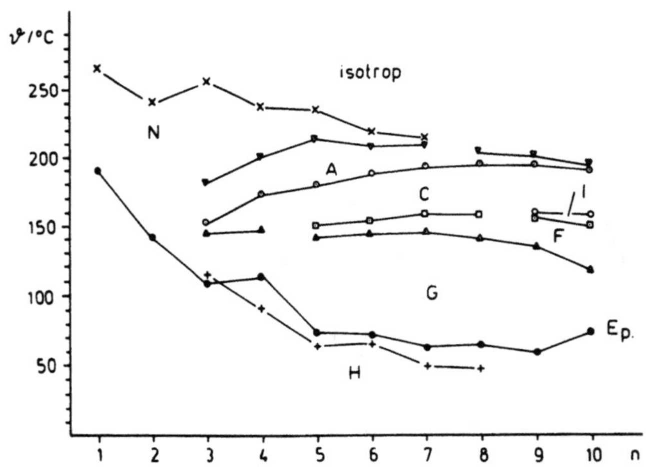

Abb. 2. Umwandlungstemperaturen in der homologen Reihe der Terephthalyliden–bis–[4-n-alkylaniline]

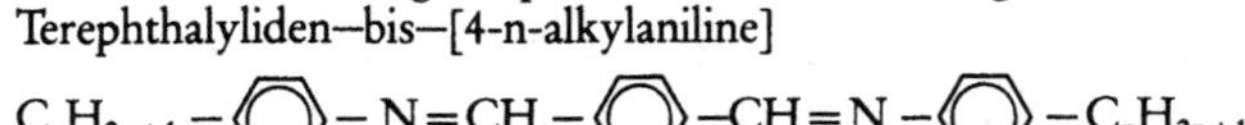

Umwandlung: x (N–is) ∇ (A–N) bzw. (A–is)
⊙ (C–A) ⊡ (F–C) bzw. (F–I)
○ (I–C) ◬ (G–C) bzw. (G–F)
+ (H–G)

H ist metastabil von n = 4 ab; siehe [2, 4]

loser Mischbarkeit verbunden, so können sie mit einem gemeinsamen Kennzeichen (Buchstabensymbol) versehen werden. Dies bedingt, daß eine lückenlose Mischbarkeit zwischen Phasen verschiedener Kennzeichnung nicht auftreten darf [2, 5, 6].

In Abbildung 3 a ist beispielhaft das Mischbarkeitsverhalten zweier Substanzen der in Abbildung 2 verwendeten homologen Reihe mit n = 5 und n = 7 dargestellt. Die sechs kr. fl. Phasen der beiden Substanzen sind in der Temperaturfolge ihres Erscheinens jeweils miteinander lückenlos mischbar. Gibt man an einer Substanz die Benennung der Phasen vor (bei n = 5: N A C F G H), so sind nach der Mischbarkeitsauswahlregel auch die Phasen der Substanz n = 7 jeweils durch diese Bezeichnungen festgelegt. Auch alle Mischphasen lassen sich in dieser Weise indizieren. Die smektischen Phasen H liegen im ganzen Konzentrationsgebiet im metastabilen, d. h. gegenüber dem kristallinen Zustand unterkühlten Phasengebiet.

In Abbildung 3 b ist ein Beispiel für auswählendes Mischbarkeitsverhalten gezeigt. Als Standardsubstanz wurde wieder die Komponente n = 5 verwendet. Als zweite Komponente wurde die Verbindung n = 9 eingesetzt. Die beobachteten lückenlosen Mischbarkeiten zeigen, daß diese Verbindung keine nematische Phase N aufweist. Dagegen besitzt sie eine Phase, welche an der Testsubstanz nicht beobachtet wird (sie wird hier mit I bezeichnet). Auch die Existenz einer metastabilen H-Phase im unterkühlten Gebiet kann nicht beobachtet werden.

Vergleicht man die kr. fl. Phasenfolge dieser drei Substanzen, so ergeben sich (die Phasenfolge ist mit steigender Temperatur aufgeschrieben) die Polymorphievarianten

n = 5	H G F C A N
n = 7	H G F C A N
n = 9	G F I C A

In den Mischphasengebieten des Systems der Abbildung 3 b kann noch eine weitere Variante be-

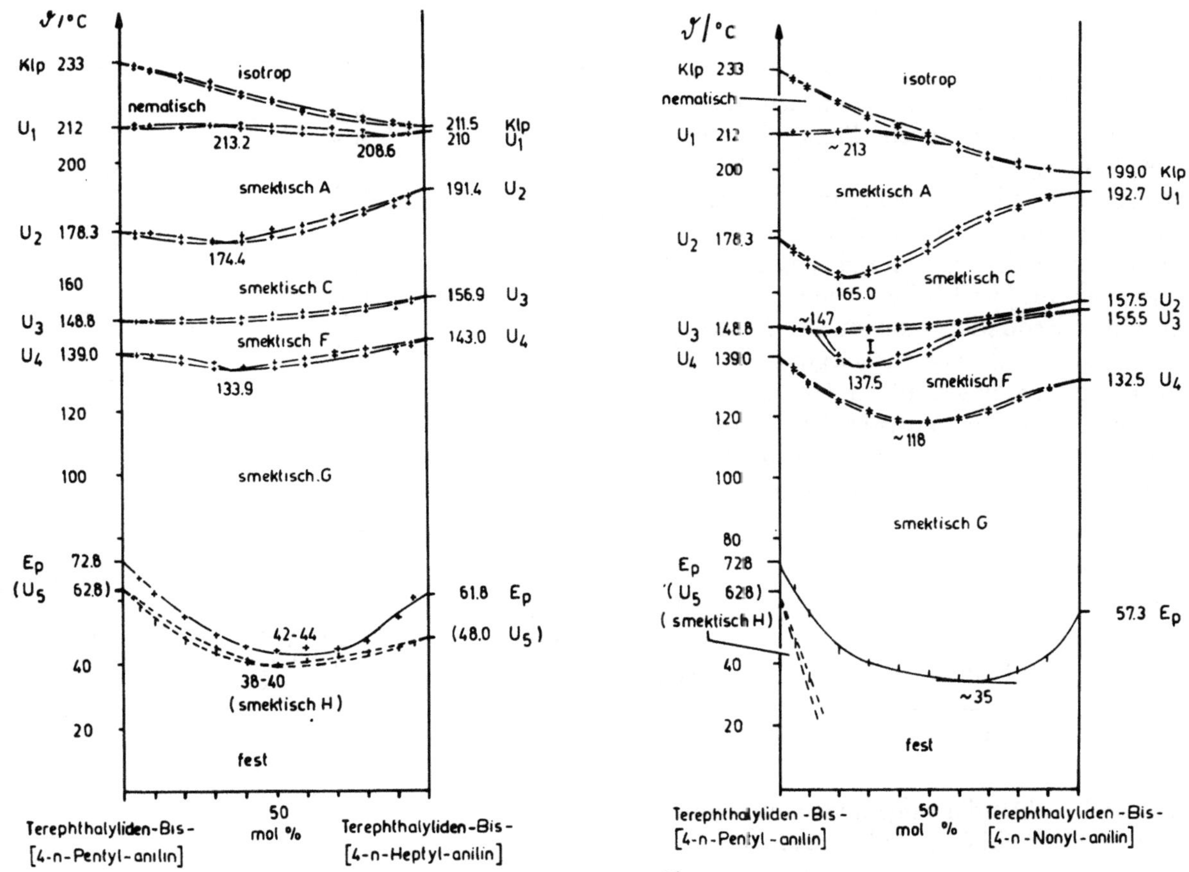

3 a 3 b

Abb. 3. Isobare Zustandsdiagramme (Temperatur – Konzentration) der Substanzen n = 5 und n = 7 (a) und n = 5 und n = 9 (b) nach [4]

obachtet werden. Bei Konzentrationen zwischen ~ 20 und ~ 60 Molprozent der Komponente n = 9 tritt die Phasenfolge

GFICAN

auf.

Die Gesamtheit der in diesem Stoffsystem vorgenommenen Untersuchungen ergibt die in Abbildung 2 enthaltene Benennung der Phasen für die einzelnen Substanzen [4].

2.3 Phasentypen und Phasensequenzen

Das in den Abschnitten 2.1 und 2.2 dargestellte Vorgehen der makroskopischen Charakterisierung kr. fl. Phasen zeigt in weitem Umfang Konsistenz von Texturbetrachtung, thermische Charakterisierung der Phasenumwandlungen in homologen Reihen und Mischbarkeit.

Die Beobachtung der Mischbarkeitsbeziehungen an einer sehr umfangreichen Substanzpalette, die nicht auf enge chemische Verwandtschaften beschränkt ist, ergibt eine Klassifikation der kr. fl. Phasen. Die individuellen Phasen können in Phasentypen zusammengefaßt werden, die mit Buchstabensymbolen gekennzeichnet sind. Gegenwärtig sind neben dem nematischen Typ N zehn Phasentypen unterschieden, die zusammenfassend als smektische Phasen bezeichnet werden: A, B, C, E, F, G, G', H, H', I. (Die Buchstabenfolge nach dem Alphabet gibt etwa die zeitliche Reihenfolge der erstmaligen Zuordnung und Benennung der Phasen wieder). G'- und H'-Phasen werden neuerdings auch mit J und K bezeichnet [7].

Neben dieser Reduktion individueller Phasen auf Phasentypen ergibt sich eine festgelegte Reihenfolge im Erscheinen dieser Phasen auf der Temperaturskala (Phasensequenz). Die an individuellen Substanzen aufgefundenen Varianten können abgeleitet werden aus zwei allgemeinen Polymorphievarianten (1 a) und (1 b):

$$\text{H H' G G' F I B C A N} \qquad (1\,a)$$

$$\text{E B C A N} \qquad (1\,b)$$

⟶ Temperatur

Die Tabelle 1 enthält die aufgefundenen Polymorphievarianten an individuellen Substanzen.

Wie ersichtlich, entstehen sie aus (1 a) oder (1 b) durch den Ausfall von Phasen. Eine Vertauschung der Reihenfolge wird nicht beobachtet.

Die isotropen Phasen sind in Tabelle 1 nicht vermerkt. Sie folgen stets auf die jeweiligen kr. fl. Hochtemperaturphasen. Auch die kristallinen Phasen sind nicht vermerkt. Im allgemeinen beenden sie die kr. fl. Sequenzen nach tieferen Temperaturen. Jedoch können eine oder mehrere kr. fl. Phasen im metastabilen, gegenüber dem Kristall unterkühlten Gebiet liegen, wie z. B. für die H-Phasen in den Abbildungen 2 und 3 gezeigt wurde (siehe auch Abb. 5).

Die E-Phasen wurden bisher nicht gefunden in Polymorphievarianten, die auch Phasen vom Typ I bis H' aufweisen. Deshalb müssen Varianten mit E-Phasen in dem Schema (1 b) zusammengefaßt werden.

Die Anzahl der Substanzen und Mischphasen in binären Substanzgemischen, an welchen die Varianten der Tabelle 1 beobachtet wurden, ist unterschiedlich. Am häufigsten wurden Varianten gefunden, die nur N-, A- und C-Phasen aufweisen.

Tabelle 1. Polymorphievarianten

monomorph	dimorph	trimorph
N	A N	B A N
A	B N	C A N
B	C N	G A N
C	G N	E A N
E	B A	G B N
I	C A	I C N
	E A	E B A
	G C	B C A
	I C	E C A
	E B	I C A
	G I	G C A
	G F	G B A
		G F A
		F I C
		G I C
tetramorph	**pentamorph**	**hexamorph**
G B A N	G B C A N	G F B C A N
E B A N	H G C A N	G F I C A N
B C A N	G I C A N	H'G'I C A N
G C A N	G F C A N	H G F C A N
I C A N	H'G' I C N	
H G C N	G F I C A	
E B C A	H G F C A	
G B C A		
G F C A		
H G C A		
G I C A		
H'I C N		

⟶ Temperatur

3. Das Struktursystem

3.1 Die smektische Schichtstrukturen

Den nematischen Phasen stehen die smektischen Phasen mit Schichtstruktur gegenüber (Abb. 1). Insbesondere durch röntgenographische Untersuchungen ist die Organisation dieser Schichtstrukturen in wesentlichen Elementen aufgeklärt worden. Sie sind in Abbildung 4 zusammengestellt [2]. Man unterscheidet zunächst zwei Gruppen von Schichtstrukturen, die in Kurzform als smektische Strukturen „ohne Ordnung" und smektische Strukturen „mit Ordnung" bezeichnet werden.

Bei den Strukturen „ohne Ordnung" liegt eine ungeordnete Verteilung der Molekülschwerpunkte innerhalb der Schichten vor. Dabei können die Moleküllängsachsen im Mittel senkrecht zu den Schichtebenen angeordnet sein (A-Typ), oder sie bilden einen Winkel mit der Schichtnormalen (C-Typ).

Die Strukturen „mit Ordnung" sind durch eine Ordnung der Molekülschwerpunkte innerhalb der Schichten ausgezeichnet. In den Strukturen von B-Typ sind die Moleküllängsachsen in einer hexagonalen Packung geordnet [8]. Sie stehen senkrecht auf den Schichtebenen. Es besteht Rotationsbewegung um die Längsachsen; sie ist aber nicht völlig frei, so daß lokale Verbände mit niederer Symmetrie vorhanden sind.

Die Strukturen von F- und I-Typ haben eine zur Schichtebene geneigte Längsachsenorientierung der Moleküle in monokliner Symmetrie. In Ebenen senkrecht zu den Achsen besteht eine hexagonale Anordnung. Wegen der Neigung der Längsachsen wird sie als pseudohexagonal bezeichnet [9, 10, 11].

Ein wesentlicher Unterschied der beiden Strukturen F und I besteht in einem unterschiedlichen Längenverhältnis der Gitterparameter a und b. Für den F-Typ gilt a/b > 1 und für den I-Typ a/b < 1: Betrachtet man das hexagonale Netz benachbarter Moleküle, so ist in einem Falle (F) die Neigungsrichtung der Längsachsen auf die Ecken der Hexagone gerichtet, im anderen Falle (I) auf die Kantenmitten derselben [12].

In den Strukturen des E-Typs sind die Längsachsen senkrecht zu den Schichtebenen orientiert [13, 14, 15]. Sie sind in einem orthorhombischen Gitter angeordnet. Im Querschnitt ergibt sich eine „herring-bone" Anordnung. Diese charakterisiert die starke Einschränkung der Rotationsbewegung um die Längsachse.

Ganz ähnlich ist die Organisation in den H- und H'-Strukturen. Die Längsachsen sind zur Schichtebene geneigt. Im monoklinen Gitter besteht eine herringbone Packung in den Schichten [16]. In Bezug auf die Parameter a und b wird a/b > 1 (H-Typ) und a/b < 1 (H'-Typ) gefunden [12].

Die Struktur des G-Typs kann zunächst als monoklines Gitter mit einer im Mittel pseudohexagonalen Packung, wie sie in den F- und I-Phasen besteht, beschrieben werden. Sie ist aber komplexer durch eine dreifache Überlagerung lokaler Gitter vom herringbone Typ in der a/b Ebene [16, 17, 18]. Der Unterschied im Parameterverhältnis a/b ergibt zwei Strukturen G (a/b > 1) und G' (a/b < 1) [12].

Damit ist eine Ebene der Strukturbeschreibung gegeben, welche die durch die makroskopischen Methoden festgelegten Phasentypen auch strukturell differenziert.

Weitergehende Strukturuntersuchungen an einzelnen Phasen der geordneten smektischen Typen haben weitere Strukturmerkmale in die Diskussion gebracht, deren strukturtypische Verallgemeinerung noch nicht endgültig beurteilt werden kann. Sie betreffen die genauere Ordnungskorrelation innerhalb der Schich-

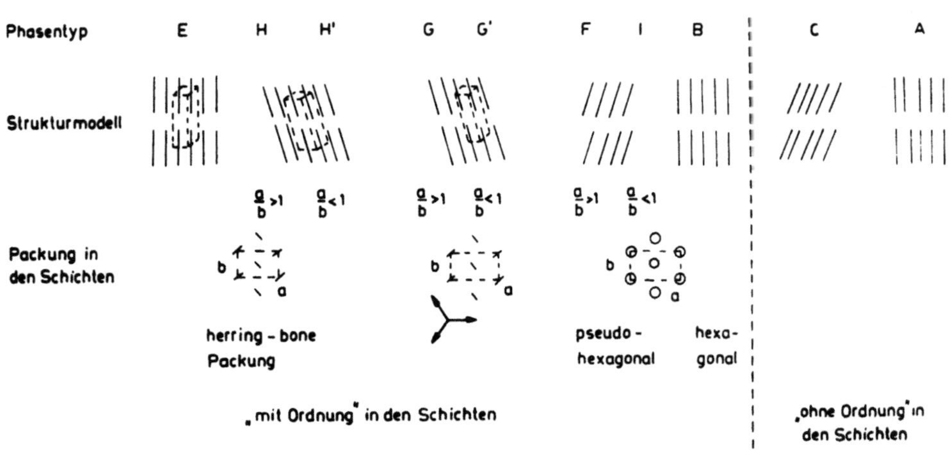

Abb. 4. Strukturelemente der smektischen Phasentypen

ten und insbesondere die Korrelation zwischen den Schichten.

Man kann davon ausgehen, daß in den E, H, H', G und G'-Phasen eine noch nicht umfassend untersuchte positionelle Fernordnung zwischen den Schichten besteht. Sie sind also 3d-Strukturen.

In B-Phasen wurden verschiedene Strukturvarianten gefunden. In einigen Phasen konnte eine 3d-Positionsfernordnung festgestellt werden. Es können auch Stapelanordnungen der Schichten auftreten mit periodischer Anordnung der Schichten aaa oder abab oder abcabc (ähnlich den dichten Kugelpackungen). Ferner wurden erhebliche Korrelationslängen in den Schichten festgestellt [19, 20, 21]. Man bezeichnet solche Strukturen als „feste" B-Phasen. Andererseits sind „hexatische" B-Phasen beschrieben worden mit einer 3dOrientierungsordnung hexatischer Schichten [22]. Als hexatische Schichten sind 2-dimensionale Strukturen benannt mit einer positionellen Nahordnung in Verbindung mit einer Orientierungsfernordnung. Da an einer Substanz ein Phasenübergang von einer „festen" in eine „hexatische" B-Struktur beobachtet wurde [23], muß eine Differenzierung der B-Phasen in zwei Phasentypen ins Auge gefaßt werden.

Auch an I- und F-Phasen liegen Untersuchungen in dieser Richtung vor, die auf hexatische Strukturen hinweisen [24, 25].

3.2 Phasensequenz, Phasenumwandlungen

Die in den beobachteten Polymorphievarianten der Tabelle 1 und in den Varianten (1 a) und (1 b) verallgemeinerte Temperatursequenz der kr. fl. Phasen erscheint nunmehr als eine Sequenz der Ordnung von Strukturen. Den stark rotationsbehinderten geordneten Phasenstrukturen vom Typ E, H und H' folgen über G- und G'-Strukturen die weniger rotationsbehinderten B-, F- und I-Strukturen. Dabei tritt auch eine Lockerung der Schichtkorrelation ein. Beim Übergang in die ungeordneten smektischen Phasen A und C wird jede Art einer Fernordnung in und zwischen den Schichten aufgehoben.

Für die Energetik der Phasenumwandlungen, die sich aus den Varianten der Tabelle 1 ergeben, liegen zahlreiche Vergleichsmöglichkeiten in Form von Umwandlungsenthalpien vor [2]. Die Gesamtheit des Materials, unabhängig von der molekularen Spezifik, läßt nur einen Umwandlungstyp hervortreten: Die Schmelzenthalpien der kristallisierten Phasen liegen meist, unabhängig vom kr. fl. Phasentyp, der beteiligt ist, um eine Größenordnung höher als die Umwandlungsenthalpien zwischen kr. fl. Phasen. Engt man jedoch die Betrachtungen auf chemisch vergleichbare Stoffe in homologen Reihen ein, so läßt sich sowohl der phasentypische als auch der molekülspezifische Einfluß erkennen (Tab. 2).

Es wird zunächst der Unterschied um eine Größenordnung für alle Phasenübergänge Kristallphase – kr. fl. Phase deutlich. Liegen genügend Vergleichsmöglichkeiten für Phasenübergänge kr. fl. – kr. fl. des gleichen Typs vor, so erkennt man (z. B. für den Übergang A–N, C–A, F–C, H–G oder auch bei Einbeziehung des isotropen Zustandes N–is, A–is) den Einfluß der Molekülparameter: Die Werte steigen (zum Teil alternierend) mit zunehmender Anzahl der Glieder der Seitenketten der Moleküle an, können jedoch auch abfallen (H–G, G–F). Bemerkenswert ist der Vergleich jener Umwandlungen, bei welchen der Typ einer der beiden Phasen geändert ist. Der Wechsel in der Größenordnung beim Vergleich der Umwandlungen G–C und G–F oder F–C und F–I zeigt die starke Änderung

Tabelle 2. Umwandlungsenthalpien (J/mol) der Terephthalyliden-bis- 4-n-alkylanilien [26, 27]

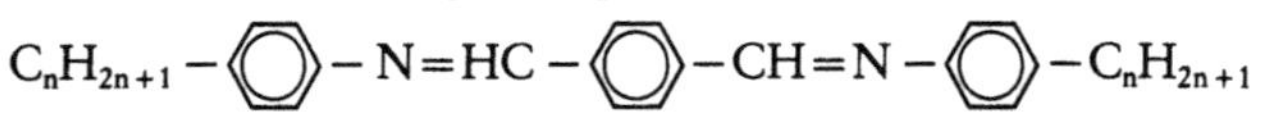

n	cr	S_5	H	G	F	I	C	A	N	is
2	· 30 100	·	—	—	—	—	—	—	· 620	·
3	· 14 300	· 1 300	· 600	· 7 000	—	—	· x	· 385	· 1 100	·
4	· 19 800	(· 410	· 1 300)	· 4 370	—	—	· x	· 590	· 1 350	·
5	· 15 100	—	(· 1 140)	· 95	· 3 660	—	· x	· 1 200	· 1 530	·
6	· 17 400	—	(· 910)	· 62	· 4 720	—	· 70	· 1 540	· 1 440	·
7	· 13 000	—	(· 990)	· 30	· 4 750	—	· 200	· 2 730	· 2 590	·
8	· 32 500	—	—	· 7	· 5 460	—	· 260	· 5 670	—	·
9	· 32 200	—	—	· 3–6	· 19	·	6 150· 550	· 6 680	—	·
10	· 50 500	—	—	· 2–4	· 18	·	6 310· 1 770	· 7 080	—	·

cr = feste Phase; S_5 noch nicht näher gekennzeichnete Phase; x Wärmekapazität zeigt einen Sprung; ( ) im metastabilen Gebiet der Ordnung der Strukturen zwischen geordneten und ungeordneten smektischen Phasen an.

Die in dieser homologen Reihe beobachteten Beziehungen wurden auch an anderen Substanzgruppen gefunden. Die vergleichende Betrachtung ergibt, daß Phasenübergänge mit großen Strukturänderungen große Enthalpiewerte haben, während bei geringen Änderungen der Struktur kleine Enthalpiewerte zu beobachten sind. Große Enthalpiewerte werden beobachtet bei Phasenübergängen zwischen geordneten smektischen Phasen einerseits und ungeordneten smektischen Phasen wie auch nematischen Phasen und auch isotropen Phasen andererseits; auch Umwandlungen zwischen ungeordneten smektischen Phasen und nematischen bzw. isotropen Phasen haben große Enthalpiewerte. Übergänge zwischen sehr ähnlichen Strukturen, etwa F–I, sind sehr klein; bei einer G–G'-Umwandlung wurden keine meßbaren Enthalpiewerte gefunden.

Es ist hier die Frage nach der Natur der Phasenumwandlungen bei kr. fl. Phasen zu stellen. Ohne auf das beträchtliche Umfeld der experimentellen und theoretischen Untersuchungen einzugehen [2], kann im Rahmen der hier interessierenden Betrachtung über das phasentypische Verhalten summarisch gesagt werden, daß für die Phasenumwandlung C–A der Umwandlungstyp 2. Ordnung angenommen werden muß. Die Umwandlungen A–N können substanzspezifisch sowohl von 1. als auch von 2. Ordnung sein. Die anderen Umwandlungen dürften von 1. Ordnung sein. Bei sehr kleinen Umwandlungseffekten, etwa bei der Umwandlung F–I, ist allerdings gegenwärtig eine Aussage nicht mögich.

Die um eine Größenordnung abgesetzten Enthalpiewerte beim Übergang Kristall – kr. fl. Phase legen die Frage nach der Abgrenzung des Systems der Flüssigen Kristalle von jenem der Kristalle vor. Im Rahmen der in 3.1 dargelegten Strukturbetrachtung kann man dem Terminus „Flüssiger Kristall" eine verbindliche physikalische Bestimmung zu geben versuchen. So kann man die A- und C-Phasen mit ihrer flüssigkeitsanalogen Schichtstruktur und den nur durch Nahordnung verbundenen Schichten als „echte" Flüssige Kristalle bezeichnen. In erweiterter Fassung kann man die hexatischen Phasen einbeziehen und nur die E-, H-, H'-, G- und G'-Phasen, sowie die „feste" B-Phase ausschließen, die eine ausgeprägte Schichtkorrelation aufweisen.

Die diskutierten Schmelzenthalpien weisen darauf hin, daß dieser Phasenübergang von einem starken Abbau eines weitgehend ungestörten Gitters begleitet

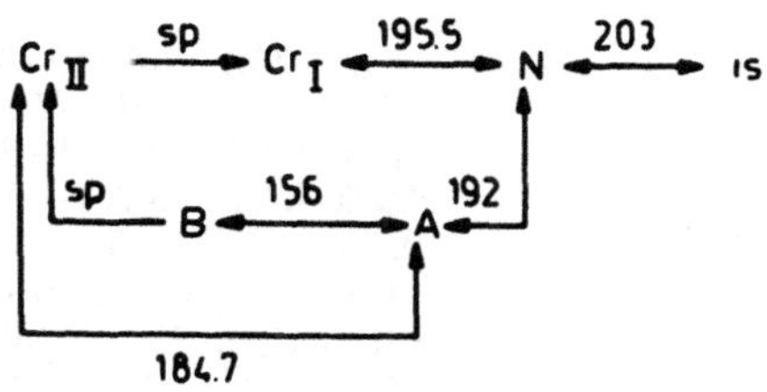

Abb. 5. Polymorphie der Substanz

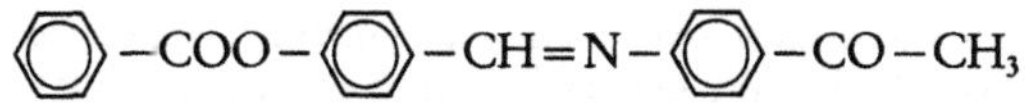 [28]

Umwandlungstemperaturen in Grad Celsius; S_p = spontane Umwandlung

ist. Auch wird eine drastische Verminderung der Röntgenreflexe beobachtet.

Es ist daher sinnvoll, die kr. fl. Phasen mit Schichtstrukturen als eine Gesamtheit den ungestörten Kristallen gegenüber zu stellen. Dies umso mehr, als eine umfassende Kenntnis der Kristallstrukturen, die an Substanzen mit kr. fl. Phasen vorkommen, noch nicht vorliegt.

Die Kristallphasen brechen die Sequenzen (1 a) und (1 b) an verschiedenen Stellen bei tieferen Temperaturen ab, wie Tabelle 1 zeigt. Gelingt eine Unterkühlung, so kann im metastabilen Gebiet eine Fortsetzung der kr. fl. Sequenz beobachtet werden (Abb. 5). Die genannte Substanz besitzt als einzige stabile kr. fl. Phase eine nematische Phase. Im unterkühlten Bereich treten zwei weitere kr. fl. Phasen auf. Eine Phase A ist metastabil in Bezug auf den Kristall cr_I, und eine B-Phase ist metastabil in Bezug auf cr_I und cr_{II}. Bezieht man den unterkühlten Bereich in die Sequenz der kr. fl. Phasen ein, dann hat diese Substanz eine trimorphe Polymorphie B A N. Auch in diesem Verhalten zeigt sich der strukturelle Zusammenhang der kr. fl. Phasen.

3.3 *Die verdrillten (twisted) Phasen*

An Substanzen mit chiralen Molekülen werden kr. fl. Phasen mit strukturellen Besonderheiten beobachtet. An Stelle von nematischen Phasen treten die bekannten cholesterinischen Phasen (N*). Der Direktor zeigt eine helikale Verdrillung (Abb. 6). In smektischen Phasen ist die Verdrillung an Strukturen mit geneigten Längsachsen gebunden. An die Stelle der smektischen C-Phase tritt eine Phase C* mit einer helikalen Verdrillung der Richtung des Neigungswinkels (Abb. 6). Die Ganghöhe der Verdrillungen (pitch) übersteigt die Moleküllänge bzw. Schichtdicke um mehrere Größenordnungen. Ferner gibt es Untersu-

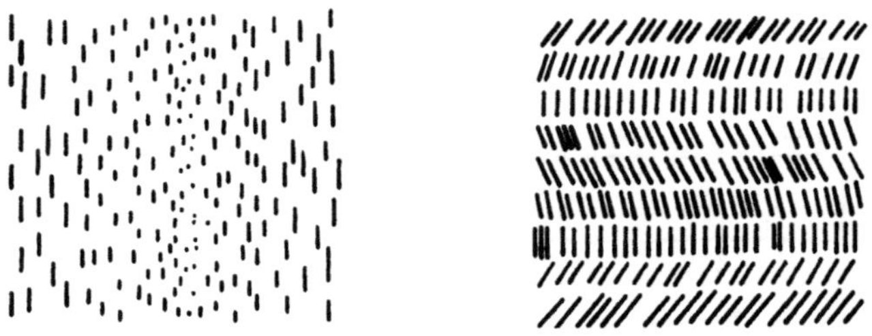

Abb. 6. Schema der Struktur der verdrillten nematischen (cholesterinischen) Phasen (links) und der verdrillten smektischen C-Phase (rechts)

chungen, die auf die Existenz verdrillter Phasen vom Typ F*, I* und G* hinweisen [29, 30, 31].

Die verdrillten Phasen treten in der Phasensequenz an die Stelle der entsprechenden unverdrillten Phasen, z. B. in den Sequenzen C* A N* oder G* C* A N.

3.4 Systeme mit nematischen reentrant Phasen, A- und C-Phasenpolymorphie

Neben den in (1a) und (1b) zusammengefaßten Temperatursequenzen kr. fl. Phasen wurde sowohl an reinen Stoffen als auch an Mischphasen ein wiederkehrendes Erscheinen der nematischen Phasen auf der Temperaturskala beobachtet [32, 33, 34]. Die Wiederholung einer Phase in einer Temperatursequenz ist an anderen Stoffsystemen lange bekannt. Bei kr. fl. Systemen wird die Bezeichnung reentrant Phasen benutzt.

So besitzt z. B. die Verbindung $C_7H_{15}O$–⌬–CH=N–⌬–⌬–CN [35] die Sequenz (Umwandlungstemperaturen in Grad Celsius):

Cr 73 E 95 B 111 N_{re} 140 A 202 N 273 is. An bisher ca. 30 Substanzen und zahlreichen Mischungen sind Varianten beobachtet worden, die in Tabelle 3 zusammengestellt sind [2]. Man erkennt, daß die N_{re}-Phasen bisher nur in Sequenzen beobachtet wurden, welche smektische A- und C-Phasen bei höheren Temperaturen aufweisen. Bei tiefen Temperaturen können A- und C-Phasen erneut auftreten. Der Klarheit wegen wird hier auf eine mögliche Bezeichnung „reentrant" verzichtet. Ein bemerkenswerter Fall mit zweifachem Wiedererscheinen der nematischen Phase ist aufgelistet [36]. An Mischphasen sind auch Sequenzen ohne nematische Hochtemperaturphasen beobachtet worden [37]. Man kann diese Hochtemperaturphasen aber bei geeigneter Wahl der Komponenten und der Konzentration auffinden, so daß die Bezeichnung N_{re} auch in diesen Sequenzen gerechtfertigt ist.

Tabelle 3. Polymorphievarianten in reentrant Systemen

N_{re} A N	C N_{re} A N I	in Mischungen beobachtet
N_{re} C A N	C N_{re} C A N I	
A N_{re} A N	N_{re} A I	
E B N_{re} A N	C N_{re} A I	
C C A N_{re} A N_{re} A N	N_{re} C A I	
	C N_{re} C A I	

Ferner wurden an reinen Stoffen und an Mischungen Phasen beobachtet, die zum Typ A gehören und durch Phasenumwandlungen direkt miteinander verbunden sind. Gegenwärtig werden vier A-Phasen unterschieden mit den Bezeichnungen A_d, A_1, A_2 und $\tilde{A}$. Ähnliches wird bei C-Phasen beobachtet [38, 39, 40, 41].

Diese Phänomene, die reentrant Polymorphie und die A-Phasen- bzw. C-Phasen-Polymorphie haben eine gemeinsame stoffliche Basis; sie wurden gefunden an Stoffen, deren Moleküle polare terminale Gruppen (in der überwiegenden Anzahl eine CN-Gruppe, aber auch die NO_2-Gruppe) aufweisen. Ein typisches Beispiel ist die in diesem Abschnitt oben genannte Verbindung.

Die strukturellen Erkenntnisse in diesen Systemen sind in Abbildung 7 zusammgefaßt. Man muß davon ausgehen, daß die CN-Gruppen und die aromatischen und aliphatischen Anteile benachbarter Moleküle eine

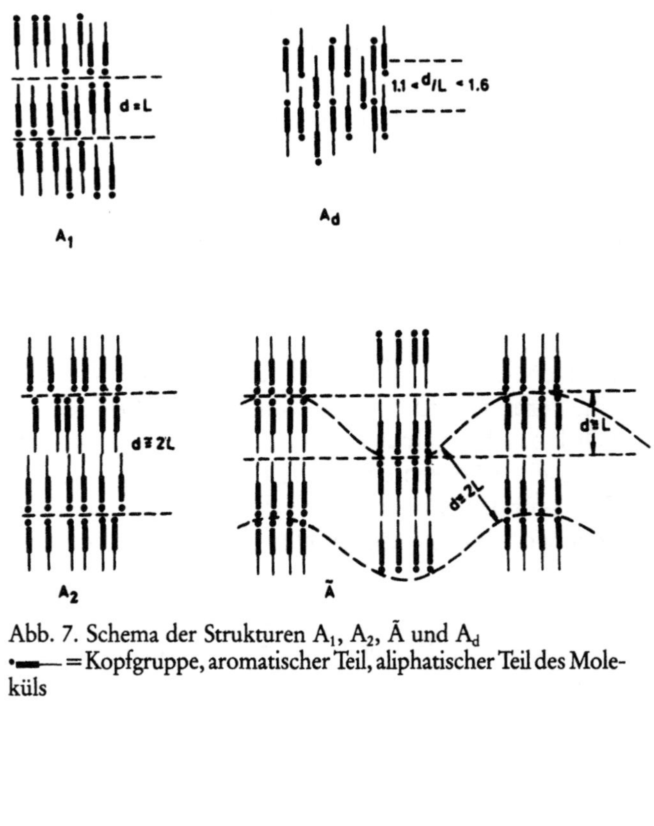

Abb. 7. Schema der Strukturen A_1, A_2, $\tilde{A}$ und A_d
•▬— = Kopfgruppe, aromatischer Teil, aliphatischer Teil des Moleküls

derartige Wechselwirkung aufeinander ausüben, daß es zu einer antiparallelen Stellung der Moleküle kommt. Diese kann in einer „Kopf-an-Kopf"-Anordnung bestehen oder in einer stärkeren Überlappung etwa in Form von „Dimeren" durch eine Wechselwirkung der CN-Gruppe eines Moleküls mit einem mehr positiven Molekülteil eines anderen.

Vergleicht man die A-Phasenstrukturen an den nicht-polaren Stoffen mit den in Abb. 7 dargestellten, so sind zunächst die unterschiedlichen Relationen d/L (d = Schichtabstand, L = Moleküllänge bei starr gestreckter Molekülanordnung) festzustellen. Bei den A-Phasen nicht-polarer Stoffe wird $d/L \leq 1$, bei denjenigen polarer Stoffe wird $d/L \geq 1$ gefunden.

In den Phasen A_2 [42, 43, 44] findet man einen Schichtabstand d, welcher der doppelten Länge der Moleküle L entspricht. Eine unregelmäßige „auf-und-ab"-Orientierung ergibt die Strukturen A_1 mit einem Monoschichtabstand [45]. Die Struktur der „Antiphase" $\tilde{A}$ ist eine Überlagerung einer Monoschicht mit einer Doppelschichtstruktur in Richtung der Schichtnormalen [46, 47]. Die Schichten der Strukturen A_d sind aus überlappenden antiparallelen Molekülen aufgebaut (d für Dimer) [2].

Gegenwärtig sind an reinen Stoffen die Übergänge A_2-A_d und $\tilde{A}$-A_1 bekannt. Die A_d-Struktur ist in einigen Hochtemperaturphasen in reentrant Systemen nachgewiesen. Als A-Phasen unterhalb der N_{re}-Phase wurden Strukturen vom Typ A_1 identifiziert. Es ist daher zu erwarten, daß die verschiedenen A- und C-Phasen der reentrant Varianten (Tab. 3) diesem Struktursystem angehören.

4. Schlußbemerkungen

Das hier dargestellte Phasen- und Struktursystem umfaßt das Hauptfeld der thermotropen Flüssigen Kristalle. Die Differenzierung der smektischen Grundstruktur bei polaren Substanzen leitet über zu kr. fl. Stoffen mit ionischen und chemisch komplizierten Kopfgruppensystemen, wie z. B. den Salzen von aliphatischen und aromatischen Carbonsäuren und den Phospholipiden. Sie organisieren sich bekanntlich in Schichtstrukturen, bei welchen die Kopfgruppen und der Kohlenwasserstoffanteil als relativ selbständig Bereiche Struktur und Eigenschaften bestimmen. Besteht in beiden Bereichen eine hohe Beweglichkeit, so entsprechen diese Schichtstrukturen den nicht geordneten smektischen Phasen, insbesondere vom A-Typ.

Diese amphiphilen Systeme sind zu differenzierten Wechselwirkungen z. B. mit Wasser befähigt, und man hat Schichtstrukturen, die dann zu den lyotropen Phasen und Strukturen gehören. Auch sie zeigen polymorphe Sequenzen. Bei tieferen Temperaturen können Schichtstrukturen entstehen, bei welchen sowohl der Kopfgruppenbereich als auch der Kohlenwasserstoffbereich eine höhere Ordnung aufweisen.

Es kann aber bei den amphiphilen Stoffen auch zu Organisationsformen kommen, bei denen die Moleküle Strukturelemente aufbauen, die in rhomboedrischen, hexagonalen oder kubischen Überstrukturen angeordnet sind.

Es soll daher ergänzend bemerkt werden, daß auch an nicht amphiphilen Stoffen neben den Schichtstrukturen kubische Strukturen beobachtet wurden. Die Gittereinheiten enthalten eine große Anzahl von Molekülen. Sie kommen in Polymorphie mit Schichtstrukturen vor, z. B. in den Sequenzen C – kubische Phase – A oder kubische Phase – C [2].

Auch die Blauen Phasen (blue phases) in thermotropen Systemen mit chiralen Molekülen müssen hier erwähnt werden. Sie treten zwischen den cholesterinischen (N^*) und den isotropen Phasen in sehr kleinen Temperaturbereichen auf [2].

Abschließend muß auf eine Gruppe von Phasen und Strukturen hingewiesen werden, die als „Scheibchenmesogene" (Disc-like Mesogens, Discotic Mesophases) bezeichnet werden. Die Moleküle besitzen eine ausgeprägte 2-dimensionale Ausdehnung meist mit einem starren Mittelteil und flexiblen Seitengruppen [48, 49]; im einfachsten Falle sind dies hexasubstituierte Derivate des Benzols. Es sind aber komplizierte Mittelstücke (core) bekannt, die aus vielfältigen Kombinationen cyclischer Kohlenstoffverbindungen, insbesondere vom aromatischen Typ, aufgebaut sind. Die scheibchenförmigen Moleküle bilden nematische Phasen oder bilden durch Stapelung Säulenaggregate, welche die konstituierenden Elemente von weiteren kr. fl. Phasen (columnar phases) z. B. vom hexagonalen Typ darstellen.

Diese Bemerkungen mögen darauf aufmerksam machen, daß das Gebiet der kr. fl. Phasen und Strukturen oder – wenn man dies bevorzugt – das Gebiet der Mesophasen ein offenes Panorama bietet, aber doch ein Thema mit Variationen darstellt.

Literatur

1. Demus D, Demus H, Zaschke H (1974) Flüssige Kirstalle in Tabellen, VEB Deutscher Verlag für Grundstoffindustrie Leipzig, ibid, 2. Band 1984, im Druck
2. Demus D, Diele S, Grande S, Sackmann H (1983) (ed) Brown GH, Polymorphism in Thermotropic Liquid Crytals, Advances in Liquid Crystals, Vol 6, Academic Press Inc

3. Demus D, Richter L (1978) Textures of Liquid Crystals, Verlag Chemie, Weinheim
4. Richter L, Demus D, Sackmann H (1981) Mol Cryst Liq Cryst 71:269
5. Sackmann H, Demus D (1966) Mol Cryst Liq Cryst 2:81, (1973) ibid 21:39
6. Sackmann H (1974) Pure Appl Chem 38:503
7. Gane PAC, Leadbetter AJ, Wrighton PG, Goodby JW, Gray GW, Tajbaksh AR (1983) Mol Cryst Liq Cryst 100:67
8. Levelut AM, Lambert M (1971) C R Hebd Seances Acad Sci 272:1018, Levelut AM, Doucet J, Lambert M (1974) J Phys (Orsay, Fr) 35:773
9. Demus D, Diele S, Klapperstück M, Link V, Zaschke H (1971) Mol Cryst Liq Cryst 15:161
10. Leadbetter AJ, Gaughan JP, Kelly B, Gray GW, Goodby JW (1979) J Phys (Orsay, Fr) 40:C3 - 178
11. Diele S, Demus D, Sackmann H (1980) Mol Cryst Liq Cryst Lett 56:217
12. Gane PAC, Leadbetter AJ, Wrighton PG (1981) Mol Cryst Liq Cryst 66:567
13. Diele S, Brand P, Sackmann H (1972) Mol Cryst Liq Cryst 17:163
14. Diele S (1974) Phys Status Solidi A 25:K 1 83
15. Doucet J, Levelut AM, Lambert M, Liebert L, Strzelecki L (1975) J Phys (Orsay, Fr) 36:C1-13
16. Doucet J, Levelut AM, Lambert M (1974) Phys Rev Lett 32:301
17. De Vries A, Fishel DL (1972) Mol Cryst Liq Cryst 16:311
18. Levelut AM, Lambert M (1971) C R Hebd Seances Acad Sci 272:1018
19. Moncton DE, Pindak R (1979) Phys Rev Lett 43:701
20. Pershan PS, Aeppli G, Litster JD, Birgeneau RJ (1981) Mol Cryst Liq Cryst 67:205
21. Leadbetter AJ, Frost JA, Mazid MA (1979) J Phys (Orsay, Fr) 40:L-325
22. Pindak R, Moncton DE, Davey SD, Goodby JW (1981) Phys Rev Lett 46:1135
23. Goodby JW, Pindak R (1981) Mol Cryst Liq Cryst 75:233
24. Gane PAC, Leadbetter AJ, Benattar JJ, Moussa F, Lambert M (1981) Phys Rev A 24:2694
25. Benattar JJ, Moussa F, Lambert M (1983) J Chim Phys (Paris) 80:53
26. Wiegeleben A, Richter L, Deresch J, Demus D (1980) Mol Cryst Liq Cryst 59:329
27. Wiegeleben A, Demus D (1982) Cryst Res Technol 17:161
28. Demus D, Sackmann H (1968) Z Phys Chem 238:215
29. Keller P, Zann A, Dubois JC, Billard J (1980) Chem Phys 11:57
30. Doucet J, Keller P, Levelut AM, Porquet P (1978) J Phys (Orsay, Fr) 39:548
31. Coates D, Gray GW (1976) Mol Cryst Liq Cryst 34:1
32. Cladis PE (1975) Phys Rev Lett 35:48
33. Cladis PE, Borgardus RR, Daniels, WB, Taylor GN (1977) Phys Rev Lett 39:720
34. Hardouin F, Sigaud G, Achard MF, Gasparoux H (1979) Phys Lett A 71 A:347
35. Gajewska B, Kresse H, Weißflog W (1982) Cryst Res Technol 17:897
36. Nguyen Huu Tinh, Hardouin F, Destrade C (1982) J Phys (Orsay, Fr) 43:1127
37. Pelzl G, Böttger U, Demus D (1981) Mol Cryst Liq Cryst Lett 64:283
38. Nguyen Huu Tinh, Hardouin F, Destrade C, Levelut AM (1982) J Phys, Paris Lett 43 L:33
39. Nguyen Huu Tinh, Hardouin F, Destrade C, Gasparoux H (1982) J Phys, Paris Lett 43 L:739
40. Hardouin F, Nguyen Huu Tinh,, Achard MF, Levelut AM (1982) J Phys, Paris Lett 43 L:327
41. Hardouin F, Levelut AM, Achard MF, Sigaud G (1983) J Chim Phys, Paris 80:53
42. Sigaud G, Hardouin F, Achard MF, Gasparoux H (1979) J Phys (Orsay, Fr) 40:3-356
43. Hardouin F, Levelut AM, Benattar JJ, Sigaud G (1980) Solid State Commun 33:337
44. Hardouin F, Levelut AM, Sigaud G (1981) J Phys (Orsay, Fr) 42:71
45. Levelut AM, Tarento RJ, Hardouin F, Achard MF, Sigaud G (1981) Phys Rev A 24:2180
46. Sigaud G, Hardouin F, Achard MF, Levelut AM (1981) J Phys (Orsay, Fr) 42:107
47. Hardouin F, Sigaud G, Nguyen Huu Tinh, Achard MF (1981) J Phys Lett (Orsay, Fr) 42:63
48. Billard J (1980) Springer Series chem Phys 11:383
49. Chandrasekhar S (1983) Phil Trans R Soc London, 309:93

Anschrift des Verfassers:

Prof. Dr. H. Sackmann
Sektion Chemie der Martin-Luther-Universität
Mühlpforte 1
GDR-4020 Halle (Saale)

New lyotropic nematic liquid crystals*)

K. Reizlein and H. Hoffmann

Lehrstuhl für Physikalische Chemie der Universität Bayreuth, Bayreuth (FRG)

Abstract: Studies of binary perfluorosurfactant/water systems are presented. The methods applied were ^{2}H- and ^{19}F-NMR spectroscopy, polarizing microscopy and viscosimetry. It is shown that several perfluorosurfactant/water systems form lyotropic nematic phases. The surfactants are rubidium, ammonium and tetramethylammonium salts of perfluorocarboxylic acids. The phase behaviour of two systems was investigated. The system $C_8F_{17}COONH_4/D_2O$ forms a nematic and a lamellar phase. The micelles in the isotropic solution seem to be spherical up to high surfactant concentrations. There is evidence for a transition to disc micelles near the isotropic/nematic phase transition. The nematic phase which consists of disc micelles aligns with its director parallel to an applied magnetic field. In the system $C_8F_{17}COON(CH_3)_4/D_2O$, the isotropic solution is viscoelastic what is probably due to rod-shaped micelles. With increasing concentration they at first grow, then shrink. The nematic phase is composed of disc micelles which also align parallel to an applied magnetic field. The alignment is not lost during the phase transition from the nematic phase to the second liquid crystalline phase, which is stable at lower temperatures.

Key words: perfluorosurfactants, lyotropic nematic LC phases, phase behaviour, NMR spectroscopy, rotational viscosity γ_1.

1. Introduction

In 1967 Lawson and Flautt [1] described the first lyotropic nematic phase consisting of an ionic surfactant, a long chain alcohol, salt and water. In the following years further ternary and quaternary nematic systems were found [2]. Further work on lyotropic nematics was done by Saupe who observed the first biaxial nematic phase [3] and by Charvolin [4]. Recently Boden [5, 6] and Lindman [7] described three binary lyotropic nematic systems. Two of them were fluorosurfactant/water systems, the surfactants being the cesium salts of the perfluorooctanoic acid and the perfluorononanoic acid respectively.

In this paper we report measurements on several other binary perfluorosurfactant/water systems exhibiting nematic behaviour.

*) Lecture given at the 31th Conference of the Kolloid Gesellschaft, Bayreuth October 11–14, 1983

KT-W 830

2. Experimental

2.1 Preparation of the surfactants

The perfluorocarboxylic acids – obtained from Riedel de Haën – contained homologous acids as impurities. They were removed by distillation over a vigreux column in the case of the perfluoroheptanoic acid and the perfluorooctanoic acid. In the case of the perfluorononanoic acid the ethylester was distilled over a spinning band column. This acid was obtained by hydrolysis of the ester with potassium hydroxide, acidification with hydrochloric acid to pH 1, extraction with ether and evaporation of the ether after drying of the ether phase by sodium sulphate.

2.2 Construction of the phase diagrams

The sequence of the liquid crystalline phases was determined by contact preparations under the polarizing microscope [8]. Then samples of defined concentrations were prepared and checked for birefringence at several temperatures. The existing liquid crystalline structures were determined by means of a polarizing microscope. For these studies a Zeiss Standard 18 Pol microscope equipped with a Zeiss hotstage and a Mettler hotstage FP 82 was used. The anisotropy and the sign of the birefringence of these liquid crystalline phases were determined by conoscopic studies using this microscope. Finally the phase boundaries were determined more exactly by NMR-measurements.

2.3 NMR experiments

For the NMR experiments, a Jeol JNM-FX 90Q Fourier Transform NMR-spectrometer was used, operating at 13.70 MHz for deuterons and 84.26 MHz for fluorine nuclei. The temperature (accuracy 1K) was read before and after each measurement by a thermometer. For ^{19}F shift anisotropy measurements of aligned liquid crystalline phases the center of the signals was evaluated.

2.4 Viscosity measurements

The viscosities were determined with the instrument LS 30 sinus from Contraves. The width of the gap of the couette system was 0.5 mm, and the shear velocity could be varied from 0 to 1.285 s^{-1}.

2.5 Measurement of the rotational viscosity γ_1

The rotational viscosity was measured in the Jeol NMR spectrometer. After a rotation of the sample tube (diameter 5 mm) about an angle of 30 degrees, ^2H NMR spectra of D_2O were recorded at intervals of 10 seconds or more and subsequently stored on a disc. The frequency range was 1000 Hz, the acquisition time 2.05 s, and the resolution 0.5 Hz.

3. Theory

3.1 Alignment and reorientation of a nematic phase in a magnetic field – rotational viscosity γ_1

A large nematic sample of dimensions of 0.1 mm or more can be aligned in magnetic fields of some kilogauss. It is thus possible to prepare single crystals of a nematic phase.

If Φ is the angle between the director of the nematic sample and the magnetic field H, then a torque Γ_m will be exerted on the sample which is

$$\Gamma_m = -\chi_a H^2 \cos\Phi \sin\Phi \tag{1}$$

per volume element. χ_a is the anisotropy of the diamagnetic susceptibility. The result of this torque is a rotation of the director of the nematic sample. The driving torque Γ_m is compensated by the frictional torque

$$\Gamma_\gamma = -\gamma_1 \frac{d\Phi}{dt} \tag{2}$$

and we obtain

$$\gamma_1 \frac{d\Phi}{dt} + \chi_a H^2 \cos\Phi \sin\Phi = 0. \tag{3}$$

After integration we obtain

$$\Phi(t) = \operatorname{arc\,tg}\left[\operatorname{tg}\Phi_0 \exp\left(-t/\tau_{\text{eff}}\right)\right] \tag{4}$$

where $\Phi = \Phi_0$ for $t = 0$ and

$$\tau_{\text{eff}} = \frac{\gamma_1}{\chi_a H^2}. \tag{5}$$

3.2 NMR spectroscopy

3.2.1 Deuterium quadrupole splitting

The nature of the quadrupole splitting of the deuteron resonance from D_2O in lyotropic liquid crystalline phases is extensively described by Johanson and Drakenberg [9]. The amount of the splitting is given by [10]

$$\Delta\nu = \frac{X_{ass}}{X_{D_2O}} \frac{3}{2} E_Q S_{OD}. \tag{6}$$

X_{ass} and X_{D_2O} are the mole fractions of the associated and the (bulk) water, respectively. E_Q is the quadrupole coupling constant, and S_{OD} is the order parameter of the O-D bond axis.

If Θ is the angle between the director of an aligned lyotropic liquid crystalline phase and the director, then the splitting is

$$\Delta\nu(\Theta) = \Delta\nu(0^\circ)\left(\frac{3}{2}\cos^2\Theta - \frac{1}{2}\right). \tag{7}$$

3.2.2 ^{19}F shift anisotropy

The ^{19}F shift anisotropy σ_a of a rotating perfluoroalkyl chain belonging to perfluorosurfactant in a lyotropic liquid crystalline phase is given by [11, 12]

$$\sigma_a = \frac{2}{3} S' (\sigma_\parallel - \sigma_\perp)\left(\frac{3}{2}\cos^2\Theta - \frac{1}{2}\right). \tag{8}$$

$\sigma_\parallel$ is the component of the screening tensor parallel to the rotation axis, and $\sigma_\perp$ is the component perpendicular to it. S′ is an order parameter describing both the movements of the lamellae or micelles of the liquid crystalline phase and of the perfluoroalkyl chains of the surfactant. Θ is the angle between the director and the magnetic field.

4. Results

We found several binary perfluorosurfactant/water systems which form a nematic phase. The surfactants were

$C_6F_{13}COORb$, $C_7F_{15}COORb$, $C_8F_{17}COORb$,
$C_6F_{13}COONH_4$, $C_7F_{15}COONH_4$,
$C_8F_{17}COONH_4$

and

$C_8F_{17}COON(CH_3)_4$.

Detailed measurements were carried out on the systems

$C_8F_{17}COONH_4/D_2O$

and

$C_8F_{17}COON(CH_3)_4/\ D_2O$.

4.1 The system $C_8F_{17}COONH_4/D_2O$

4.1.1 Phase diagram

Figure 1 shows a section of the phase diagram for the system $C_8F_{17}COONH_4/D_2O$. The nematic phase is situated between the isotropic solution and the lamellar phase. Below 25 °C the surfactant crystallizes.

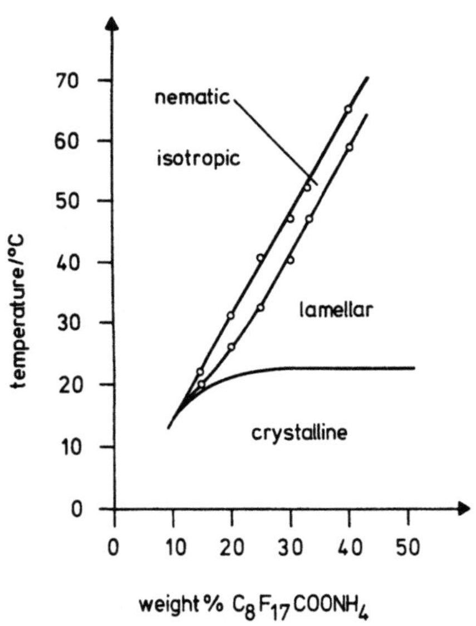

Fig. 1. Section of the phase diagram of the system $C_8F_{17}COONH_4/D_2O$

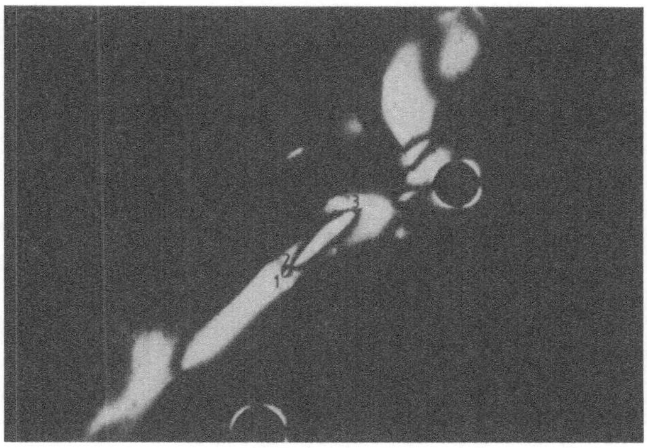

a

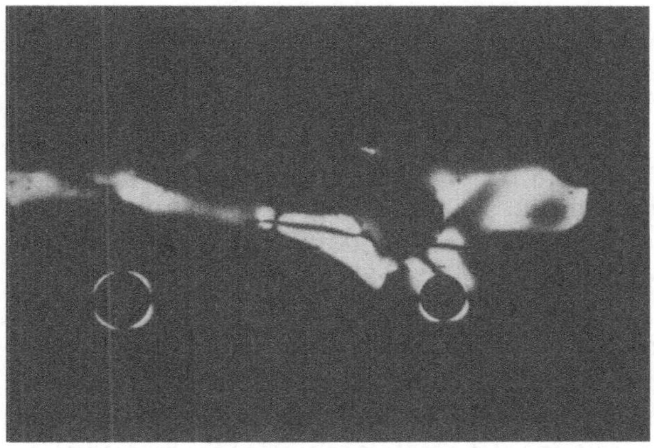

b

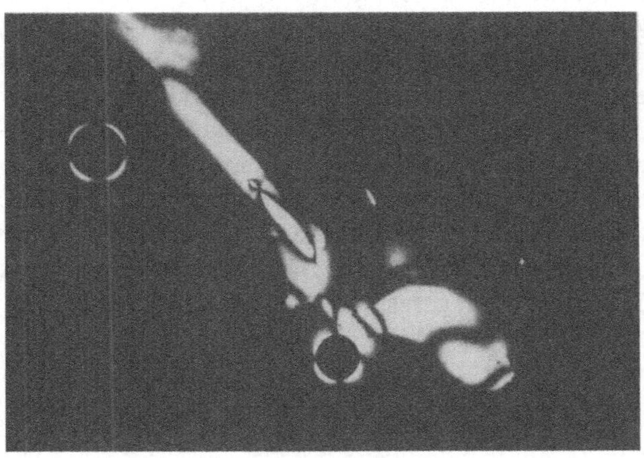

c

Fig. 2. Nematic schlieren texture: two singularities with $|s| = 1/2$ (points 1 and 3) and one singularity with $|s| = 1$ (point 2) and pseudoisotropic texture (25 weight percent $C_8F_{17}COONH_4/D_2O$; 36 °C; × 30; cell thickness 0.1 mm). Sample was rotated clockwise about 45° between a and b and between b and c

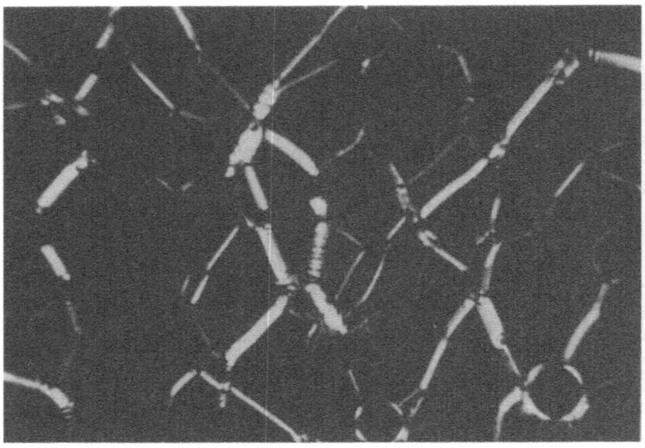

Fig. 3. Pseudoisotropic lamellar texture with oily streaks (20% $C_8F_{17}COONH_4/D_2O$; 22 °C; × 30; cell thickness 0.1 mm)

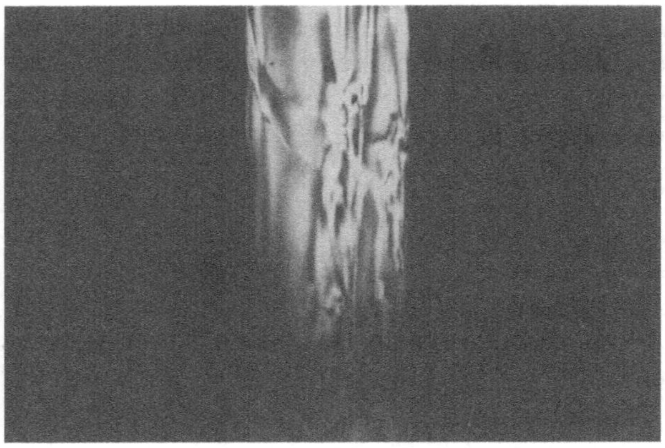

a

b

Fig. 4. Nonaligned nematic phase in a 5 mm NMR tube; chromatic polarization (30% $C_8F_{17}COONH_4/D_2O$; 45 °C): a) crossed polarizers, b) parallel polarizers

4.1.2 Polarizing microscopy

The cell used for investigations with the polarizing microscope had a thickness of 0.1 mm. Because of wall effects the nematic phase aligned with its director perpendicular to the cell windows. The result was a pseudoisotropic – non-birefringent – texture. Only disturbances of this alignment were visible between crossed polarizers. A schlieren texture produced by shearing such a sample disappeared after the end of the shear. Figures 2a–c were taken at three different orientations of the sample relative to the crossed polarizers. The sample had been turned clockwise about 45 and 90 degrees respectively. The schlieren meet at three fixed points corresponding to singularities. A singularity with $|s| = 1$ (point 2 in fig. 2a) lies between two singularities with $|s| = 1/2$ (points 1 and 3). This ensemble is similar to the inversion wall described by Nehring and Saupe [13]. Figure 3 shows a pseudoisotropic lamellar texture with "oily streaks" belonging to the lamellar phase of the same system. Conoscopic studies show that both the nematic and the lamellar phases are uniaxial and the sign of the birefringence is positive in both cases.

4.1.3 Alignment in magnetic fields – NMR spectroscopy

Figure 4 shows a nonaligned nematic sample in a 5 mm NMR tube between crossed and parallel polarizers respectively, viewed in white light. The sample appears coloured because of the occurrence of chromatic polarization [14]. In a magnetic field of several kilogauss the nematic phase aligns with its director parallel to the magnetic field. The birefringence vanishes in this direction (fig. 5a). Perpendicular to the magnetic field, strong birefringence can be seen if the incident light is polarized neither parallel nor perpendicular to the magnetic field (fig. 5b).

In figure 6 the splitting of the deuteron resonance from D_2O of aligned nematic and lamellar phases is plotted versus the temperature for several concentrations. In the lamellar region the temperature dependence of the splitting is linear, but not in the nematic region. At the phase transition nematic/isotropic which takes place within 0.5 to 1.0 K both phases coexist. The splitting vanishes at a certain temperature characteristic of each concentration.

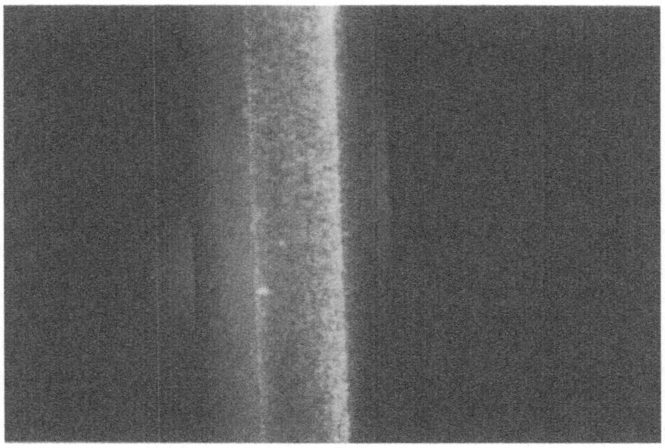

a

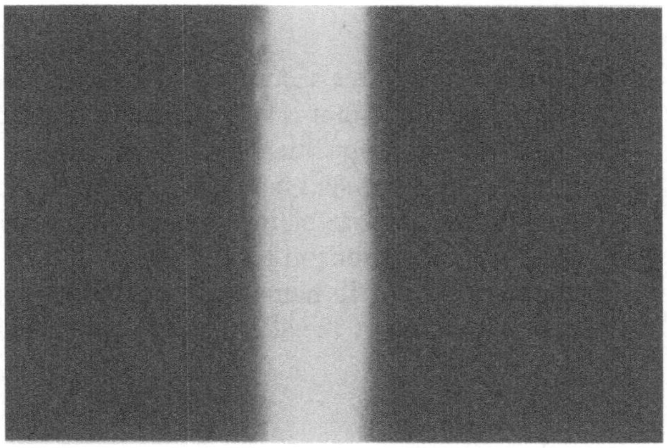

b

Fig. 5. Nematic phase aligned by a magnetic field (same conditions as in fig. 4); a) viewed parallel to the field, b) viewed perpendicular to the field

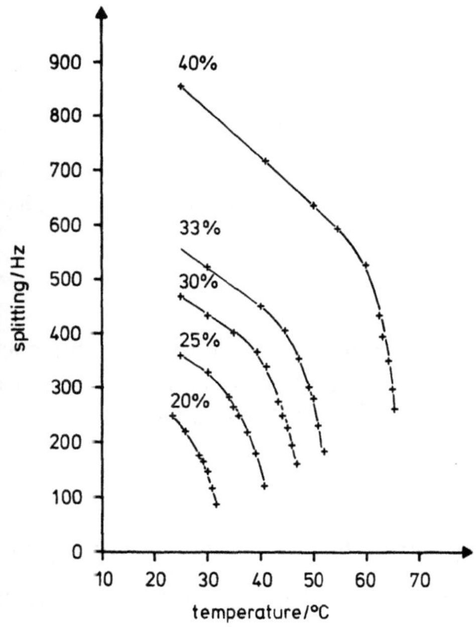

Fig. 6. Temperature dependence of the deuterium splitting of several aligned $C_8F_{17}COONH_4/D_2O$ systems

Although the lamellar phase aligns in a magnetic field in the course of several days or weeks, while the nematic phase takes several minutes or seconds, the lamellar/nematic phase transition can, however, be studied by observing the deuteron resonance from D_2O which gives information of the alignment. Nonaligned liquid crystalline samples were put into the magnetic field of the NMR spectrometer, and spectra were taken repeatedly over several hours. Spectra of such nonaligned polycrystalline samples are called Pake spectra. Each crystallite of the sample causes a pair of lines whose separation corresponds to its orientation in the magnetic field. The resulting spectrum comes about by superposition of the spectra of all crystals where each orientation has the same statistical weight in the case of a totally disordered sample. If the Pake spectrum did not vanish, then the phase had to be lamellar (fig. 7). If the final spectrum consisted of two sharp peaks (fig. 8), then the phase was nematic.

By rotating an aligned lamellar sample in the NMR spectrometer we could show that the splitting is indeed proportional to $(3/2 \cos^2\Theta - 1/2)$ where Θ is the angle between the magnetic field and the director.

The ^{19}F NMR signals of the system $C_8F_{17}COONH_4/D_2O$ are not broadened in the isotropic region away from the phase boundary. Broadening does take place in the immediate neighbourhood of the isotropic/nematic phase boundary.

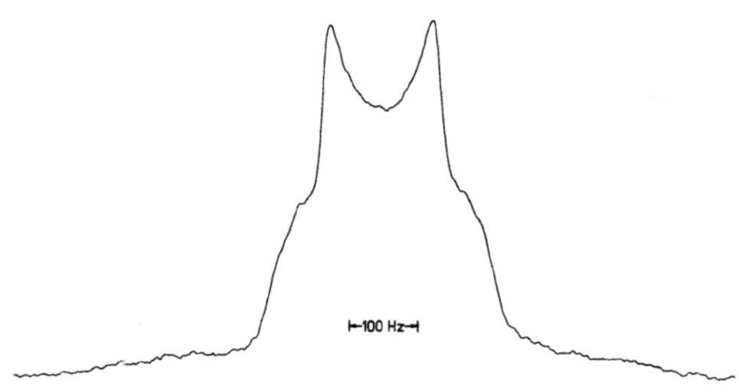

Fig. 7. Deuterium NMR spectrum of a nonaligned lamellar phase – Pake spectrum (25 % $C_8F_{17}COONH_4/D_2O$; 33.3 °C)

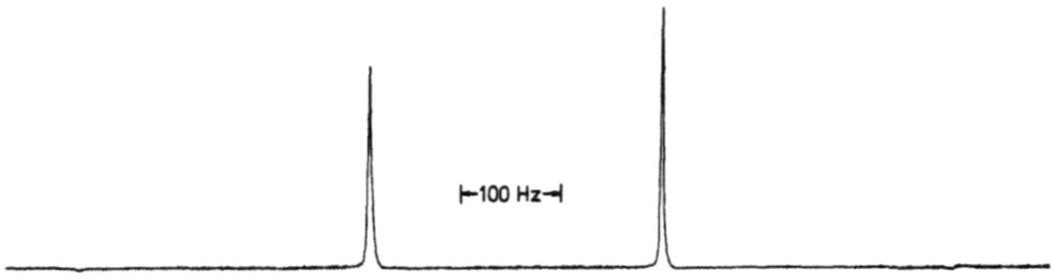

Fig. 8. Deuterium NMR spectrum of an aligned nematic phase (25 % $C_8F_{17}COONH_4/D_2O$; 34.5 °C)

In figure 9 the temperature dependence of the chemical shift anisotropy is plotted versus the temperature for a solution with 25 weight percent surfactant content.

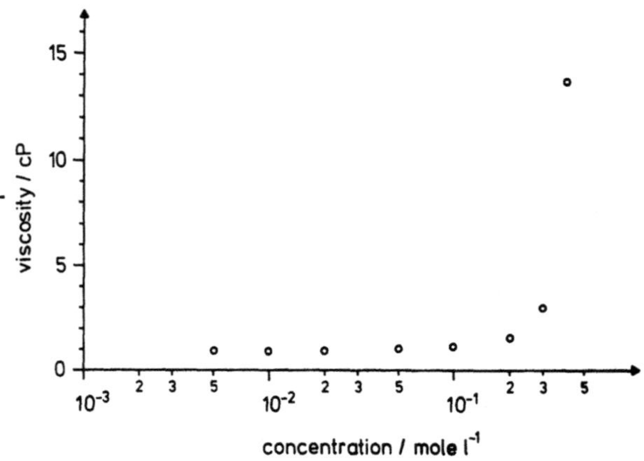

Fig. 10. Concentration dependence of the viscosity of solutions of $C_8F_{17}COONH_4$ in D_2O at 30 °C

4.1.4 Viscosity measurements – rotational viscosity γ_1

Figure 10 shows the viscosity of surfactant solutions in the isotropic region at 30 °C. It is nearly constant up to concentrations of 0.1 M, and then it increases asymptotically until reaching the isotropic/nematic phase boundary just above 0.5 M.

Figure 11 shows the temperature dependence of the viscosity of the system $C_8F_{17}COONH_4/D_2O$ with a surfactant content of 20 weight percent. In the nematic and in the isotropic region the viscosity decreases very fast with increasing temperature. The nematic/isotropic phase transition is accompanied by a viscosity maximum if the surfactant is very pure, but not if the surfactant contains impurities. Such a maximum of thermotropic liquid crystals has been known since the beginning of this century (cf. the papers of Miesowicz [15] and Porter and Johnson [16]).

As the alignment of the nematic phase in the couette system is undefined at the shear rates applied here,

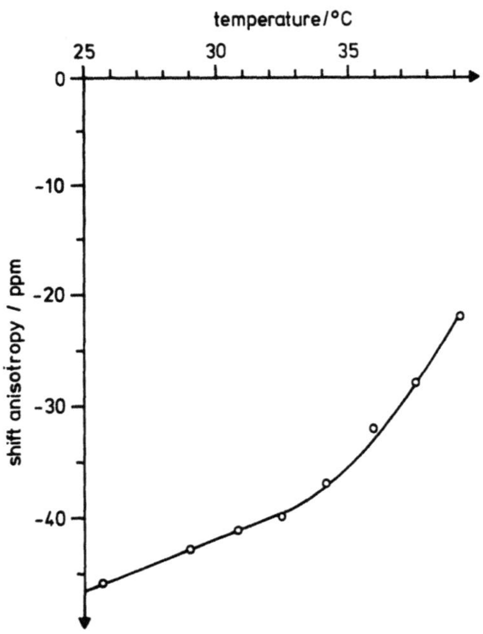

Fig. 9. Temperature dependence of the chemical shift anisotropy of the ^{19}F signal of the perfluoroalkyl chain of the aligned system 25 % $C_8F_{17}COONH_4/D_2O$

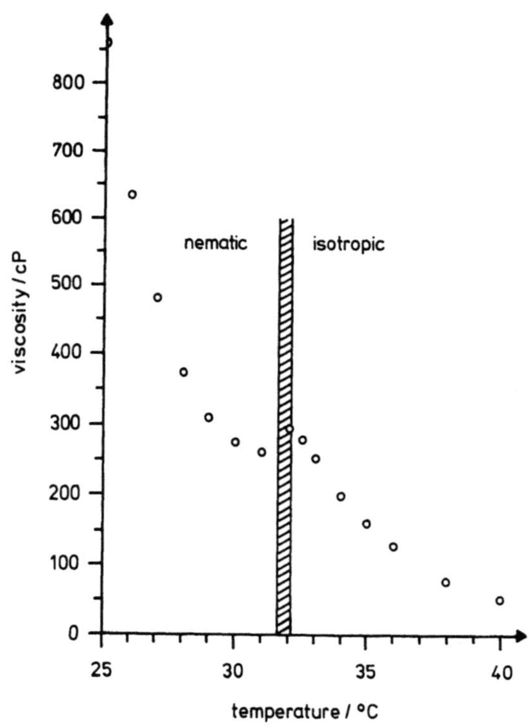

Fig. 11. Temperature dependence of the viscosity of the system 20 % $C_8F_{17}COONH_4/D_2O$

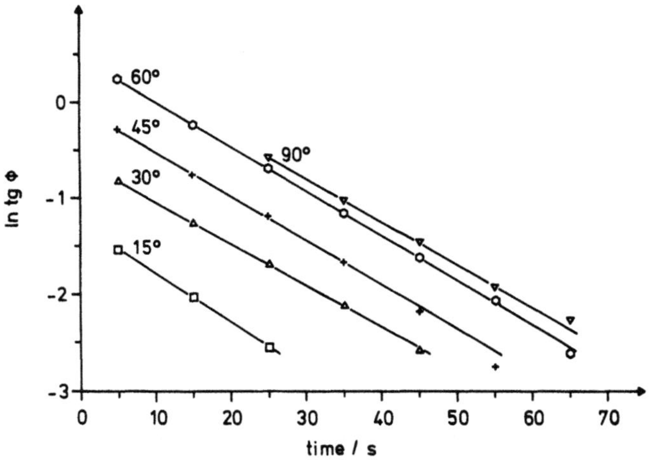

Fig. 12. Plot of ln tg Φ versus the time for several starting angles (20 % $C_8F_{17}COONH_4/D_2O$; 31.6 °C)

measurements of this kind in the liquid crystalline region cannot be interpreted quantitatively [17, 18].

Finally we measured the ratio γ_1/χ_a. The relaxation time τ_{eff} was shown to be independent of the sample diameter and the starting angle (fig. 12). The homogeneity of the alignment was destroyed at starting angles greater than 60 degrees (fig. 13 a and b). Therefore a starting angle of 30 degrees was chosen. Figure 14 shows the temperature dependence of γ_1/χ_a.

4.2 The system $C_8F_{17}COON(CH_3)_4/D_2O$

4.2.1 Phase diagram

In figure 15 a section of the phase diagram of the system $C_8F_{17}COON(CH_3)_4/D_2O$ is shown. When this

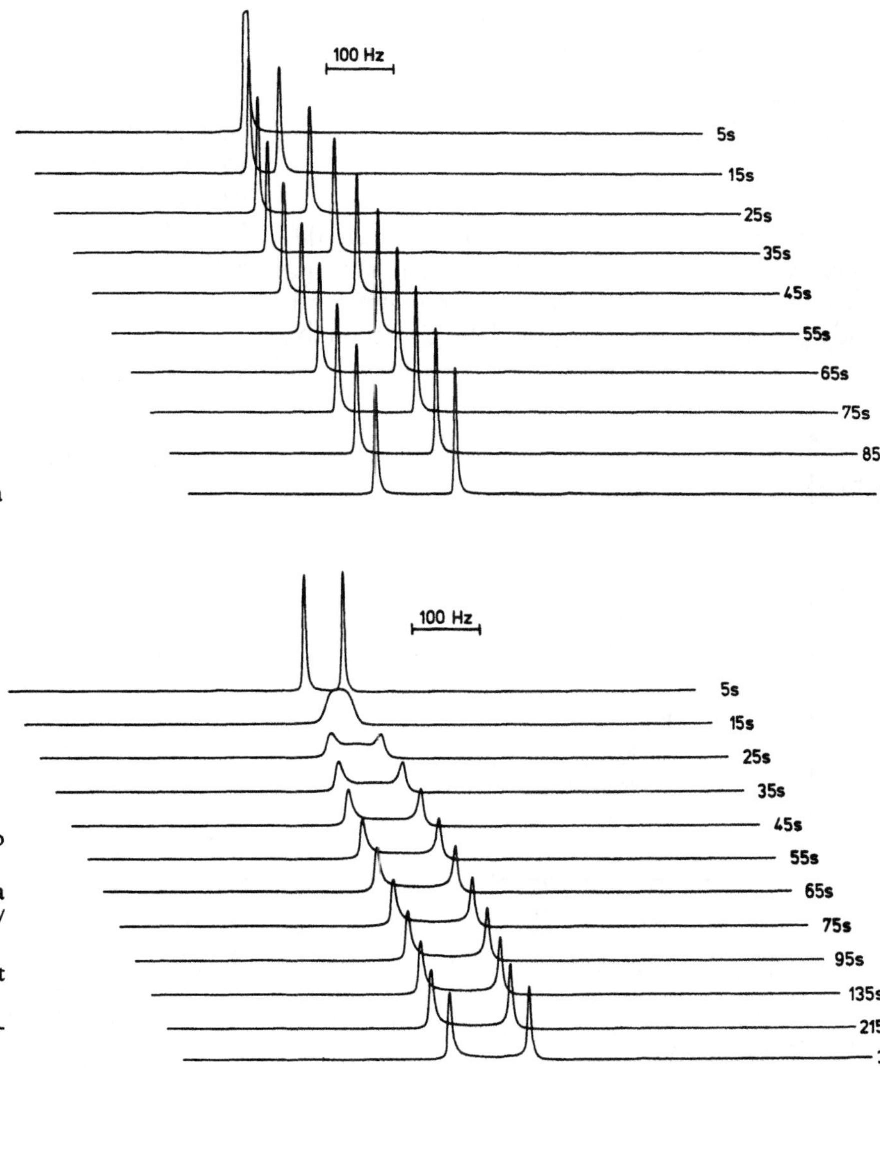

Fig. 13. Rotation of a nematic phase in a magnetic field (20 % $C_8F_{17}COONH_4/D_2O$; 31.6 °C)
a) homogeneity of the alignment not destroyed (starting angle 60 degrees)
b) homogeneity of the alignment destroyed (starting angle 90 degrees)

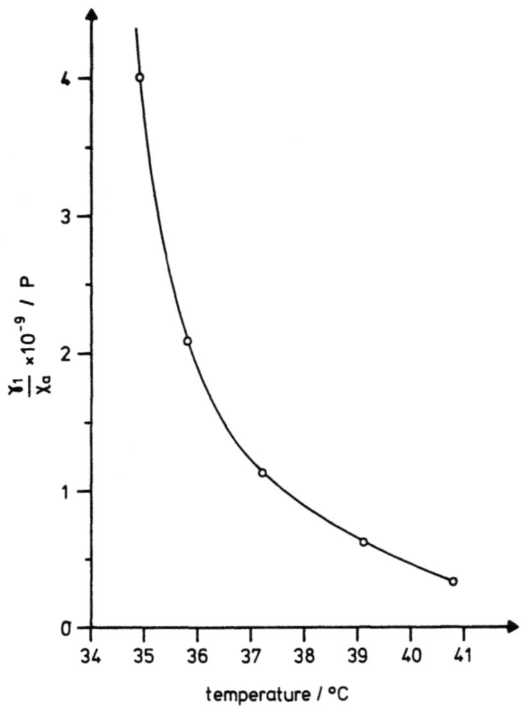

Fig. 14. Temperature dependence of the ratio γ_1/χ_a of the system 25% $C_8F_{17}COONH_4/D_2O$

system is compared with the preceding one, three differences are noticed at once: firstly, even at 10 °C no surfactant crystals are formed (the ammonium salt already precipitates between 20 and 25 °C). Secondly, the nematic range lies at higher temperatures. Thirdly, at 25 °C, the solutions are viscous at concentrations of 0.06 M and higher while the samples of the ammonium salt are flowing readily in the whole concentration range studied.

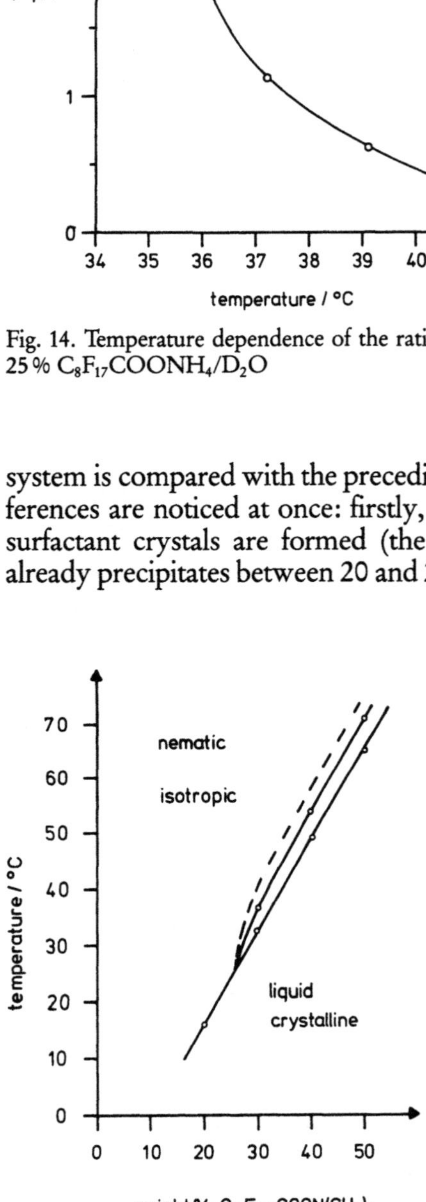

Fig. 15. Section of the phase diagram of the system $C_8F_{17}COON(CH_3)_4/D_2O$

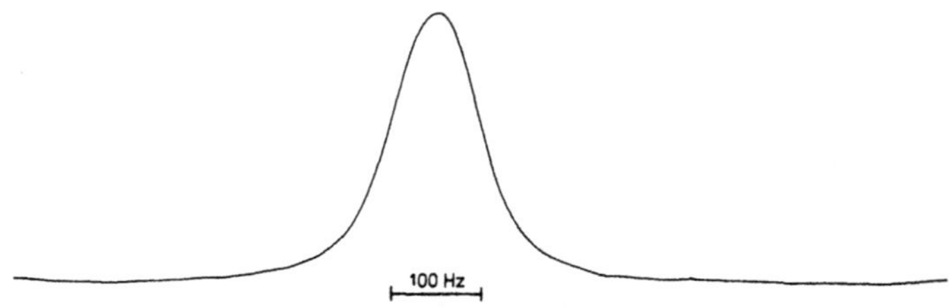

Fig. 16. Deuterium NMR spectrum of a nonaligned sample of the system 40% $C_8F_{17}COON(CH_3)_4/D_2O$ at 45 °C

4.2.2 *The isotropic region*

Solutions of $C_8F_{17}COON(CH_3)_4$ in D_2O are viscoelastic at concentrations of 0.03 M and higher at room temperature and show birefringence if an electric field is applied. The viscoelasticity is accompanied by a line broadening which has its maximum at 0.1 M and decreases again at higher concentrations.

4.2.3 *The liquid crystalline region*

The deuteron spectra from D_2O of a nonaligned liquid crystalline phase of this system consist of a single very broad signal (fig. 16). The corresponding spectra of the ammonium system are Pake spectra. Upon warming such a sample, a splitting into two sharp lines appears in the nematic region which disappears during the nematic/isotropic phase transition. Upon cooling the aligned nematic phase in the magnetic field, the splitting does not vanish (fig. 17), i.e., the alignment is not destroyed. The alignment does not disappear even if the sample is left standing for several weeks at room temperature (ca. 20 °C) without an applied magnetic field.

The angle dependence of the splitting is proportional to $3/2\cos^2\Theta - 1/2$ where Θ is the angle between the director of the liquid crystalline phase and the magnetic field.

The sign of the ^{19}F shift anisotropy is the same as it is for the ammonium system; however the amount is reduced by approximately the factor two. So it is −28

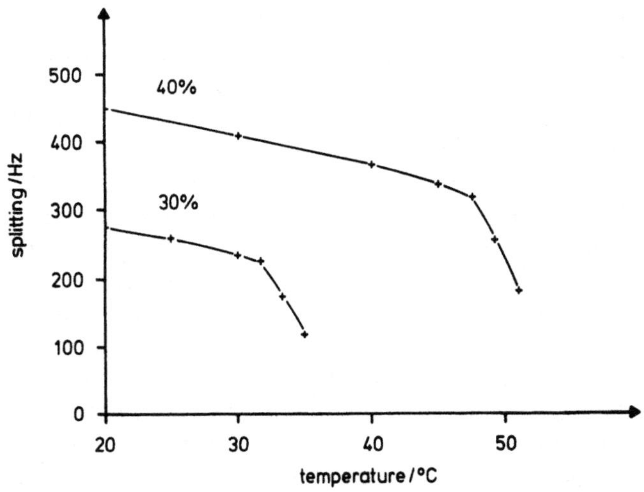

Fig. 17 Temperature dependence of the deuterium splitting for aligned $C_8F_{17}COON(CH_3)_4/D_2O$ systems

ppm for a 30 percent solution at 25 °C. It is − 46 ppm for a 25 percent solution of the ammonium salt at the same temperature.

5. Discussion

Before this study only three binary surfactant/water systems with nematic phases were known. We could now show that the existence of nematic phases is not very exceptional. The surfactants were pure rubidium, ammonium and tetramethylammonium salts of perfluorocarboxylic acids.

5.1 The system $C_8F_{17}COONH_4/D_2O$

5.1.1 The liquid crystalline region

The present system has an uniaxial nematic phase and an uniaxial lamellar phase. Recently Fontell and Lindman published a phase diagram of this system without a nematic phase [7].

The existence of the nematic phase could be demonstrated firstly by the occurrence of a schlieren texture with point singularities with $|s| = 1/2$ (fig. 2a–c) known from nematic schlieren textures exclusively [13] and secondly by the fact that the phase could be readily aligned in a magnetic field.

As the nematic phase adjoins a lamellar phase, and the alignment is not lost during the nematic/lamellar phase transition, the nematic phase should consist of disc micelles. There are experiments which prove this assumption.

Lyotropic nematic phases consisting of disc micelles (N_D-phases) align in a cell with a thickness of 0.1 mm or less with the optical axis perpendicular to the cell windows [19]. The result is a pseudoisotropic texture. This behaviour is also shown by the nematic phase of this system. This alignment is stable: a schlieren texture, obtained by shear of a nematic phase with pseudoisotropic texture, disappears within a few minutes after of the end of the shear.

The observed angle dependence of the deuterium splitting which satisfies the equation (7) shows that the nematic phase aligns with its director parallel to an applied magnetic field. For the examination of the ^{19}F shift anisotropy values (fig. 9), the equation (8) and the $\sigma_\parallel$ and $\sigma_\perp$ values for aligned crystalline teflon were used ($\sigma_\parallel = -69.5$ ppm and $\sigma_\perp = 34.7$ ppm) [12]. The sign and the amount of the shift anisotropy show that the perfluoroalkyl chains of the surfactant molecules are aligned parallel to the magnetic field. These results can be explained only by the existence of disc micelles which align with their short axis parallel to the magnetic field.

With this model the angle dependence of the birefringence can also be explained. There was no birefringence parallel to the magnetic field at all, and it vanished perpendicular to the field if the incident light was polarized either parallel or perpendicular to the field, i. e. if the light propagated along the alkyl chains or was polarized parallel or perpendicular to them.

The isotropic/nematic phase transition can be observed by deuteron NMR spectra and by inspection of the samples through crossed polarizers. In the transition region there is a two phase region whose temperature interval can be very small in the case of a very pure surfactant. This phase transition seems to be a first order phase transition. During the nematic/lamellar phase transition a two phase region can be observed by deuteron NMR. This can be a result of temperature gradients in the probehead, however, this must be eliminated before the question of the order of the phase transition can be decided unambiguously.

The temperature dependences of both the deuterium splitting and the ^{19}F shift anisotropy are the same. Therefore we can conclude that they are proportional to the order parameter S. The temperature dependence of S is then nonlinear in the nematic range, just as it is in the case of thermotropic nematics, and linear in the lamellar range.

5.1.2 The isotropic region

The absence of a Kerr effect and of a line broadening of the perfluoroalkyl chain ^{19}F signals and the low viscosity up to concentrations of 0.1 M indicate that, at a temperature of 30 °C, the micelle shape must be spherical or can have only small deviations from a spherical shape. At higher concentrations the NMR line-width as well as the viscosity increase. The reason can be a micelle growth as well as an increase of the intermicellar interaction [20].

5.5 The system $C_8F_{17}COON(CH_3)_4/D_2O$

5.2.1 The liquid crystalline region

The nematic phase adjoins a second liquid crystalline phase which could not be clearly identified by Fontell and Lindman [7]. They had not observed the nematic phase in this system.

The deuterium splitting and the ^{19}F shift anisotropy show that the nematic phase – just like the nematic phase of the system $C_8F_{17}COONH_4/D_2O$ – consists of disc micelles which align with their minor axis parallel to an applied magnetic field. Upon cooling, the alignment remains. Therefore the structure of the second liquid crystalline phase should be lamellar. Recently obtained neutron scattering results support this assumption [21].

5.2.2 The isotropic region

The increase of the line broadening of the perfluoroalkyl chain ^{19}F NMR signal with the concentration up to 0.1 M and the following decrease can be explained by a corresponding increase and decrease of the micellar rotational correlation time which is influenced both by the micellar size and the micellar interaction [20]. A possible interpretation is the following: at low concentrations the micelles grow with increasing surfactant concentrations; at higher concentrations they shrink again – or a transition from rod micelles to disc micelles takes place.

5.3 Mechanism of the alignment in a magnetic field

Some years ago Charvolin discussed the effects of the diamagnetic anisotropy of alkyl chains and of the magnetic shape anisotropy of homogeneous prolate and oblate objects on the alignment of a lyotropic nematic phase [19]. Surfactant molecules with hydrocarbon chains whose diamagnetic anisotropy is negative align with their chains perpendicular to an applied magnetic field. Homogeneous oblate objects orient with their minor axis perpendicular, prolate with their major axis parallel to an applied uniform magnetic field. In the case of surfactant molecules with hydrocarbon chains, such as sodium dodecylsulphate, the effects of the molecular and the shape magnetic anisotropy converge, and Charvolin, who used such surfactants, could not decide which effect was the dominating one. In the following years, Boden, Radley and Holmes carried out experiments with cesium perfluorooctanoate which with water forms a lyotropic nematic phase consisting of disc micelles [6]. The diamagnetic anisotropy of perfluoroalkyl chains is positive. The nematic phase was aligned by a magnetic field with its director parallel to the field, i. e. with the perfluoroalkyl chains of the surfactant parallel to the magnetic field. This was the first example with opposing molecular and shape magnetic anisotropy, and the molecular anisotropy was the dominating one. Apparently the diamagnetic anisotropoy of the perfluoroalkyl chains also dominates the alignment of the systems we have presented here.

Acknowledgments

We gratefully acknowledge financial support of this work by the Deutsche Forschungsgemeinschaft and the Fond der Chemischen Industrie.

References

1. Lawson KD, Flautt TJ (1967) J Amer Chem Soc 89:5489
2. Forrest BJ, Reeves LW (1981) Chem Rev 81:1
3. Yu LY, Saupe A (1980) Phys Rev Lett 45:1000
4. Hendrikx Y, Charvolin J (1981) J Physique 42:1427
5. Boden N, Jackson PH, McMullen K, Holmes MC (1979) Chem Phys Lett 65:476
6. Boden N, Radley K, Holmes MC (1981) Mol Phys 42:493
7. Fontell K, Lindman B (1983) J Phys Chem 87:3289
8. Tiddy GJT (1980) Phys Rep 57:1
9. Johansson A, Drakenberg T (1971) Mol Cryst Liq Cryst 14:23
10. Haven T, Radley K, Saupe A (1981) Mol Cryst Liq Cryst 75:87
11. Saupe A (1968) Angew Chem 80:99
12. Garroway AN, Stalker DC, Mansfield P (1975) Polymer 16:161
13. Nehring J, Saupe A (1971) J C S Faraday Trans 61:1
14. Szivessy G (1928) in: Handbuch der Physik 19:917
15. Miesowicz M, Bull Acad Pol Sci A 1936:228
16. Porter RS, Johnson JF (1967) in: E R Eirich: Rheology, New York 4:317

17. de Gennes PG (1974) The Physics of Liquid Crystals, Oxford Univ Press, London
18. Leslie FM, Brown GH (1979) Advances in Liquid Crystals 4:1
19. Charvolin J, Levelut AM, Samulski ET (1979) J Physique – Lettres, 40:L587
20. Ulmius J, Wennerström H (1977) J Magnet Res 28:309
21. Hoffmann H, Kalus J, Thurn H, unpublished results

Authors' address:

H. Hoffmann
Lehrstuhl für Physikalische Chemie
der Universität Bayreuth
Universitätsstraße 30
D-8580 Bayreuth

Progress in Colloid & Polymer Science Progr Colloid & Polymer Sci 69:94–99 (1984)

Investigations on pretransitional phenomena of the isotropic-nematic phase transition of mesogenic materials by means of electrically induced birefringence*

M. Eich, K. Ullrich, and J. H. Wendorff

Deutsches Kunststoff-Institut, Darmstadt, (F. R. G.)

Abstract: Pretransitional phenomena were characterized for low molecular weight liquid crystals as well as for a mesogenic side chain polymer by means of electric briefringence studies. The magnitude of the induced birefringence was found to diverge in the isotropic phase as a hypothetical second order phase transition into the nematic phase is approached.

This behaviour also holds for the relaxation time of correlated "mesogenic groups" in the case of the polymeric sample. The results are interpreted in terms of the Landau-de Gennes theory of the nematic-isotropic phase transition. Additionally, an example of an isotropic-smectic phase transition is given, where no critical fluctuations of the nematic order parameter occur.

Key words: Low molecular weight liquid crystals, liquid crystalline polymer, electrically induced birefringence, pretransitional phenomena, Landau-de Gennes model, order parameter fluctuations, critical slowing down.

Introduction

In most atomic or molecular systems the transition from the crystalline solid state to the isotropic melt takes place in one step at the melting temperature. This transition is characterized by the simultaneous destruction of the positional and the orientational long range order. Consequently only a short range order exists in the high temperature phase.

In certain cases, however, the transition from the crystalline to the isotropic fluid state happens stepwise involving one or more intermediate phases. Molecules, which are characterized by elongated molecular shapes, may exhibit liquid crystalline phases i. e. nematic, cholesteric or smectic phases, including specific modification, of these phases. In the following we will only be concerned with the nematic phase. It is characterized by a short range positional order while displaying at the same time a long range orientational order. The orientational order is expressed in terms of the scalar order parameter S defined as

$$S = \frac{1}{2} \cdot \langle 3 \cos^2 \theta - 1 \rangle \tag{1}$$

where θ is the angle between the locally preferred direction and the molecular long axis. The order parameter is zero in the isotropic fluid state, it increases discontinuously to non-zero values at the transition from the isotropic to the nematic phase. Thus a first order phase transition occurs.

This transition, however, has some of the features of a second order phase transition. For this reason this particular transition has often been described as being only weakly first order. This indicates that the discontinuities in the first derivative of the Gibbs free energy at the transition are small in comparison to the corresponding values of the isotropic to crystalline transition. Furthermore pretransitional effects occur.

We know that a characteristic property of a second order phase transiton is the presence of strong fluctuations of certain thermodynamic parameters such as the relevant order parameter and the divergence of certain

---

* Lecture given at the 31th Conference of the Kolloid Gesellschaft, Bayreuth October 11–14, 1983.

KT-W 831

thermodynamic properties in its neighborhood. Pretransitional effects related to order parameter fluctuations have been observed in the isotropic fluid state close to the nematic transition for low molecular weight systems [1].

Efffects of this kind have recently met with considerable interest also in the case of the fluid state of chain molecules. It has been assumed, for instance, that due to the anisotropic shape of chain molecules strong orientational correlations may occur in the melt of chain molecules, indicative either of the presence of a nematic phase or of a hypothetic low temperature nematic phase occurring in a temperature range below the actual crystallization temperature. In fact, results from the depolarized light scattering of several n-alkane melts have successfully been interpreted in terms of a low temperature nematic phase [2]. The experimental data at higher temperatures were interpreted in terms of pretransitional effects [3].

For this reason, we became interested in such pretransitional effects in general and specifically in polymers, known to show a nematic-isotropic phase transition [4]. We selected a side chain polymer, the side chain of which carries a mesogenic group and a spacer group. The pretransitional effects were studied both with respect to static and dynamic properties, using the electric birefringence technique (Kerr Effect). The data obtained will be discussed in relation to data available for low molecular weight mesogenic systems.

Theoretical considerations

The nematic-isotropic transition as well as pretransitional effects are usually treated in terms of the Landau-de Gennes model, which is based on the simple phenomenological Landau theory of phase transitions [5]. It is generally accepted that this theory is an adequate qualitative description, while being quantitatively less reliable. According to this model, the free energy density $F(T)$ in the vicinity of the phase transition can be expressed as a function of the order parameter S as [6]:

$$F(T) = F_o(T)_{S=0} + \frac{1}{2} \cdot a(T)\, S^2 - \frac{1}{3} \cdot b(T)\, S^3 + \frac{1}{4} \cdot c(T)\, S^4 + \dots \tag{2}$$

In this model, the coefficient of the term quadratic in S, $a(T)$ is taken to be $a = a_o \cdot (T - T^*)$, where T^* is a hypothetical second order phase transition temperature. The cubic term b has to be included in the case of the isotropic-nematic transition since the free energy for $S < 0$ is unequal to the free energy for $S > 0$. The inclusion of b has the consequence, that in a zero field a first order phase transition will take place at a temperature T_{NI}, at which the order parameter S will increase discontinuously from zero to a positive value. Below this transition, a second order phase transition at a temperature T^* is hidden.

The order parameter fluctuations increase very strongly, already in the isotropic state, as this temperature is approached. The expansion coefficients, a, b, c are related to thermodynamic parameters such as the discontinuous jump of the order parameter ΔS_{NI}, the heat of fusion ΔH_{NI} and the first order transition temperature T_{NI} relative to the hypothetical second order phase transition:

$$\Delta S_{NI} = \frac{2 \cdot b}{3 \cdot c} \tag{3}$$

$$\Delta H_{NI} = \frac{2 \cdot a_o \cdot b^2}{9 \cdot c^2} \cdot T_{NI} \tag{4}$$

$$T_{NI} = T^* + \frac{2 \cdot b^2}{9 \cdot a_o \cdot c}. \tag{5}$$

In the presence of an electric field, the average order parameter S will be non-zero already in the isotropic phase. It will increase very strongly as the second order phase transition is approached. By taking into account the influence of the electric field E on the free energy density of the total system,

$$F(T) = F(T)_{E=0} - \frac{1}{3} \cdot \Delta\varepsilon^o \cdot \varepsilon_o \cdot E^2 \cdot S \tag{6}$$

we obtain the result that the induced order parameter S_E takes a value of

$$S_E = \frac{\Delta\varepsilon^o \cdot \varepsilon_o \cdot E^2}{3 \cdot a_o \cdot (T - T^*)} \tag{7}$$

where ε_o = influence constant
and $\Delta\varepsilon^o$ = anisotropy of static dielectric constant of the ideally ordered nematic state ($S = 1$) [7].

The macroscopically observed birefringence is given by

$$\Delta n_o = \Delta n^o \cdot S_E \tag{8}$$

where Δn^o is the birefringence of the perfectly ordered molecules.

The Kerr constant B, defined as

$$B = \frac{\Delta n_o}{\lambda \cdot E^2} \tag{9}$$

(λ = wavelength of light)

is thus given by

$$B = \frac{\Delta n^o \cdot \Delta \varepsilon^o \cdot \varepsilon_o}{3 \cdot \lambda \cdot a_o \cdot (T - T^*)}. \tag{10}$$

B is predicted to diverge as T^* is approached, and with a critical exponent $\gamma = 1$, according to the Landau theory.

Up to now, only the static response has been considered. In the following, the dynamical properties will also be treated. It is known that fluctuations in the neighborhood of a critical point show a critical slowing down of the characteristic relaxation time, due primarily to a strong increase of the size of the correlation lenght of the fluctuations. According to de Gennes' treatment, the relaxation time of the electric birefringence is governed by the dynamical properties of the order parameter S [8].

Neglecting the expansion terms of third and fourth order in S (η = viscosity), the result for the rise and decay are as follows:

$$S_R(t) = A \cdot (1 - e^{-\frac{t}{\tau}}) \tag{11}$$

$$S_D(t) = A \cdot e^{-\frac{t}{\tau}} \tag{12}$$

or

$$\Delta n_R(t) = \Delta n_o \cdot (1 - e^{-\frac{t}{\tau}}) \tag{13}$$

$$\Delta n_D(t) = \Delta n_o \cdot e^{-\frac{t}{\tau}} \tag{14}$$

with

$$A = \frac{2 \cdot \Delta \varepsilon^o \cdot \varepsilon_o \cdot E^2 \cdot \tau}{9 \cdot \eta} = \frac{\Delta \varepsilon^o \cdot \varepsilon_o \cdot E^2}{3 \cdot a}. \tag{15}$$

The relaxation time τ is given by

$$\tau = \frac{3 \cdot \eta(T)}{2 \cdot a_o \cdot (T - T^*)}. \tag{16}$$

Thus the relaxation times diverges as T^* is approached; the critical exponent being 1. $\eta(T)$ represents the temperature dependence of the viscosity of the polymer, which in turn influences the temperature dependence of the reorientational relaxation time of the molecular groups.

Results and discussion

1. Low molecular weight samples

Different low molecular weight liquid crystals have been studied. In the case of the isotropic phase of the smectic sample (fig. 1), the temperature dependence of the Kerr constant varies comparable to normal liquids because fluctuations of the relevant order parameter of the smectic phase could not be detected by electric birefringence [9].

Thus, no pretransitional behaviour is observed.

In the case of nematics, critical order parameter fluctuations appear in the isotropic phase (fig. 2), resulting in a strongly increasing Kerr constant. The sign of the Kerr constant depends on the anisotropy of permittivity, because the birefringence of the perfectly ordered phase is always positive (eq. (10)).

Dynamical properties were not resolutable with our experimental set-up, because the relaxation times of the low molecular weight samples are of the magnitude of some hundred ns.

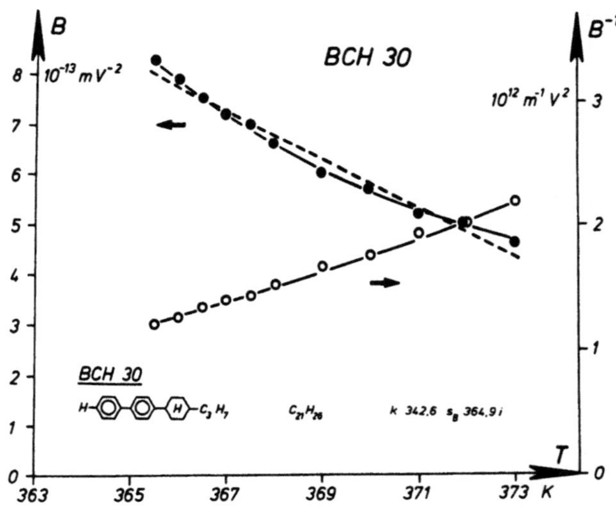

Fig. 1. Kerr constant B versus temperature

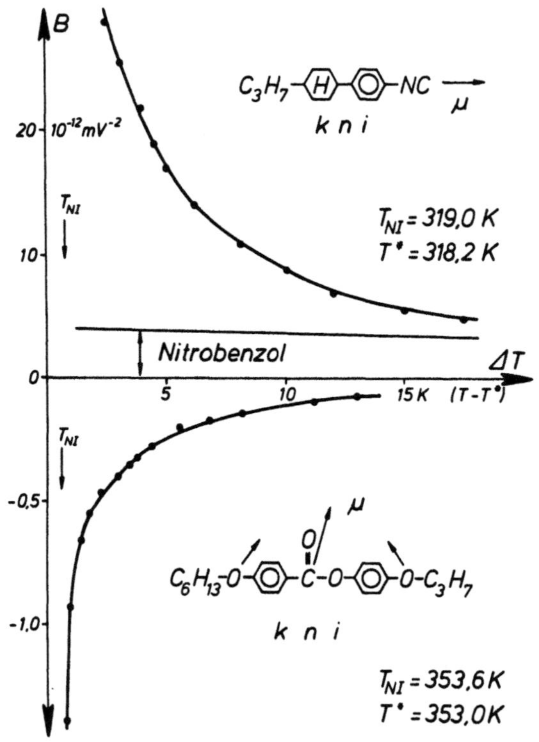

Fig. 2. Kerr constants B versus temperature

2. *Polymeric sample P 1*

From figure 3 it is already apparent that the magnitude of the Kerr constant depends very strongly on the temperature. The Kerr constant increases with decreasing temperature as the nematic transition temperature is approached. The temperature dependence of the Kerr constant in the temperature range $T_{NI} \leq T \leq (T_{NI} + 25\ \mathrm{K})$ is displayed in figure 3.

It is obvious that the Kerr constant diverges and that it approaches infinity at a specific fictive temperature T^*, which may be obtained from the plot of B^{-1} versus T. This plot is also given in figure 3. The fictive temperature T^* was found to be 405.6 K in the case of the samples studied here. It is thus located about 1.2 K below the actual transition temperature from the isotropic to the nematic phase. Thus the Kerr constant behaves, as a function of the temperature, to a very good approximation as

$$B \sim (T - T^*)^{-\gamma} \tag{17}$$

with $\gamma \approx 1$.

In the case of the dynamic experiments the rise and decay time of the electrically induced birefringence was determined as a function of the temperature.

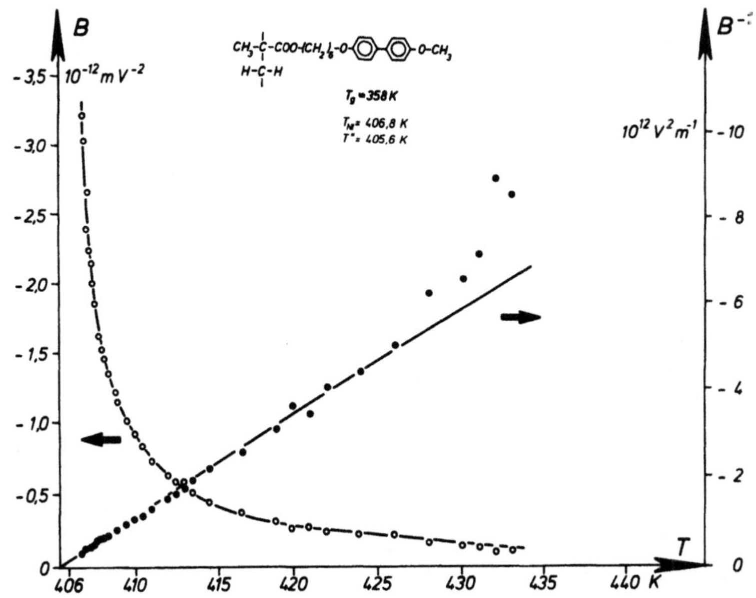

Fig. 3. Kerr constant B versus temperature of polymeric sample P 1

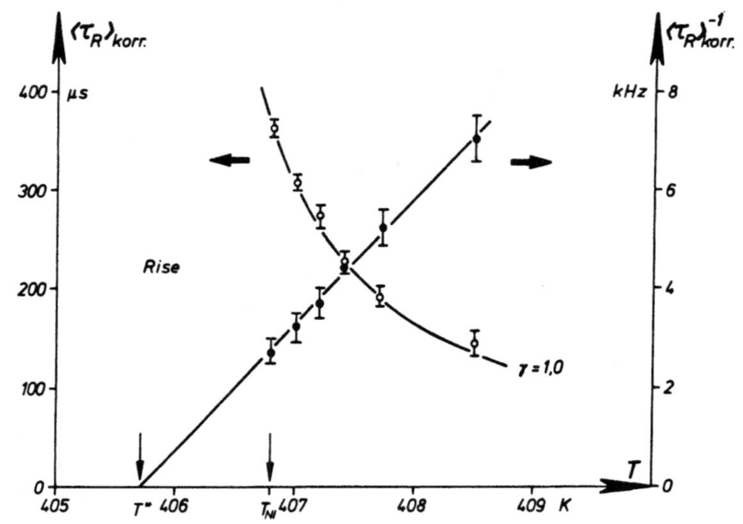

Fig. 4. Relaxation times versus temperature of polymeric sample P 1

As displayed in figure 4, the relaxation times show similar behaviour to the Kerr constant when cooling down to the isotropic-nematic phase transition, known as the "critical slowing down".

The experimentally observed relaxation times have to be corrected for the temperature dependence of the viscosity of the polymer according to equation (16) [11]. We did this by using the WLF equation for calculating the temperature shift factor. The justification for the use of the WLF equation is that the temperature range in which we performed the experiments was close to the glass transition temperature of about 358 K [12].

After these corrections the characteristic relaxation time of the correlated motions was found to diverge according to

$$\tau \sim (T - T^*)^{-\gamma} \tag{18}$$

as the characteristic temperature T^* is approached, in agreement with the theory. The exponent was thus found to be close to $\gamma \approx 1$. The characteristic temperature was $T^* = 405.6$ K, as in the case of the static experiments. In order to obtain the approximate magnitude of the relaxation times for the rise and decay, we calculated the time after which the normalized response attained the value $1/e$ and $(1 - 1/e)$ for the decay and rise respectively.

Based on the experimental results on T^* and T_{NI} as well as on the heat of transition $\Delta H_{NI} = 8.14\,\mathrm{J/cm^3}$ and assuming that at T_{NI} the order parameter changes discontinuously from zero to approximately $S_{NI} \approx 0.35$ [13], we are able to calculate the characteristic constants of the phenomenological Landau-de Gennes theory. The values are

$$a_o = 0.33\,\mathrm{J/cm^3\,K};\quad b = 3.36\,\mathrm{J/cm^3};\quad c = 6.41\,\mathrm{J/cm^3};$$

$$\Delta\varepsilon^o_{\mathrm{cal.}} \approx -0.8;\quad \varrho_{\mathrm{isotr.}} = 1.1\,\mathrm{g/cm^3}\ [10].$$

In the case of an electric field of 1 MV/m and a temperature of $T = 408.6$ K corresponding to $\Delta T = 3$ K, we calculated an induced order parameter in the isotropic melt of about

$$|S_E| \approx 5 \cdot 10^{-6}.$$

The values given above correspond closely to the values known for low molecular weight liquid crystals, for instance for n-heptylcyanobiphenyl [14, 15]. This indicates that the thermodynamic properties of low molecular weight mesogenic materials and polymeric mesogenic materials are similar. In the case of the dynamical studies we found, however, that the time scale of the characteristic relaxation is quite different from the corresponding time scale of low molecular weight liquid crystals [16, 17]. This may be attributed to the very large values of the viscosity of the polymer melt ($\eta \approx 1000$ Poise). It is larger by a factor $10^3 - 10^4$, as compared to the corresponding values of low molecular weight liquid crystals such as MBBA [18]. The viscosity is of the same order of magnitude as observed for main chain mesogenic polymers such as copolymers of PET and PHB, which are able to display nematic melts [19].

Abbreviations: MBBA (Methoxybenzylidene butyl aniline); PET (Polyethylene terephthalate); PHB (p-hydroxy-benzoesäure)

References

1. Kelker H, Hatz H (1980) Handbook of Liquid Crystals, Verlag Chemie, Weinheim
2. Dettenmaier M (1979) Prog Colloid Polym Sci 66:169
3. De Gennes PG (1969) Phys Lett A 30:454
4. Blumstein A (1978) Liquid Crystalline Order In Polymers, Academic Press, New York, San Francisco, London
5. Landau L, Lifshiz E (1958) Statistical Physics, Pergamon Press Ltd, Headington Hill Hall, Oxford
6. Senbetu I, Woo C (1982) Mol Cryst Liq Cryst 84:101
7. Maier W, Meier G (1961) Z Naturforschung 16 a:262
8. De Gennes PG (1974) Physics of Liquid Crystals, Oxford Univ Press, London
9. Krause S (1981) Molecular Electro-Optics, Plenum Press, New York, London, 435
10. Hahn B, Wendorff JH, Portugall M, Ringsdorf H (1981) Colloid Polym Sci 259:875
11. Stinson III TW, Litster JD (1970) Phys Rev Lett 25:503
12. Hahn B (1980) Diplom-Arbeit, TH Darmstadt
13. Finkelmann H, Benthack H, Rehage G, Journal de Physique Chimie, in press
14. Dunmur DA, Tomes AE (1981) Mol Cryst Liq Cryst 76:231
15. Poggi Y, Filippini J, Aleonard R (1976) Phys Lett 57 A:53
16. Coles HJ (1978) Mol Cryst Liq Cryst Lett 49:67
17. Yamamoto R, Ishihara S, Hayakawa S, Morimoto K (1978) Phys Lett 69 A, 4:276
18. Stinson III TW, Litster JD (1972) Journal de Physique C1:69
19. Jerman RE, Baird DG (1981) J Rheol 25, 2:275

Authors' address:

M. Eich
Deutsches Kunststoff-Institut
Schloßgartenstraße 6 R
D-6100 Darmstadt

Progress in Colloid & Polymer Science Progr Colloid & Polymer Sci 69:100–105 (1984)

Formation of lyotropic liquid crystals of metal dodecyl benzene sulphonates*)

Ð. Težak, F. Strajnar, O. Milat**), and M. Stubičar**)

Department of Physical Chemistry, Faculty of Natural Sciences and Mathematics, University of Zagreb, Zagreb and **) Institute of Physics, University of Zagreb, Zagreb, Yugoslavia

Abstract: The formation of uni-, bi-, and trivalent metal ion dodecyl benzene sulphonate (DBS) precipitates was investigated using light scattering, low angle X-ray diffraction and polarizing microscopy at 293 K. The thermodynamic equilibrium constants were calculated, improving some of our previous experimental results and applying a fitting program to the theoretical curve for association and solubility product equilibrium involving ionic strengths. The appearene of liquid crystals formed in a particular precipitation region is characterized by relationships of the reacting components. The question of the structural model of the metal ion – DBS liquid crystalline mesophase remains to be solved in future investigations.

Key words: lyotropic liquid crystals, solubility products.

Introduction

Investigations of solubility/precipitability in detergent solutions are particularly interesting because of detergent accumulation in natural waters. No data were found in the literature on the solubility/precipitability equilibria and the precipitate structure of Na, K, Cs, Mg, Ca, Sr, Ba, Cu, La or Al -dodecyl benzene sulphonate (DBS), although such data exist for dodecyl sulphates (DS) [1]. The identification of liquid crystalline mesophases of metal ion – DBS in the present work was done at 293 K, since DS liquid crystalline phases were formed only at higher temperatures [2]. Our previous paper [3] described experimental work on the precipitation of alkaline earth metal ions and DBS. It was proved that this detergent forms lamellar mesophases at room temperature. In the present paper the numerical values for equilibria constants have been experimentally corrected and an attempt at computer fitting of experimental results to the theoretical curve has been made. The thermodynamic constants were obtained by taking into account the ionic strengths. Similar behavior was found in some uni-, bi-, and trivalent metal – DBS systems in aqueous solutions using tyndallometry, low angle X-ray diffraction and polarizing microscopy.

*) Lecture given at the 31th Conference of the Kolloid Gesellschaft, Bayreuth October 11–14, 1983.

KT-W 828

Experimental

Materials

The p. a. chemical used were as follows: HDBS from Prva Iskra Barić, Beograd, Na, K, Mg, Ca, Sr, Ba, Cu, La, Al nitrates from Kemika, Zagreb, and $CsNO_3$ from Merck, Darmstadt; all were used without further purification. Molar solutions of HDBS and metal nitrates were prepared with double distilled water. $NaNO_3$, KNO_3 and $CsNO_3$ solutions were standardized using an ion exchange method. $Mg(NO_3)_2$ and $Ca(NO_3)_2$ were standardized by complexometric titrations with EDTA using Eryochrome Black *T* as indicator, $Cu(NO_3)_2$ was standardized with EDTA using Chromazurol *S* as indicator, $Sr(NO_3)_2$ and $Ba(NO_3)_2$ were not standardized. $La(NO_3)_3$ was standardized potentiometrically, $Al(NO_3)_3$ was standardized by precipitation from homogeneous solution as 8-hydroxychinolate.

Methods

Precipitation diagrams for the anion – cation systems were obtained by light scattering measurements using a Zeiss tyndallometer connected to a Pulfrich photometer. Since the precipitation processes in the examined systems were fast, the precipitation systems could be considered as being totally equilibrated, i. e. the precipitation limits became constant after a time. The experimental systems were prepared by the usual method described earlier [4]. The solubility/precipitability limits were determined by the criterion of the „first clear system", i. e. the sample which did not show any turbidity even 24 hours after preparation in comparison to water as a reference system.

The method of X-ray low angle diffraction was used in order to obtain the interplanar spacings (D) of M^{+n} – DBS, as described earlier [3].

Optical micrographs were made using the Leitz Wetzlar optical microscope.

All measurements were performed at 293 K.

Interpretation of experimental data

If we assume that HDBS and $M(NO_3)_2$ were totally dissociated:

$$HDBS \rightarrow H^+ + DBS^-$$

and

$$M(NO_3)_2 \rightarrow M^{+2} + 2NO_3^-$$

then the equilibrium at the solubility limit is described by:

$$M^{+2} + 2DBS^- \rightarrow M(DBS)_2\ (s) \tag{1}$$

(the region of solubility product),
and

$$M^{+2} + 2DBS^- \rightarrow [M(DBS)_2]\ (ass) \tag{2}$$

(the region of associates).

The numerical values for equilibrium constants and solubility products for Mg, Ca, Sr, Ba – DBS were more accurate than those obtained previously [3] since precise measurements were made in experiments which were repeated several times. Experimental data were subjected to nonlinear curve fitting. The results for the other metal ions estimated have to be considered as concentration parameters of solubility.

The solubility product (eq. (1)) is therefore:

$$K_s^o = a(M) \cdot a^2(D) \tag{3}$$

where $M = M^{+2}$, and $D = DBS^-$, hence

$$K_s^o = y(M) \cdot c(M) \cdot y^2(D) \cdot c^2(D)\,. \tag{4}$$

Assuming that the activity coefficient of M is

$$y(M) = y^4(D) = y^4$$

we can simplify the equation into

$$K_s^0 = y^6 \cdot c(M) \cdot c^2(D) \equiv S\,. \tag{5}$$

The association equilibrium (eq. (2)) is described by

$$K_{ass} = \frac{c(MD_2)}{S} \equiv K\,. \tag{6}$$

Under the solid/liquid equilibrium conditions at the solubility limit the concentrations of ions in solution are equal to the total concentrations of ions excluding those bonded in the solid phase. Therefore we can write:

$$c(M) = [M] - c(MD_2)$$

and

$$c(D) = [D] - 2c(MD_2)\,.$$

By substitution equation (5) becomes

$$S = (M - KS) \cdot (D - 2KS)^2 y^6\,. \tag{7}$$

Equation (7) was used for nonlinear curve fitting.

Results and discussion

Precipitation phenomena in aqueous systems of Na, K, Cs, Mg, Ca, Sr, Ba, Cu, La and Al nitrates and HDBS were examined, and the summarized results indicating solubility limits are presented in figure 1. The solubility limits show that the valencies of precipitation components determine the slopes of straight lines indicating the equilibria of solubility products, and that in our systems the slopes are related to the stoichiometric relationships 1 : 1 for univalent, 1 : 2 for bivalent and 1 : 3 for trivalent metal ions. The only exception is the Al – DBS system, which shows a 1 : 1 relation, because of a nonreliable pH region, i. e. hydrolysis of aluminium. In future it will be necessary to have pH conditions controlled.

Precipitation following the equilibrium of solubility product of univalent metal ions when these are in great

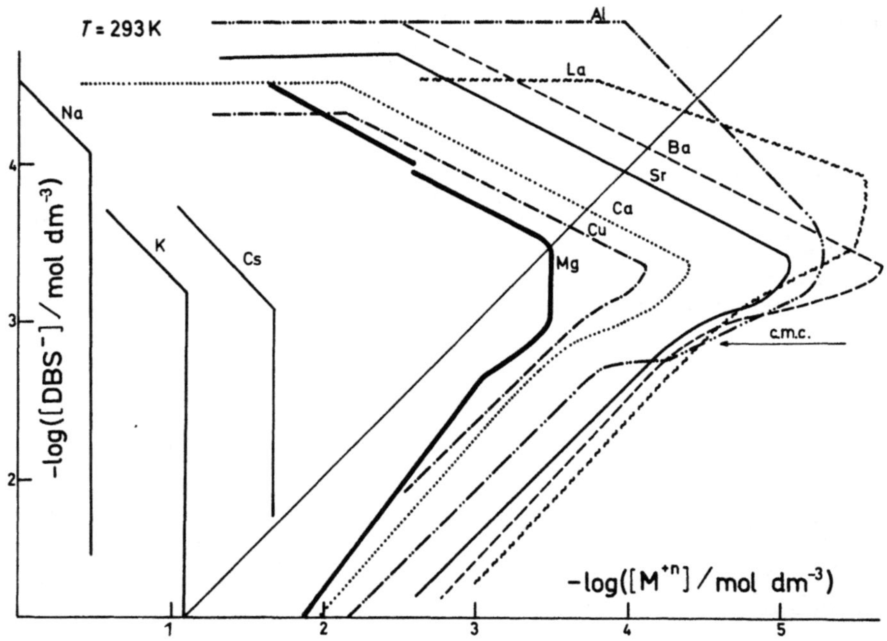

Fig. 1. The contours of solubility boundaries in precipitation diagrams of metal ion – DBS systems

excess, was found to occur very far from the equivalency region. This effect was caused by the formation of associates – aquacomplexes in the great excess of metal ions indicated by the solubility limit being parallel to the ordinate. The associates were formed in the Na – DBS system and K and Cs – DBS systems from the concentration point of HDBS where the formation of dimers and premicelles was found, respectively.

Bi- and trivalent ions exibit the same type of precipitation diagram, i. e. the solubility product equilibrium, and associates – aquacomplexes in the excess of metal ion concentration being parallel to the abscissae. The equivalent precipitation occurs in the equimolar region and complex ions are formed in excess of HDBS. The cessation of predominant equilibrium processes, where the calculation of solubility product is possible, is shown by the sudden change in slopes of straight lines of solubility limits, which is exibited in the CMC, of HDBS for bi- and trivalent cations.

The solubility of HDBS in water is limited by aggregation processes. The critical dimerization (CDC) and premicellization concentrations (CPC) found previously [3] can be related to the limiting association concentration (LAC) introduced by L. Mandell [5].

The concentration solubility parameters defined as $[M^{+n}] \cdot [DBS^-]^n$ were calculated for Na, K, Cs, Cu, La and Al – DBS salts and are listed in table 1.

Table 1. The concentration solubility parameters for $M(DBS)_n$ at 293 K

Cation	pK_s	c.c.
Na^+	4.66 ± 0.07	3×10^{-1}
K^+	4.01 ± 0.07	8×10^{-2}
Cs^+	4.86 ± 0.04	2×10^{-2}
Cu^{+2}	10.78 ± 0.15	8×10^{-5}
[a]) $(Al\,L_2)^+$	8.99 ± 0.08	6×10^{-6}
La^{+3}	17.44 ± 0.19	3×10^{-6}

[a]) L = ligand is considered to be $(OH)^-$ (pH at boundary from 4 to 5).

c.c. = critical concentration = $[M^{+n}]$ in mol dm^{-3} at which boundary changes slope.

$K_s = [M^{+n}] \cdot [DBS^-]^n/(\text{mol dm}^{-3})^{n+1}$.

We have used a computer program to analyze the experimental results for Mg, Ca, Sr, Ba – DBS presented in figure 1. Activity points at solubility limits were calculated from concentration data using the Debye-Hückel theory. Figure 2 shows the best fitting of these points to the theoretical straight lines of slope -0.5.

The inadequate fitting of results for Mg^{+2} – DBS equilibria is also presented. The conversion of concentrations into activities also gave straight lines. The straight line obtained by applying the mean square root calculation passing through all the experimental points is presented by the dashed line.

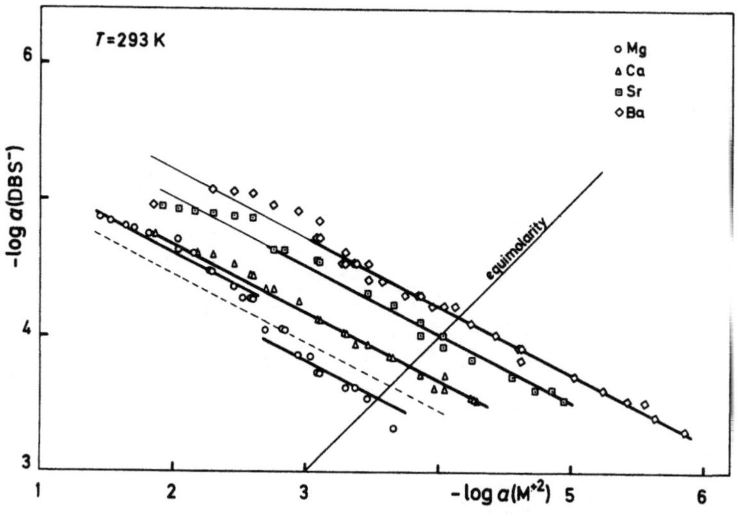

Fig. 2. Thermodynamic solubility limits of Mg, Ca, Sr, Ba – DBS systems. Log. activities were calculated from concentration values so as to fit a theoretical straight of the slope −0.5

Figure 2 shows a very good agreement of experimentally estimated points with the theoretical straight lines. In this paper we wish to include the equilibrium processes appearing in precipitation diagram by introducing all ionic species under CMC of HDBS into the equation describing the solubility curve. For the curve fitting program the slope of the straight lines was theoretical (−0.5). There were no significant differences in standard deviations of K_s^0 when the value of K_{ass} was taken as 0 or as the value obtained from the fitting program. For this reason the question of associates – aquacomplexes formation remains open for future experimental and theoretical considerations. Table 2 presents thermodynamic constants K_s^0 obtained by applying the fitting program. Solubility products K_s^0 were calculated from experimental points lying on the heavy lines in figure 2.

The formation of liquid crystalline mesophases in all cases of the examined precipitation systems was observed above CMC of HDBS in the region between the concentration excess of metal ions and equimolarity. Low angle X-ray diffraction analyses have shown that these are smectic A lamellar mesophases with interplanar sacings (D) listed in table 3. It should be mentioned that the length of the HDBS molecule was calculated to be 26.5 Å. As the depth of a double layer of DBS micelle arrangement was found to be about 30 Å (D experimentally obtained), the tail of HDBS molecules must be distorted. It seems that small differences in D values (table 3) are not indicative enough to be considered as real phenomena.

The micrographs from the polarizing microscope showing lyotropic mesomorphism are presented in figures 3 and 4. When the photographs were compared under crossed polarizers and without them various patterns appeared in macroscopic forms of liquid crystalline structures. Spherulites of various sizes indicated the formation of chains always having the same size of vesicle. The „oily streaks" pattern [6] and the coacervation in detergent solutions [7] were described. It is very clearly seen in the micrograph of the Mg – DBS system that the chains are joined in double helix ordering, causing a greater layer thickness of material and in particular swelling sites at the highest optical density.

Table 2. The solubility products for $M(DBS)_2$ at 293 K

Cation	pK_s^o	Slope
Mg^{+2} (I)	11.22 ± 0.98	−0.5
Mg^{+2} (II)	10.65 ± 0.80	−0.5
Ca^{+2}	11.37 ± 0.41	−0.5
Sr^{+2}	12.04 ± 0.79	−0.5
Ba^{+2}	12.47 ± 0.28	−0.5

$K_s^0 = a(M^{+2}) \cdot a^2(DBS^-)$

Table 3. Interplanar spacings (D) for $M(DBS)_n$

Cation	D/Å
Na^+	30.9 ± 1.1
Mg^{+2}	30.6 ± 0.9
Ca^{+2}	30.9 ± 1.9
Sr^{+2}	28.8 ± 0.4
Ba^{+2}	30.8 ± 1.3
Al^{+3}	29.6 ± 0.7
La^{+3}	28.9 ± 1.0

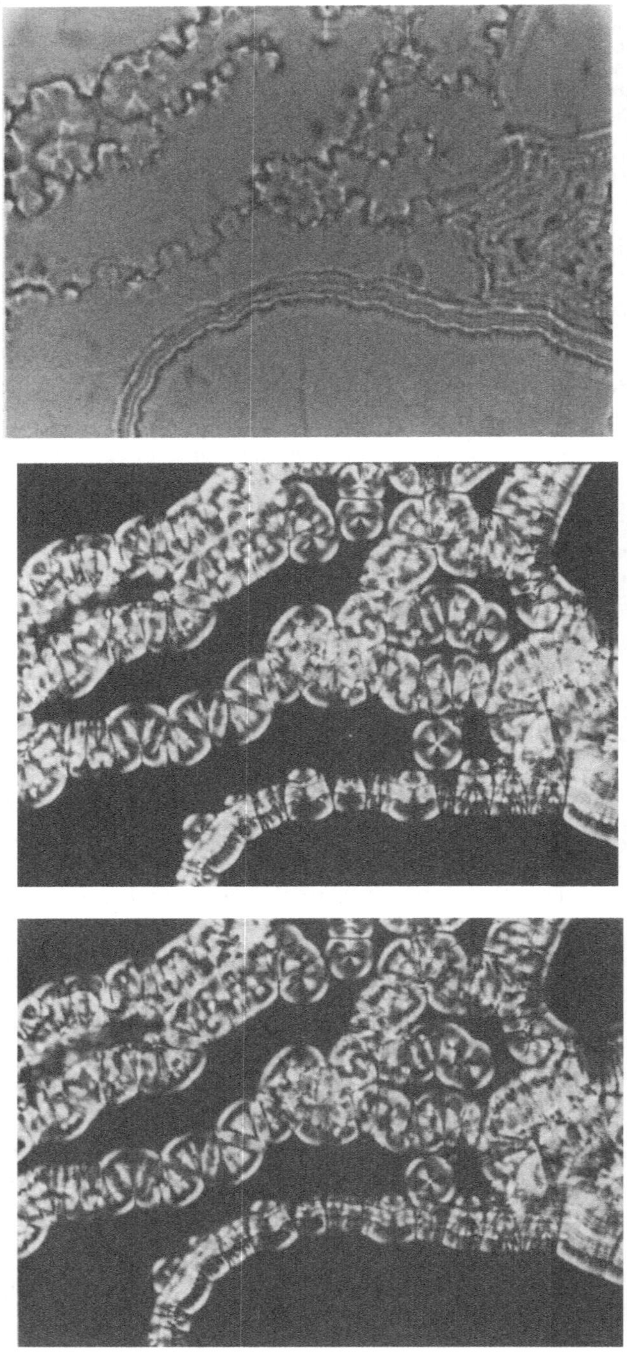

Fig. 3. The optical pattern of $Mg(DBS)_2$ liquid crystals. The concentrations in mol dm^{-3}: $[Mg(NO_3)_2] = 6 \times 10^{-2}$, $[HDBS] = 6 \times 10^{-2}$. Total magnification 200 ×; (a) without, (b) with crossed polarizers, (c) with crossed polarizers and λ plate

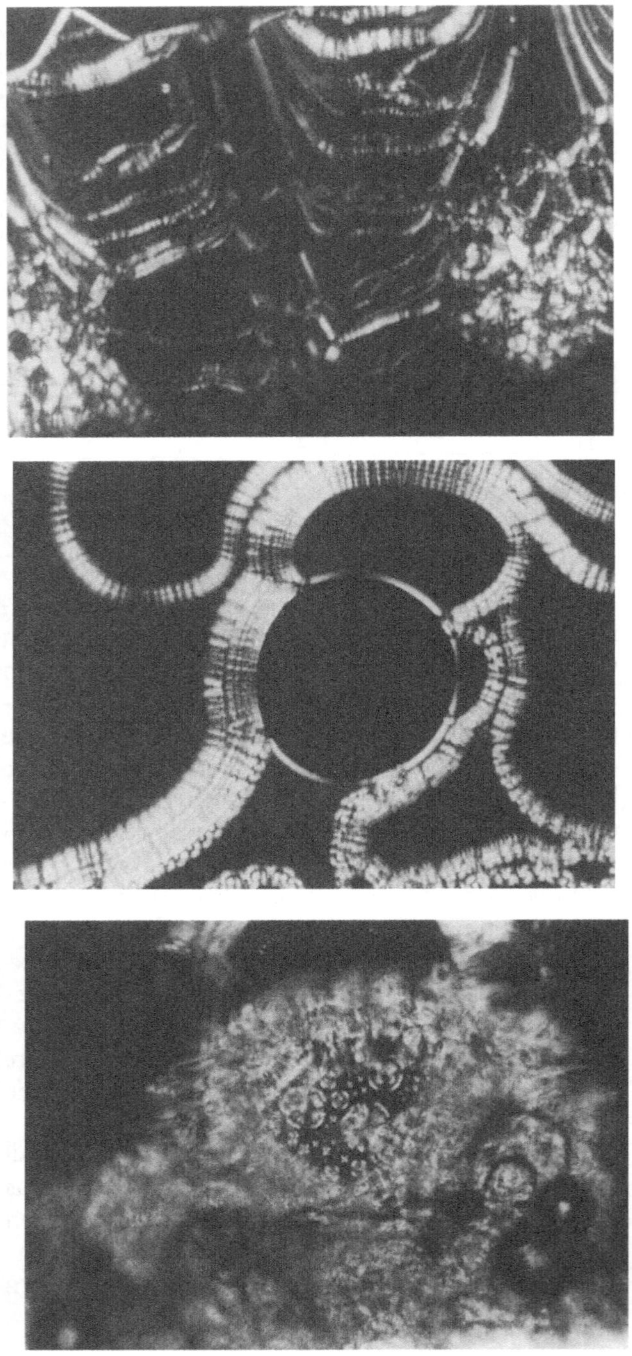

Fig. 4. The optical patterns under crossed polarizers of: (a) KDBS; $[KNO_3] = 2.5 \times 10^{-1}$, $[HDBS] = 1.5 \times 10^{-1}$; (b) $Mg(DBS)_2$; $[Mg(NO_3)_2] = 6 \times 10^{-2}$, $[HDBS] = 6 \times 10^{-2}$; (c) $La(DBS)_3$; $[La(NO_3)_3] = 4 \times 10^{-2}$, $[HDBS] = 4 \times 10^{-2}$. All concentrations in mol dm^{-3}. Total magnification 200 ×

For this reason they show a thin curved line which is recognizable in the micrograph taken in nonpolarized light. This work raises questions of the micro and macro structure of HDBS micelles, of formation of lamellar smectic layers into vesicles, and then of the connection of vesicles into chains and chains into the double helix structure. We hope to resolve some of these questions in our future investigations.

Acknowledgement

The authors thank Dr. N. Kallay for helpful discussions, and Mr. D. Babić for a computer fitting program.

References

1. Božić J, Krznarić I, Kallay N (1979) Colloid & Polymer Sci 257:201
2. Leigh ID, McDonald MP, Wood RM, Tiddy GJT, Trevethan MA (1981) J Chem Soc, Faraday Trans 1, 77:2867
3. Težak Đ, Strajnar F, Šarčević D, Milat O, Stubičar M (1984) Croat Chem Acta 57:93
4. Težak B, Matijević E, Schulz K (1951) J Phys Colloid Chem 55:1558
5. Mandell L, Ekwall P (1968) Acta Chem Scand 22:1
6. El-Nokaly MA, Ford LD, Friberg SE (1981) J Colloid and Interface Science 84:228
7. Mukhayer GI, Davis SS (1978) J Colloid and Interface Science 66:110

Authors' addresses:

Đ. Težak and F. Strajnar
Department of Physical Chemistry
Faculty of Natural Sciences and Mathematics
University of Zagreb
Marulićev trg 19/II
41001 Zagreb, P.O. Box 163
Yugoslavia

O. Milat and M. Stubičar
Institute of Physics
University of Zagreb
Bijenička cesta 46
41001 Zagreb, P.O. Box 304
Yugoslavia

Progress in Colloid & Polymer Science Progr Colloid & Polymer Sci 69:106–108 (1984)

Kritische Eigenschaften von Lipid-Doppelschichten am Hauptphasenübergang*

F. S. Rys**

Fritz-Haber-Institut der Max-Planck-Gesellschaft, Berlin (West)

Zusammenfassung: Das Verhalten von Lipid-Doppelschichtmembranen am Hauptphasenübergang wird im Rahmen der allgemeinen Theorie der kritischen Phänomene diskutiert. Bei der Auswertung experimenteller Daten müssen allgemeine Ergebnisse der Theorie berücksichtigt werden. Eine Molekularfeld-Analyse liefert i. a. falsche Resultate.

Abstract: The behaviour of lipid double-layer membranes at the main phase transition is discussed within the framework of the general theory of critical phenomena. For the evaluation of experimental data general results of the theory must be considered. A mean-field analysis yields false results in general.

Key words: Phase transition, critical phenomena, membranes.

Für das Studium der komplexen Vorgänge in biologischen Membranen ist es notwendig, die kollektiven Eigenschaften von künstlichen Membranen, z. B. von Phospholipid-Doppelschichten, quantitativ zu beschreiben. Das kooperative Verhalten der Lipidmolekül-Ketten und -Kopfgruppen läßt sich geeignet an Phasenübergängen untersuchen. In neueren experimentellen Arbeiten wurde über das thermische Verhalten von Membranen, insbesondere von DPPC in wässriger Lösung, ausführlich berichtet [1, 2, 3, 4].

In der Nähe von T_c zeigen verschiedene statische und dynamische Größen ein anomales Verhalten: Die spezifische Wärme (Abb. 1 in [1]), die Ultraschall-Absorption (Abb. 1 in [2], und -Relaxation (Abb. 2 in [3]), und verschiedene Relaxationszeiten und Amplituden in Temperatursprung-Experimenten (Abb. 1 in [4]), usw. zeigen eine ausgeprägte Spitze als Funktion von T bei $T = T_c$, während z. B. die Ultraschall-Geschwindigkeit einen scharfen Abfall bei T_c zeigt (Abb. 1 in [2]). Diese Tatsachen stellen ein gewichtiges Indiz für die Existenz von kritischen Fluktuationen am Phasenübergang dar, und lassen T_c als kritischen Punkt erscheinen.

Im folgenden wird eine kurze Zusammenfassung des heutigen Standes der Theorie gegeben. Im Rahmen der Theorie der kritischen Phänomene [5] von Phasenübergängen 2. Ordnung bestimmen die *ano-*

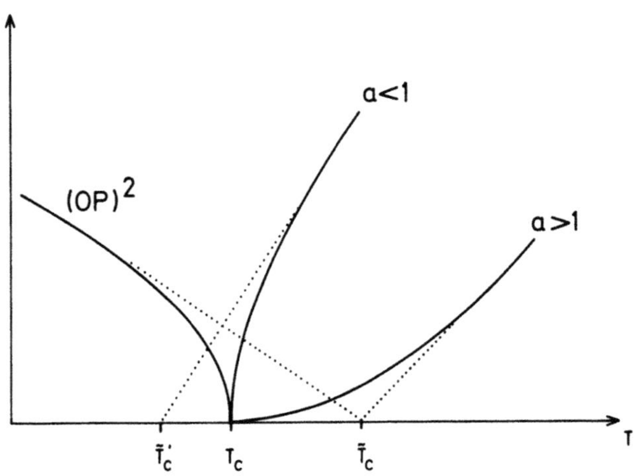

Abb. 1. Kritisches Verhalten des Ordnungsparameters (OP) ~ $(T_c - T)^\beta$, $\beta < \frac{1}{2}$, und einer divergierenden Größe ~ $(T - T_c)^{-a}$ für $a > 1$ und $a < 1$, reziprok gegen T aufgetragen (ausgezogene Linien). Extrapolationen aus dem Molekularfeldbereich führen zu falschen Werten $\tilde{T}_c$ für die kritischen Temperaturen (punktierte Linien)

* Vortrag, gehalten auf der 31. Hauptversammlung der Kolloid-Gesellschaft, Bayreuth 11. bis 14. Oktober 1983.

** Früher: Institut für Theoretische Physik, Freie Universität Berlin, Arnimallee 14, 1000 Berlin 33 (West)

KT-W 819

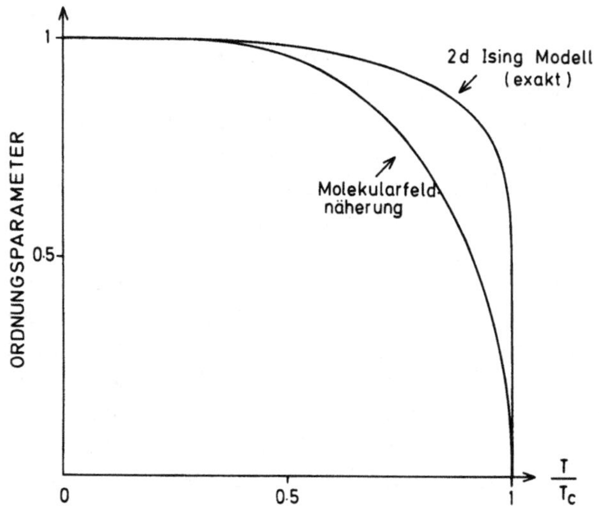

Abb. 2. Temperaturverlauf des Ordnungsparameters (OP) für das 2-dimensionale Ising-Modell (exakte Lösung nach Onsager, vgl. Bd III, [5]) und für die Molekularfeld-Näherung (Landau-Theorie)

malen Fluktuationen des Ordnungsparameters das kritische Verhalten der statischen und dynamischen Größen. Die Korrelationsfunktion ist charakterisiert durch eine bei T_c unendliche Korrelationslänge ξ. Dadurch ergibt sich eine für $T = T_c$ gültige Skaleninvarianz bezüglich einer Transformation der Längenskala. Zahlreiche Größen, wie die spezifische Wärme, der Ordnungsparameter, die Kompressibilität, die Relaxationszeit, usw. zeigen bei T_c ein Potenzverhalten als Funktion von $(T - T_c)$ bzw. $(T_c - T)$. Verschiedene Größen divergieren bei T_c (negativer Exponent oder logarithmisches Verhalten), einige andere (z. B. der Ordnungsparameter) fallen zu Null ab (positiver Exponent).

Aus der Skaleninvarianz ergeben sich allgemein gültige Beziehungen zwischen den Exponenten (*„Skalengesetze“*), die für alle Theorien gültig sind und die Rolle von Zustandsgleichungen am kritischen Punkt spielen. Ferner führt die Skaleninvarianz zum sog. Renormierungsgruppen-Formalismus, einer sehr nützlichen Rechenvorschrift, die sämtliche kritische Exponenten für ein konkretes Model in guter Näherung zu berechnen erlaubt. Dabei zeigt sich, daß sehr verschiedenartige Modelle dasselbe kritische Verhalten, d. h. dieselben kritischen Exponenten haben können. Man spricht dann von einer *Universalitätsklasse*. Verschiedene Klassen sind durch wenige Charakteristika gekennzeichnet: Die Dimensionalität des Systems, die Symmetrie des Ordnungsparameters, die Reichweite der Kräfte. So ist z. B. die Ising-Universalitätsklasse gekennzeichnet durch einen einkomponentigen Ordnungsparameter, kurzreichweitige Kräfte und die Dimension $d (= 2$ bzw. $3)$, während die Klasse der Molekularfeldtheorie oder Landau-Theorie nur für extrem weitreichende Kräfte oder unphysikalisch hohe Dimensionszahl ($d = 4$) gültig ist. Allerdings zeigen explizite Modellrechnungen, daß weit außerhalb des kritischen Gebietes (um T_c), also außerhalb des eigentlichen Gültigkeitsbereiches der Theorie der kritischen Phänomene eine Landau-Theorie durchaus brauchbare Werte liefert. Man beobachtet in diesem Zusammenhang einen sog. *“cross-over”* vom *“klassischen”* (d. h. Molekularfeld-)Verhalten in ein kritisches Verhalten bei Annäherung an den kritischen Punkt. Eine einfache Extrapolation vom Molekularfeld-Bereich in den kritischen Bereich ist unzuläßig; sie würde im allgemeinen zu völlig falschen Resultaten führen (vgl. Abb. 1). Aus exakten Lösungen von 2-dimensionalen Modellen ist bekannt, daß Systeme mit kurzreichweitigen Kräften sehr stark von der Landau-Theorie abweichen. So fällt der Ordnungsparameter des 2d Ising-Modells wie $(T_c - T)^{1/8}$ ab, während er in der Landau-Theorie wie $(T_c - T)^{1/2}$ gegen Null geht (vgl. Abb. 2). Eine oberflächliche Analyse experimenteller Daten im Landau-Rahmen könnte einen Isingartigen stetigen Abfall fälschlicherweise als einen 1. Ordnungssprung interpretieren.

Für endliche (d. h. kleine) Systeme wachsen divergierende Größen bei Annäherung an T_c zwar zunächst an, flachen jedoch nahe bei T_c zu einem endlichen „Buckel“ ab. Zudem ist der kritische Punkt gegenüber dem Wert für große Systeme verschoben. Dieser „finite-size cross-over“ eines 2d Modells ist in Abbildung 3 illustriert.

Aus diesen Bemerkungen lassen sich folgende Schlüsse ziehen:

1. Die Auswertung der experimentellen Daten muß im Rahmen der allgemeinen Theorie der kritischen Phänomene erfolgen. Größen, die bei T_c divergieren (spezifische Wärme, Ultraschall-Absorption, Relaxationszeit, usw.) bzw. zu Null abfallen (Ordnungsparameter, Ultraschall-Geschwindigkeit, usw.), ergeben, als Funktion von $|T - T_c|$ doppelt logarithmisch aufgetragen, Informationen über den Wert des betreffenden kritischen Exponenten nahe bei T_c, sowie einen eventuell auftretenden “cross-over” bzw. eine “finite-size” Abrundung [4]. Dabei ist, insbesondere nahe bei T_c, eine genaue Temperaturmessung von T und T_c erforderlich, da ein Fehler ΔT im Ausdruck log $|T - T_c|$ stark in Erscheinung tritt.

Eine Analyse, die auf einer Landau-Entwicklung beruht, wird im allgemeinen keine korrekten Resultate ergeben (vgl. Abb. 1).

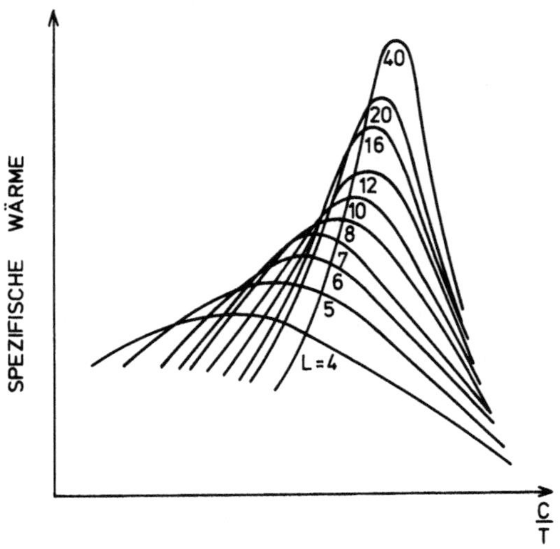

Abb. 3. Illustratives Beispiel des Temperaturverlaufs der spezifischen Wärme eines 2-dimensionalen Gittermodells für endliche Gittergrößen $N = L^2$, aufgetragen gegen die reziproke Temperatur (in willkürlichen Einheiten). Mit wachsender Systemgröße wächst der Maximalwert an, während sich dessen Lage (in diesem Beispiel zu höheren Temperaturen) verschiebt

2. Aus den verschiedenen experimentellen Daten, den geometrischen Verhältnissen, den (angenommenen oder gemessenen) Kräften zwischen den Molekülen, usw. wird ein mikroskopisches Modell formuliert, welches das beobachtete kritische Verhalten beschreibt. Wegen der komplexen Natur der Lipid-Moleküle ist dies eine schwierige Aufgabe, die heute noch nicht gelöst ist. Die Kenntnis des kritischen Verhaltens ist hierfür eine wichtige Voraussetzung.
Der Einfluß der (kritischen) Fluktuationen der Lipide auf die in einer biologischen Membran eingelagerten Proteinmoleküle bzw. deren innere Konformationen führt auf weitere interessante Probleme des kritischen Verhaltens von gekoppelten Systemen.

3. Die Frage nach einem überlagerten 1. Ordnungssprung, z. B. in der spezifischen Wärme, läßt sich bis heute noch nicht schlüssig beantworten. Weder läßt der oben skizzierte Rahmen die Beschreibung von 1. Ordnungsübergängen (außer für trikritische Punkte) bis heute zu, noch können aus experimentellen Daten einwandfreie Resultate über die Existenz eines möglichen Sprungs erhalten werden. Möglicherweise sind 1. Ordnungsphänomene an sehr schnelle Prozesse im ns-Bereich gekoppelt, während die langsameren Prozesse als kritische Phänomene in Erscheinung treten [6]. Ein theoretisches Verständnis hierzu fehlt jedoch noch weitgehend.

Literatur

1. Albon N, Sturtevant JM (1978) Proc Natl Acad Sci, USA 75:2258
2. Mitaku S (1981) Mol Cryst Liq Cryst 70:21
3. Mitaku S, Jippo T, Kataoka R (1983) Biophys J 42:137
4. Holzwarth JF, Rys FS (1984) Progr Colloid Polym Sci, dieser Tagungsband
5. Domb C, Green MS (1972–1976) (eds) Phase Transitions and Critical Phenomena, Bände I, II, III, V a, V b, VI, Acad Press, London
6. Jähnig F, private Mitteilung

Anschrift des Verfassers:

Franz S. Rys
Fritz-Haber-Institut der
Max-Planck-Gesellschaft
Faradayweg 4–6
D-1000 Berlin 33 (FRG)

Progress in Colloid & Polymer Science Progr Colloid & Polymer Sci 69:109–112 (1984)

Beobachtung einer kritischen Trübung und Verlangsamung am Hauptphasenübergang von Phospholipid-Membranen, bestimmt mit der Laser-Temperatursprungmethode*

J. Holzwarth und F. Rys**

Fritz-Haber-Institut der Max-Planck-Gesellschaft, Berlin (West)

Zusammenfassung: Kinetische Daten aus Trübungsgrad-Messungen mit der Laser-Temperatursprungmethode an unilamellaren DPPC-Vesikeln mit einem mittleren Durchmesser von 60 nm in 10^{-3}molarer wässriger Lösung werden in der Nähe des Hauptphasenüberganges als Funktion der Temperatur analysiert. Dabei lassen sich 5 verschiedene Prozesse mit Relaxationszeiten im ns-, 100 ns-, 10 µs, 100 µs und ms-Bereich unterscheiden. Die Analyse der letzteren 3 Prozesse zeigt ein charakteristisches kritisches Verhalten („kritische Opaleszenz", „critical slowing-down" und „finite-size cross-over"), das gut in den Rahmen einer allgemeinen Theorie kritischer Phänomene passt.

Abstract: The dynamics of the main phase transition in unilamellar vesicles of DPPC have been investigated using the „Iodine-Laser Temperature-Jump" technique with turbidity as detection parameter. As a result we could observe 5 well separated relaxation processes between 4 ns and 10 ms. The analysis of the three processes between 5 µs and 10 ms showed a critical slowing down of the relaxation processes which represent the phase transition. This behaviour can be treated using a theory of critical phenomena.

Key words: Temperature jump, phospholipids, phase-transition, kinetics, turbidity.

Die empfindliche Methode der Trübungsgrad-Messung nach einem Laserblitz-induzierten Temperatursprung von 0.5 bis 1 Grad erlaubt eine genaue Bestimmung des zeitlichen Ablaufs von Relaxationsphänomenen auf verschiedenen Zeitskalen von 10^{-10} bis 1 *s* [1, 2, 3, 4]. Für den Fall verdünnter Phospholipid-Lösungen von Dipalmitoylphosphatidylcholin (DPPC) und Dimyristoylphosphatidylcholin (DPMC) wurde ein charakteristisches Verhalten des Trübungsgrades, A, beobachtet [2], welches durch 5 verschiedene Relaxationszeiten, τ_i, beschrieben werden kann:

$$A = \sum_{i=1}^{5} A_i \, e^{-t/\tau_i} \qquad (1)$$

In der Nähe des Hauptüberganges, $T = T_c \pm \Delta T (\Delta T$ klein) tritt eine Verlangsamung der Relaxationsprozesse ein, wobei die dazugehörigen Amplituden A_i anwachsen.

Für den Fall von DPPC liegen genügende Meßresultate vor, die eine erste numerische Analyse erlauben. In Abbildung 1 sind die Daten des Trübungsgrades für eine ~ 0.001 molare Lösung von DPPC zusammengefaßt.

Ein typisches nicht-lineares Anwachsen von A_i und τ_i bei Annäherung an T_c wird für die Prozesse $i = 2, \dots$

* Vortrag, gehalten auf der 31. Hauptversammlung der Kolloid-Gesellschaft, Bayreuth 11. bis 14. Oktober 1983.

** Früher: Institut für Theoretische Physik, Freie Universität Berlin, (West)

KT-W 818

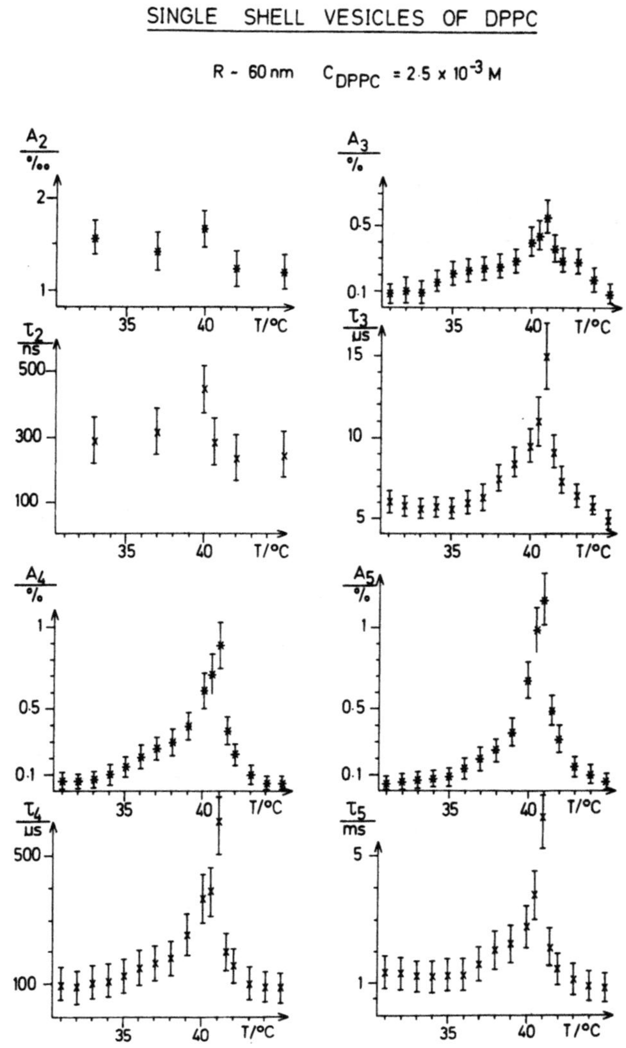

Abb. 1. Temperaturverlauf der Amplituden $A_i^{\pm}$ und der Relaxationszeiten $\tau_i^{\pm}$ $(i = 2, \ldots 5)$ für unilamellare DPPC Vesikel (Ø =120 nm)

5, nicht aber für den schellsten Prozeß $\tau_1 \simeq 4$ ns beobachtet.

Im folgenden konzentrieren wir uns auf die Analyse der Relaxationsprozesse $i = 3, 4$ und 5, da für den Prozeß Nr. 2 vergleichsweise wenige Daten vorliegen.

Zunächst fällt auf, daß sowohl die Amplituden A_i als auch die Relaxationszeiten τ_i weit weg von der Übergangstemperatur T_c ($\Delta > 5\,°C$) innerhalb der Fehlergrenzen fast dieselben Werte oberhalb und unterhalb von T_c annehmen, wobei A_3^+ und, weniger deutlich, A_4^+ ein abweichendes Verhalten zeigen, was auf Meßfehler bei der Bestimmung der sehr kleinen Amplitude zurückzuführen sein könnte. Aus den dynamischen Trübungsgradmessungen innerhalb der betrachteten Zeitskalen lassen sich daher keine markanten Unterschiede zwischen der kristallinen ($T < T_c$) und der fluiden Phase ($T > T_c$) feststellen.

In der Nähe von T_c steigen sowohl die Relaxationszeiten τ_i als auch die entsprechenden Amplituden A_i (für $i = 3, 4, 5$) beim Annähern an T_c zu einer ausgeprägten Spitze an, wobei der Anstieg von der kristallinen Seite her ($T < T_c$) weniger steil ist als jener von der fluiden Seite ($T > T_c$). Generell kann das Verhalten von τ_i als kritische Verlangsamung *(„critical slowing down")* bei T_c interpretiert werden, die als dynamisches kritisches Phänomen am kritischen Punkt von kontinuierlichen Phasenübergängen auftritt. Der Hauptphasenübergang von verdünnten Lösungen von Phospholipid-Doppelschichten erscheint daher im Lichte dieser Trübungsgradmessungen als kritischer Punkt eines *kontinuierlichen* Phasenübergangs zweiter Ordnung [5].

Ferner läßt sich der kritische Anstieg der Amplituden A_i als Analogon zur kritischen Opaleszenz von einfachen Flüssigkeiten am kritischen Punkt deuten. In beiden Größen is *kein* dem kritischen Anwachsen überlagerter endlicher Sprung feststellbar, wie er bei einem Phasenübergang 1. Ordnung zu erwarten wäre.

Um diese Interpretation im Rahmen einer modernen Theorie dynamischer kritischer Phänomene [5] zu quantifizieren, haben wir die experimentellen Daten auf ein Potenzverhalten hin untersucht, wie es die Skalentheorie ("dynamical scaling") annimmt:

$$A_i = \begin{cases} A_{io}^{+}\,(T - T_c)^{a+} & (T > T_c) \\ A_{io}^{-}\,(T_c - T)^{a-} & (T < T_c) \end{cases},$$
$$\tau_i = \begin{cases} \tau_{io}^{+}\,(T - T_c)^{t+} & (T > T_c) \\ \tau_{io}^{-}\,(T_c - T)^{t-} & (T < T_c) \end{cases}, \qquad (2)$$

In Abbildung 2 sind die Relaxationszeiten τ_i und die Amplituden A_i in doppelt-logarithmischer Darstellung gegen log $|T - T_c|$ aufgetragen. Ein reines Potenzgesetz (Glg 2) erscheint als lineare Abhängigkeit. Die Auftragung ist sehr empfindlich bezüglich der genauen Wahl von T_c. Da dieser Wert experimentell nicht genügend genau bestimmt wurde, haben wir mit dem Wert:

$$T_c = 41{,}25\,°C \text{ (DPPC)} \qquad (3)$$

eine gute Anpassung an die experimentellen Werte erzielt, die mit den Werten aus der Literatur gut übereinstimmt [6] und die Gleichheit der Exponenten $a_i^- = a_i^+$, $t_i^- = t_i^+$ oberhalb und unterhalb von T_c in etwa ergibt. Die verschieden steilen Flanken der Kurven für A_i und τ_i werden daher durch verschiedene Werte der Vorfaktoren beschrieben: $A_{io}^- \neq A_{io}^+$, $\tau_{io}^- \neq \tau_{io}^+$.

Ein weiteres Merkmal, das für sämtliche analysierte Kurven beobachtet wurde, ist ein Abknicken des linearen Verlaufs der log-log Auftragung, bei ca. $|T - T_c| \sim 1 \div 2\,°C$. Dieses Verhalten ("cross-over") könnte durch die Endlichkeit (d. h. Kleinheit) der Vesikelgröße bedingt sein. Es wäre daher interessant, auch wesentlich größere Vesikel zu untersuchen, für die der Knick dann näher bei T_c liegen müßte.

Zusammenfassend kann gesagt werden, daß eine Analyse der hier vorliegenden experimentellen Daten des Trübungsgrades nach einem Laserstrahl-induzierten kleinen Temperatursprung an verdünnten Phospholipid-Membranen Evidenzen für ein *kritisches Verhalten* am Hauptphasenübergang ergeben hat, bei dem *kein* 1. Ordnungssprung festgestellt wurde. Diese Ergebnisse passen gut in den allgemeinen Rahmen einer dynamischen Skalentheorie von endlich ausgedehnten Systemen, da ein *"critical slowing down"*, eine *kritische Opaleszenz* und ein *"finite-size cross-over"* beobachtet wurden. Eine befriedigende quantitative Beschreibung des Lipidmembran-Hauptphasenüber-

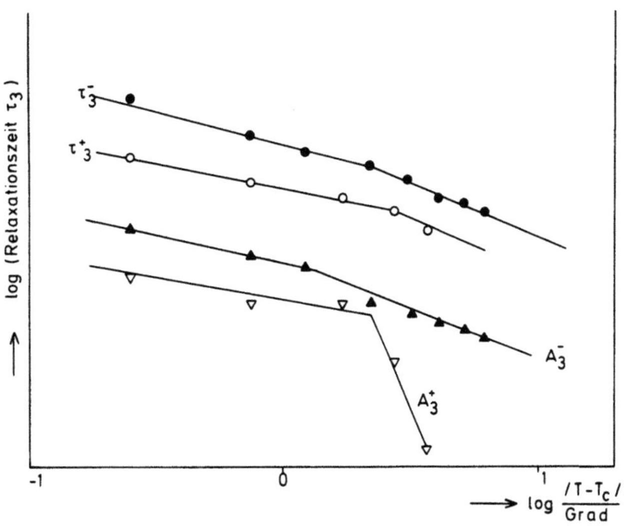

Abb. 2a

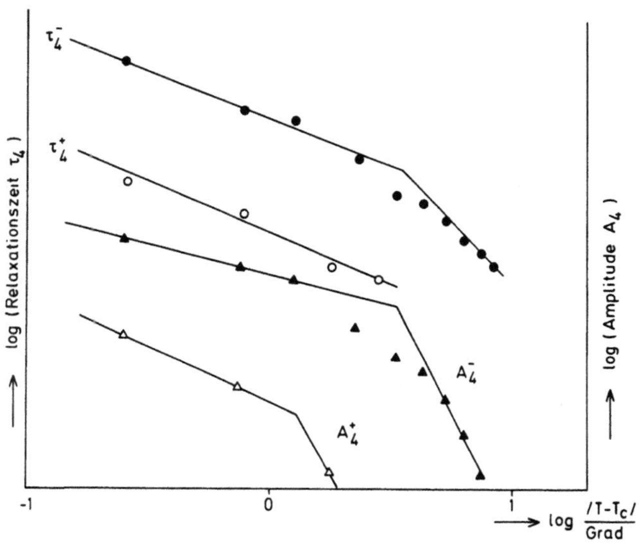

Abb. 2b

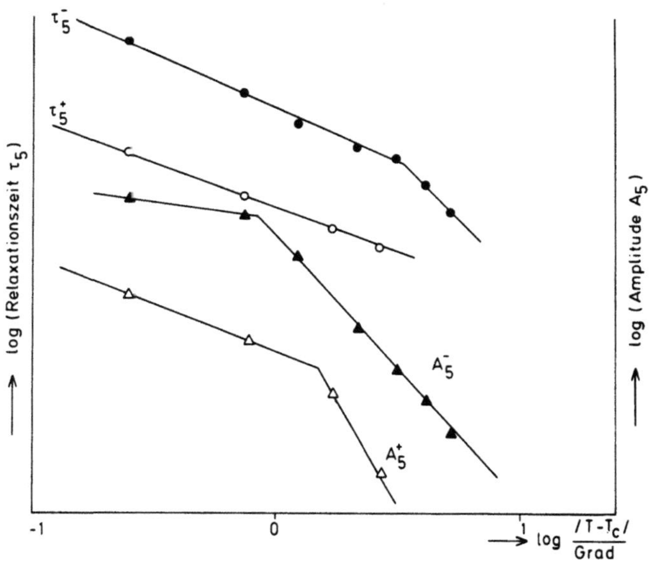

Abb. 2c

Abb. 2a–2c. Doppelt-logarithmische Auftragung von $A_i^\pm$ und $\tau_i^\pm$ aus Abbildung 1 als Funktion von $\log|T - T_c|$, mit dem Wert der kritischen Temperatur: $T_c = 41{,}25\,°C$

ganges, an dem offensichtlich mehrere Mechanismen beteiligt sind, wird sich an diesem Rahmen zu orientieren haben. Durch weitere dynamische Messungen mit noch kleineren Temperatursprüngen und unterschiedlich großen Vesikeln sollen die hier angesprochenen Fragen weiter bearbeitet werden.

Literatur

1. Holzwarth JF (1979) (ed) Gettins WJ, Wyn-Jones E, In: Techniques and Applications of Fast Reactions in Solutions, Reindel D, Publ Comp Dordrecht (NL) and Boston (USA), 47:69
2. Holzwarth JF, Frisch W, Gruenewald B (1982) (ed) Robb JD, In: Microemulsions, Plenum Publ Corp NY 185:205
3. Genz A, Holzwarth JF (1984) Progr Colloid and Polym Sci, dieser Tagungsband
4. Eck V, Holzwarth JF (1984) (ed) Mittal KL, Surfactants in Solution, Lund, Plenum Publ Corp NY 2059:2079
5. Rys FS (1984) Progr Colloid and Polym Sci, dieser Tagungsband
6. Mitaku S, Jippo T, Kataoka R (1983) Biophys J 42:137

Anschrift der Verfasser:

J. F. Holzwarth
Fritz-Haber-Institut der
Max-Planck-Gesellschaft
Faradayweg 4–6
D-1000 Berlin 33

On the interplay of microscopic order and macroscopic properties in solvent-saturated lipid films*

H. Wendel and P. M. Bisch**

Abteilung für Biophysik, Universität Ulm, Ulm (Donau), F. R. G.

Abstract: We review a recent theory of ours on the hydrodynamic behaviour of thin fluid films formed from short chain hydrocarbon molecules. For this we first a sketch semi-phenomenological model of steric repulsion in these films. The statistical mechanical formulation of the van der Waals theory of simple liquid-vapour interfaces is employed. The model is based on the picture that upon diminishing the film thickness, the adjacent interfacial layers of surface adsorbed molecules start overlapping in the film center. The overlap raises the degree of orientation of the hydrocarbon segments present there, thus causing a mutual repulsion of the two film surfaces. The general order parameter profile is found to depend on two parameters which are fixed by fitting the results of an experiment on the steric disjoining pressure in solvent-containing mono-oleate films. We further discuss the film tension and the film elasticity in terms of our model. At long wavelengths, the sign of these quantities determines the film stability against bending and squeezing deformations, respectively. In both cases, sterics is found to counteract the destabilizing van der Waals attraction and electrostatics. We establish a stability diagram displaying the regimes of film stability as a function of film thickness and electric cross film potential. The film is found to be stable against thickness fluctuations from a critical thickness down (of the order of 60 Å), the thickness depending slightly on the applied voltage. Further, voltages above 170 mV trigger bending of low tension films (single surface tension: 0.05 dyn/cm) just below their critical squeezing thickness. At these voltages, bending of the film is prevented only by increasing the disjoining pressure or by diminishing the film thickness, respectively.

Key words: Amphiphiles, liquid films, disjoining pressure, film tension, film elasticity, film stability.

1. Introduction

On the occasion of the 30th Meeting of the Kolloid-gesellschaft we introduced a model to describe the hydrodynamic behaviour of short chain hydrocarbon or lipid films, respectively, with a thickness of several thousand Angstroms (coloured state) [1]. At the same time we sketched roughly how the theory could be extended to the regime of black lipid films (LF) displaying a thickness of the order of 60 Å. In the present communication we would like to to discuss this extension in greater detail.

LF are formed [2–4] by introducing a small amount of lipid solution onto the opening of a hydrocarbon support immersed in aqueous solution. First the film appears coloured implying a film thickness of the order of 0.1–1 μm. Its bulk phase consists of a hydrocarbon solution sandwiched between two layers of surface adsorbed lipids. The surfactant layers take advantage of the amphiphilic character of the lipid molecules: their polar head groups face the aqueous phase while their hydrocarbon chains point into the hydrophobic film interior.

Under suitable conditions, the coloured film begins to drain while colour bands indicate varying thickness across the film. The bands increase in breadth and width until finally the appearance of "black spots" signals the beginning of the formation of the black film (cf. fig. 1 of ref. [1]). Then the chain ends of lipids belonging to adjacent surface layers start interacting (indirectly)

---

* Lecture given at the 31th Conference of the Kolloid Gesellschaft, Bayreuth October 11–14, 1983.

** Permanent address: Centro Brasileiro de Pesquisas Fisicas – CBPF, Rio de Janeiro, Brazil.

KT-W 805

with each other effecting a stabilization of the LF at the quasi bimolecular thickness of the order of 60 Å.

The large scale dynamics of LF has been the research community's step-child for a long time and only scarce data exist on this topic to date [5–7]. On a first glance the situation is somewhat surprising since LF have been employed amply to simulate the socalled lipid bilayer membrane (LBM), a film of biological relevance [8]. Yet, the experimental and theoretical research interest for the macroscopic behaviour of pure lipid films has been exhausting itself mainly in establishing their phase diagrams and studying the corresponding phase transitions [9]. As a matter of fact, experimental results obtained by photon correlation spectroscopy on LF have been reported only recently [5, 6]. In view of this development the vistas are open now for a hydrodynamic theory of LF which goes beyond the interpretation of the light scattering experiments on the basis of a single interface picture [6].

A large number of theoretical papers on the hydrodynamics of thin films up to the present [10–12] has employed the DLVO picture [13] in which film stability relies on the presence of overlapping diffuse layers in the film. The diffuse layers of LF extend into the disjoint aqueous bulk phases while the hydrocarbon bulk is free of any charge. Thus within the framework of the theories mentioned above, LF would collapse or break up into bubbles, respectively, under the action of the van der Waals forces, which become more effective with a thinning of the film. This is at variance with an experiment which encounters stable LF of a small thickness. Thus a repulsive mechanism is called for which stabilizes black LF, while the references above apply only to coloured lipid films.

Other hydrodynamic theories have dealt with biological viscoelastic membranes [14, 15]. These models have aimed high in trying to describe actual membranes whose detailed morphology remains unknown. Thus in our opinion these models had to fall short; they introduced only additional adjustable parameters into the problem without granting further insight. Also none of the models above is suited to treat coloured and black LF on the same footing. They thus generate an artificial distinction which is not justified according to the observed experimental fact of a continuous transition from coloured to black films [2].

Recently, Gallez [16], following the lines of reference [1], has proposed an alternative purely phenomenological model to treat steric repulsion. Unfortunately the microscopic origin of the repulsive forces cannot be traced in her theory. Finally, a static analysis for thickness fluctuations of lipid films was performed by Hladky and Gruen, but they also used a purely phenomenological model without any wavelength dependence of fluctuating forces [17].

The system under consideration is a very complex multi-body system. The host of microscopic studies of the statics of lipid systems supports the impression that it is hard imagining any realistic bilayer model to be accessible to an analytical treatment without excessive computer processing. This must hold true even more for the dynamics. By choosing length and time scales sufficiently coarse (hydrodynamics) the theoretical effort may be reduced. Modelling the dynamics still remains a formidable task. On the other hand a purely phenomenological treatment with adjustable parameters involves the danger that a subsequent microscopic model may not be able to assign a strict physical meaning to the parameters involved.

Thus recently we have developed an analytical *semiphneomenological* theory of repulsion [1, 18]. In view of our critics noted above with respect to various models in existence we emphasize: i) Our model accounts for our present microscopic understanding of LBM; ii) it unifies the hydrodynamics of coloured and black lipid films from a theoretical point of view, too in the following we will review this model.

In detail, in Section 2 we will recall two experiments and their theoretical interpretation which will help us in formulating our model. We will also introduce an interpretation from the point of view of interfacial physics, thus anticipating verbally the formal theory to be developed in consequence. In Section 3 we formulate our model of "steric" repulsion. Based on the present development of microscopic theories we show how the complex problem can be reduced to the problem of solving for the density profile of a binary liquid film. With the assumption of constant hydrocarbon density in the film our formulation of steric repulsion effectively comes down to fixing the density profile between two interacting liquid vapour interfaces. In Section 4 this correspondence is exploited within our model to derive the steric disjoining pressure. While doing so we disentangle the general confusion which arose in the past concerning the term disjoining pressure. The steric disjoining pressure for solvent-saturated LF was measured. The experimental results will serve for fixing two parameters in our theory. In Section 5 we briefly recall that at long wavelengths and low frequencies the dynamic stability of the film depends on two modes called the bending and squeezing mode, respectively. Stability against these deformations depends on the sign of the film tension and film elasticity, respectively. We will define these quantities in general and within the framework of our theoretical model. We further employ the model results in order to establish a diagram depicting the areas of film stability as a function of film thickness and of the strength of an electric potential applied across the film. Section 6 concludes our communication.

2. Real and computer experiments; their hints [19, 20, 22]

In this Section we want to discuss two types of real experiments [19, 20] which will lead us to propose a specific theoretical model for steric repulsion, in Section 3. This model will exhibit two parameters which are fixed by means of the quantitative results of reference [19].

In their experiment, Andrews, Manev and Haydon [19] applied an electrical potential across different types of optically black films, thus subjecting them to a large compressive force. They report that most of the films became significantly thinner presumably due to a squeezing-out process of solvent naturally present when forming the films. The authors measured the specific capacitance C of the film as a function of the applied potential V related in a simple manner to the electrically active film thickness δ_e. Thus they could derive the magnitude of the steric repulsion employing the following scheme: at equilibrium, the forces per unit area (pressure) due to the London-van der Waals interaction (F_L), the applied cross film potential (F_e), and the steric interaction (F_s) balance each other, i. e.,

$$F_L + F_e + F_s = 0 \tag{1}$$

with

$$F_L = -H/6\pi h^3 \tag{2a}$$

$$F_e = -CV^2/2\delta_e . \tag{2b}$$

Here H is the Hamaker constant and h is the total film thickness given by $h = (\Delta + \delta_e)$ nm with Δ being the diameter of the polar head group of the lipid forming the film. The specific capacitance C is related to δ_e via $C = \varepsilon/4\pi\delta_e$ where ε is the dielectric constant of the hydrocarbon bulk ($\varepsilon \sim 2.1$).

All the quantities appearing in equations (2 a) and (2 b) are accessible to experiment. The steric repulsion pressure is thus determined by

$$F_s = (H/6h^3) + (CV^2/2\delta_e) . \tag{3}$$

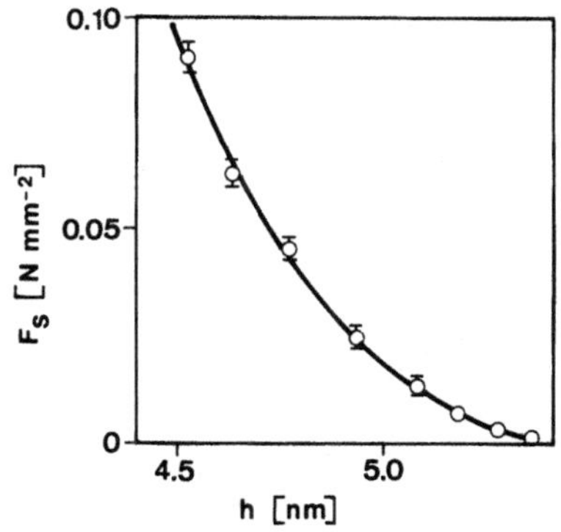

Fig. 1. Steric repulsive force F_S per unit area of lipid film surface versus film thickness as measured by Andrews, Manev and Haydon for glyceryl mono-oleate in n-decane [19]

A typical experimental result for F_s as a function of h is shown in figure 1. We note the steep rise of the steric repulsion, even more pronounced in some other types of film. The important conclusion of the authors of this experiment is: a very small overlap of lipid chains belonging to adjacent monolayers ("surfactant" layers) at very small segment concentration generates a repulsive force sufficient to stabilize the film.

In the foregoing experiment the integral static behaviour of LF was probed. The exact nature of the lipid conformation and dynamics was immaterial. In recent years, however, experiments emphasized the significance of phospholipid conformation for the organization of biological membranes [20].

The method of deuterim nmr [20] contributed fundamentally to our modern view of the individual behaviour of a lipid molecule or parts of it, respectively, in the evironment provided by a membrane.

In this experiment, deuterium is introduced at a specific site of the phospholipid molecule either via chemical synthesis or by means of chemical incorporation. Selected substitution of ^1H by ^2H does not perturb the natural arrangement of the molecule in the membrane and the small natural abundance of ^2H (0.06 %) assigns the ^2H nmr signal immediately to the deuterium labelled site.

The characteristic quantity measured in this experiment is the deuterium quadrupole splitting $\Delta\nu_Q$. The splitting is proportional to the deuterium order parameter S_{CD}, i. e.,

$$\Delta\nu_Q \sim S_{CD} = \frac{1}{2}\langle 3\cos^2\theta - 1\rangle . \tag{4}$$

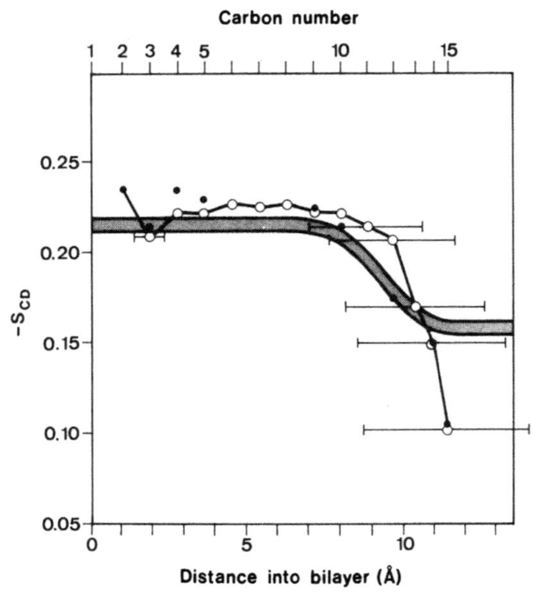

Fig. 2. Comparison of experimental (black dots) and theoretical (open circles connected by solid line) deuterium order parameter S_{CD} as a function of carbon atom label (upper abscissa) for one half of a lipid bilayer membrane. Also shown is the order parameter as a function of position in the lipid bilayer (lower abscissa, screen dotted stripe) [4]

Here θ denotes the instantaneous angle between the CD-bond of a selectively deuterized lipid hydrocarbon chain with the direction of the applied magnetic field, usually aligned in the direction of the bilayer normal.

Figure 2 shows a typical order profile of a lipid bilayer (solid dots) in the fluid state above the transition temperature T_c,i. e., it depicts the order parameter S_{CD} as a function of the number of the C atom involved in the CD-Bond. Apart from a zig-zag structure at the carbon atoms 2–4, the order parameter is found to be relatively constant in a large region at the beginning of the chain (carbon atoms 2–9) and decreases subsequently towards the methyl terminal. Also shown is the theoretical calculation (circles) based on a mean field picture [21] agreeing well with experiment. In addition to these order profiles down the chain, the figure also displays the profile across the bilayer (smooth continuous curves), i. e., as a function of the position in the bilayer. Qualitatively, though less pronounced, a similar behaviour is found: the order prevailing close to the membrane surface and continuing into an extended plateau region decays (slowly) towards the center of the membrane.

We point out the experiment and theory of figure 2 hold strictly only for bilayers without solvent. The theory, however, was recently extended to bilayers containing hydrocarbon solvent [22] and a behaviour of the order parameter qualitatively similar to that in the solvent-free film was predicted (fig. 3). The only major change was reported for the film center where,

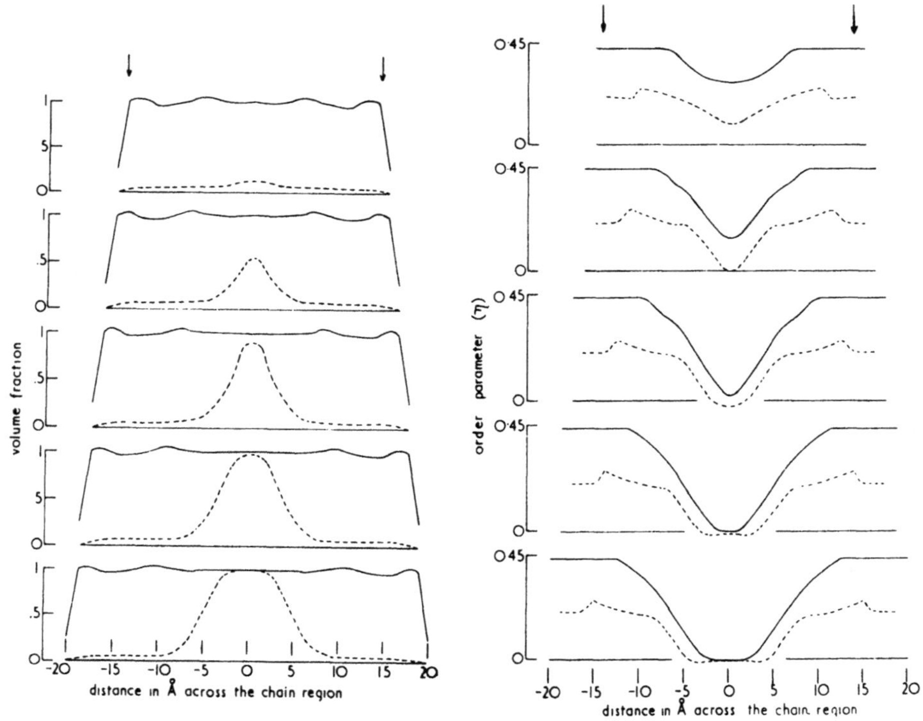

Fig. 3. Volume fractions and order parameter profiles across butane-containing bilayers at different thicknesses (increasing thickness from top to bottom). (---) volume fraction and order parameter, respectively, of butane bilayer. (——) volume fraction and volume weighted order parameter, respectively, for all chains [22]

depending on the type of solvent and on the film thickness, in most cases nearly complete disorder (order parameter identically zero) was met.

The picture which arose for black LF [22] is as follows: in a solventless LBM the lipid chains are packed closely to reduce chain-water contact. They are preferentially oriented along the normal of the plane of the bilayer in which plane the chains exert a lateral pressure. In a membrane containing a very small amount of alkane which is chemically similar to the lipid chains, the volume fraction of lipid is high throughout the membrane. Hence, the alkane chains present are subject to considerable lateral pressure and pack with similar properties to the surrounding lipid chains. In a LF containing a substantial volume fraction of alkane the only lipid chains present in the center are those which are fully extended. In the fluid state only a small fraction of the lipid chains is fully extended. The alkane chains in the outer regions of the LF line up essentially parallel to the lipid chains, while those in the center are almost completely disordered. Thus, at the center of such a film, the alkane chains sit in an environment approaching that of bulk liquid alkane. The lateral pressure acting upon them must be smaller than for the chain segments nearer the polar interfaces and they will be subject to forces similar to those in bulk liquid alkane. Altogether the chains (both lipid and alkane) are in an essentially liquid state with no well defined interfaces between opposing monolayers.

Let us analyse the problem from the point of view of interface chemical physics: a single interface represents a highly inhomogeneous transition region between two different homogeneous bulk phases. The anisotropic interactions together with the non-uniform composition in the interface give rise to a spatially varying free energy density which beyound a certain "decay length" assumes the constant values of the homogeneous bulk phases on either side. A (planar) symmetric film comprises two identical interfaces. As long as the decay length of the interfacial inhomogeneity is small compared to the film thickness the van der Waals attraction of the external phases across the film causes the film thickness to shrink. From a thickness of the order of twice the decay length on, the interfacial inhomogeneities start overlapping in the film center thus raising the value of the free energy density above its homogeneous value in the thick film. This fact manifests itself in the appearance of repulsion between the two film surfaces.

From our preceding discussion it is obvious that, for constant packing density of the hydrocarbon segments, in our system the interfacial inhomogeneity in the film phase is related to the hydrocarbon order: for thick films, the order close to the film surface (head groups) "decays" into the isotropic segment orientation prevailing in bulk hydrocarbon. Upon approaching the film surfaces the order in the film center increases (cf. fig. 3 in ref. [1]) reducing the film entropy and increasing the short range van der Waals interaction; thus the free energy is raised, which fact finds its expression in the repulsive force measured in reference [19].

In the next Section we model this intuitive picture.

3. Density functional formalism and orientational order profile [1, 18]

For complex interfaces as in LF there is no simple statistical mechanical theory relating the molecular properties to macroscopic interfacial forces. Nevertheless, here we will take advantage of a molecular theory of interfacial phenomena in multicomponent liquids [23–25]. On its basis we will argue how the problem of "steric" repulsion between two coupled interfaces (= film) reduces to the problem of solving for the density profiles of a binary liquid film comprised of "oriented" and "non-oriented" particles.

The system formed by the hydrocarbon chains of lipid and solvent molecules is described by a classical Hamiltonian of the general form

$$H_N = \sum_{\alpha=1}^{N} E_\alpha + V(\vec{r}_1, \ldots, \vec{r}_N, \theta_1, \ldots, \theta_N)\,. \tag{5}$$

Here N is the total number of hydrocarbon monomers, E_α is the intrinsic energy of segment α, and V is the potential energy of interaction which depends on the positions $\{\vec{r}_\alpha\}$ and orientations $\{\theta_\alpha\}$ of the segments. The angles $\{\theta_\alpha\}$ are referred to an axis chosen arbitrarily. According to the basic ideas of the density functional formalism, the macroscopic properties of the system are determined completely by the configuration averages $<\hat{n}(\vec{r}, \theta)>$ where

$$\hat{n}(\vec{r}, \theta) = \sum_{\alpha=1}^{N} \delta(\vec{r} - \vec{r}_\alpha)\,\delta(\theta - \theta_\alpha) \tag{6}$$

is the local density of monomers with orientation θ. Equations (5) and (6) represent the basic ingredients of a general statistical description of a system of interacting particles with an orientational degree of freedom.

For our purposes, however, this description can be simplified in the light of current mean field theories [21, 22, 26].

i) In these theories, the angle θ_α is measured with respect to the average direction of alignment of the segments, i. e., the bilayer normal. Further, the orientational interaction of the Mayer-Saupe type [27] is only a function(al) of the segment order ($3/2 \cos^2 \Theta_\alpha - 1/2$); it is described in the density functional formalism by the microscopic density

$$\hat{n}_1(\vec{r}) = \int d\theta \left(\frac{3}{2}\cos^2\theta - \frac{1}{2}\right)\hat{n}(\Theta, \vec{r}) = \sum_\alpha \left(\frac{3}{2}\cos^2\theta_\alpha - \frac{1}{2}\right)\delta(\vec{r} - \vec{r}_\alpha). \quad (7)$$

ii) The other relevant parameter in these theories is the local packing of the chains, or equivalently, the microscopic density of monomers,

$$\hat{n}(\vec{r}) = \int d\theta\, \hat{n}(\vec{r}, \theta). \quad (8)$$

Without adopting any explicit form, we suppose the interaction potential to be a function(al) of $\hat{n}_1(\vec{r})$ and $\hat{n}(\vec{r})$ only. Thus the macroscopic system is characterized completely by the mean values $\langle \hat{n}_1(\vec{r}) \rangle$ and $\langle \hat{n}(\vec{r}) \rangle$. These assumption include especially all the microscopic mean field theories put forward up to now.

Introducing arbitrary external fields $U(\vec{r})$ and $U_1(\vec{r})$, which couple to $\hat{n}(\vec{r})$ and $\hat{n}_1(\vec{r})$, respectively, the total Hamiltonian reads

$$H_o = H_N + \int d^3r\, U(\vec{r})\,\hat{n}(\vec{r}) + \int d^3r\, U_1(\vec{r})\,\hat{n}_1(\vec{r}). \quad (9)$$

Within the assumptions discussed above, the grand partition function for the system at temperature T is given by

$$e^W = Tr_{cl}\{e^{-\beta(H_o - \mu_\theta \bar{\eta} N - \mu N)}\} \quad (10)$$

with

$$N = \int d^3r\, \hat{n}(\vec{r}) \quad (11)$$

and

$$\bar{\eta} = \frac{1}{N}\int d^3r\, \hat{n}_1(\vec{r}) = \frac{1}{N}\sum_\alpha \left(\frac{3}{2}\cos^2\theta_\alpha - \frac{1}{2}\right). \quad (12)$$

The classical trace, Tr_{cl}, is performed over the molecular configurations, and $\beta = 1/K_B T$, K_B being the Boltzmann constant. The chemical potentials μ and μ_θ are associated with the monomer density and with the segmental order, respectively. For maximum segment order, i. e., $\cos^2 \Theta_\alpha = 1$ for any α, equation (12) gives $\bar{\eta} = 1$, and for complete segment disorder, $\bar{\eta} = 0$. The chemical potential μ_θ, is thus the energy (per monomer) required to orient segments. Consequently the quantity $\bar{\eta} N = N_1$ can be interpreted as an effective number of oriented segments while there are $N_2 = N - N_1$ non-oriented monomers. On the microscopic level this interpretation corresponds to employing the densities $\hat{n}_1$ and $\hat{n}_2 = \hat{n} - \hat{n}_1$, instead of $\hat{n}_1$ and $\hat{n}$. The corresponding grand potential is defined via

$$e^W = Tr_{cl}\{e^{-\beta(H_o - \mu_1 N_1 - \mu_2 N_2)}\} \quad (13)$$

instead of equation (10) with $\mu_1 = \mu_\theta + \mu$ and $\mu_2 = \mu$. The local monomer densities, n_1 and n_2, are given by

$$n_1(\vec{r}) = \langle \hat{n}_1(\vec{r}) \rangle = \frac{\delta}{\delta u_1(\vec{r})} W$$

$$n_2(\vec{r}) = \langle \hat{n}_2(\vec{r}) \rangle = \frac{\delta}{\delta u_2(\vec{r})} W \quad (14)$$

with $u_1(\vec{r}) = U_1(\vec{r}) + U(\vec{r}) + \mu_1$ and $u_2(\vec{r}) = U(\vec{r}) + \mu_2$.

The total density is

$$n(\vec{r}) = n_1(\vec{r}) + n_2(\vec{r}), \quad (15)$$

and the macroscopic order parameter is [20–22]

$$\eta(\vec{r}) = \left\langle \int d\theta \left(\frac{3}{2}\cos^2\theta - \frac{1}{2}\right)\hat{n}(\vec{r}, \theta)/n(\vec{r}) \right\rangle = n_1(\vec{r})/n(\vec{r}). \quad (16)$$

Thus the problem of determining the density of the segment order is reduced to the problem of solving for the density profiles of a binary system constituted of oriented and non-oriented "particles". Henceforth we are able to employ the general theory of classical non-uniform systems [23–25]. H_N need not be specified. The only requirement is that the interaction potential be short ranged.

For $n(\vec{r}) = \text{const.}$, which assumption has been discussed in detail in reference [18], our problem simplifies even further allowing us to employ the general ideas of the theory of simple liquid interfaces [23, 25]. The spatially varying density in this theory corresponds to the concentration of the oriented particles in our map; it is numerically identical to the value of the order parameter density as measured in experiment [20] or calculated numerically [22].

In the theory of non-uniform one-component fluids, the Helmholtz free energy at fixed temperature is expressed as a definite integral of a local free energy density,

$$F = \int d^3r f(\vec{r}) \tag{17}$$

with

$$f(\vec{r}) = f_o(\eta(\vec{r})) + \frac{1}{2} A(\eta(\vec{r})) \cdot (\vec{\nabla}\eta(\vec{r}))^2 . \tag{18}$$

In equation (18) it is assumed that i) the (order parameter) density $\eta(\vec{r})$ is sufficiently slowly varying in its spatial dependence on a scale set by the interaction range, and ii) there is no external field coupling to $\eta(\vec{r})$. $f_o(\eta(\vec{r}))$ is the local free energy density of a uniform equilibrium system displaying density $\eta(\vec{r})$. $A(\eta)$ is related to the direct correlation function $C(\vec{r}, \eta)$ of a uniform system having density $\eta = \eta(\vec{r})$, by

$$A(\eta) = \frac{kT}{6} \int d^3r\, r^2\, C(\vec{r}, \eta(\vec{r})) . \tag{19}$$

The Ornstein-Zernike direct correlation function is derived from the interaction part of the free energy,

$$C(\vec{r} - \vec{r}'; \eta) = \frac{1}{kT} \frac{\delta^2 (F_{\text{ideal}} - F)}{\delta\eta(\vec{r})\, \delta\eta(\vec{r}')} \tag{20}$$

where F_{ideal} is the free energy of the system in the absence of interactions.

The equilibrium density $\eta_o(\vec{r})$ arises from the minimum condition for equation (17) plus the condition of constant total particle number. The corresponding Euler-Lagrange equation reads

$$\mu = \mu_o(\eta(\vec{r})) - A(\eta) \cdot \vec{\nabla}^2 \eta(\vec{r}) - \frac{1}{2} \frac{\delta A(\eta)}{\delta\eta(\vec{r})} \cdot (\vec{\nabla}\eta(\vec{r}))^2 . \tag{21}$$

μ is the chemical potential of the inhomogeneous system. $\mu_o(\eta(\vec{r}))$ represents the chemical potential a homogeneous liquid of (constant) density $\eta = \eta(\vec{r})$ would have, i. e.,

$$\mu_o(\eta(\vec{r})) = \frac{\delta f_o(\eta)}{\delta\eta(\vec{r})} . \tag{22}$$

The stress tensor derives from the condition of mechanical equilibrium and reads [25]

$$\vec{\vec{\sigma}}(\vec{r}) = -p(\vec{r}) \cdot \vec{\vec{I}} + \vec{\vec{\Pi}}(\vec{r}) \tag{23}$$

with

$$p(\vec{r}) = f_o(\eta(\vec{r})) - \mu \cdot \eta(\vec{r}) \tag{24}$$

and

$$\vec{\vec{\Pi}}(\vec{r}) = A(\eta) \cdot \left\{ \vec{\nabla}\eta(\vec{r})\, \vec{\nabla}\eta(\vec{r}) - \frac{1}{2} (\vec{\nabla}\eta(\vec{r}))^2 \cdot \vec{\vec{I}} \right\} \tag{25}$$

with $\vec{\vec{I}}$ being the 3d-unit tensor.

In principle the quantities $f_o(\eta(\vec{r}))$, $A(\eta)$, $\delta f_o / \delta\eta(\vec{r})$ etc. can be calculated given appropriate microscopic potentials, and equation (9) can be solved for $\eta(\vec{r})$ and the stress tensor equation (23) can be calculated.

Based on equation (21) we now want to derive the order parameter profile for the film system. For that we assume the film normal to be parallel to the z-axis of our Cartesian coordinate system. Further, as in the experiment, our black LF is thought to be in equilibrium with bulk alkane (cf. fig. 4). Because of thermodynamic equilibrium, the chemical potential μ in the film is equal to that in the homogeneous bulk phase, i. e.,

$$\mu = \mu_o(\eta_b) . \tag{26}$$

Here η_b is the constant value of the density in the homogeneous bulk phase. For film densities η different from η_b but such that $\hat{\eta} = (\eta - \eta_b) \ll 1$, the first term on rhs of equation (21) can be expanded:

$$\mu_o(\eta) = \mu_o(\eta_b) + \left.\frac{\partial \mu_o(\eta)}{\partial \eta}\right|_{\eta_b} \cdot \hat{\eta} + 0(\hat{\eta}^2) . \tag{27}$$

Combination of equations (21), (26), and (27) while keeping only terms of order $\hat{\eta}$ yields

$$\left(\frac{d^2}{dz^2} - \beta^2 \right) \hat{\eta} = 0 . \tag{28}$$

Here

$$\beta^2 = \left.\frac{\partial \mu_o(\eta)}{\partial \eta}\right|_{\eta_b} / A(\eta_b) . \tag{29}$$

The parameter β^2 is related to the competitive mechanisms leading to an inhomogeneous profile of the order parameter: the direct correlation $C(\vec{r})$ between different (neighbouring) points tends to increase the order, and the change in internal energy due to the variations of $\eta \left(\frac{\partial \mu_o}{\partial \eta} \right.$, positive partial compressibility$\Big)$

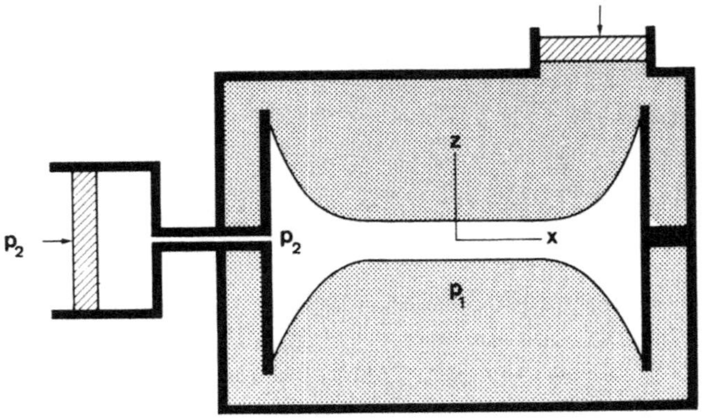

Fig. 4. Schematic film set-up. The film in the interior of the chamber is in equilibrium with its meniscus, i.e. with the bulk phase of the film forming material. The isotropic pressure p_2 in this phase differs in general from the pressure p_1 in the "gas" phase. This difference is called disjoining pressure [28]

is unfavourable for an increase of order. The length β^{-1} measures then an effective range of direct correlations as compared to the increase of internal energy due to variations of η.

Equation (15) represents the "asymptotic" form of equation (9), i.e., it holds for regions in which i) η is close to η_b, and ii) $\frac{\partial A}{\partial \eta} \cdot \left(\frac{d}{dz}\hat{\eta}\right)^2 \ll \left|\frac{d^2}{dz^2}\hat{\eta}\right| \cdot A(\eta)$. For symmetry reasons the latter conditions is always satisfied in the center of a symmetric film independently of its thickness. So there the validity of equation (28) is only restricted by condition i) or, equivalently, to the treatment of films with small overlap of the interfacial inhomogenities. This limitation for its part then gurantees equation (28) to also hold further away from the film center.

We note that for our purpose it sufficies to model well the order parameter in and close to the film center where it depends most sensitively on the film thickness [22].

The general solution of equation (15) writes

$$\eta(z) = B \cdot \cos h(\beta z), \tag{30}$$

where B is fixed by the boundary conditions either as

$$B = \hat{\eta}(0) = \eta_F - \eta_b, \tag{31}$$

with η_F being the actual value of the order parameter in the film center, or as

$$B = \eta_s / \cos h(\beta h/2), \tag{32}$$

where η_s is the order parameter at the film suface. η_s need not correspond to the real density prevailing at the film surface; it rather leads to the correct "asymptotic" solution close to the film center. As already mentioned, the exact density profile or the interfacial forces close to the surface, respectively, are irrelevant for the aspect of repulsion since they contribute to the surface tension which will be defined in Section 5.

Having established the order parameter profile (eqs. (28), (29)), in principle we could quantify the state of stress due to the repulsion (eqs. (23–25)) if we knew the values of the quantities β and $\hat{\eta}(0)$ or $\hat{\eta}_s$, respectively. Lacking a microscopic theory for calculating these quantities we have to look for some other means to fix them. We regress to the macroscopic experiment discussed in Section 2. For that we first introduce the notion "disjoining pressure" [28].

4. Steric repulsion and steric disjoining pressure [18, 29, 30]

We consider a flat lipid film in thermal, chemical, and mechanical equilibrium with its film forming phase (cf. fig. 4). The film is surrounded by two aqueous bulk electrolytes and subjected to an electric cross film voltage of variable strength.

In the film, the intermolecular forces acting on an element of volume are different from those acting on an element of identical volume situated in the interior of the bulk phase and far away from a surface of discontinuity. Further, in the thin film, the two surfaces are so close to each other that no part of the thin film is out of range of the surface forces. Hence the properties of thin films are described in terms of a stress tensor (cf. eq. (23)). It has, however, become customary in the literature to discuss the properties of thin films in terms of the disjoining pressure [31–33].

In mechanical equilibrium, the disjoining pressure is defined as being the difference between the pressure in the external phase, p_1, and the pressure p_2 in the bulk phase forming the film, i.e., in the meniscus (cf. fig. 4). For equilibrium thickness, studied by Andrews et al. [19], the repulsive force nearly compensates for the attractive forces of electrical and van der Waals origin; the *total* disjoining pressure is negligibly small. Thus the pressure difference arising exclusively in the hydrocarbon part of the film system is equal to the steric repulsion force per unit area, F_s, measured by Haydon and collaborators [19].

The pressure distribution $p(\vec{r})$ in the film system (film plus meniscus) follows from the *conditions of mechanical equilibrium*, i. e.,

$$\vec{\nabla} p(\vec{r}) = -\varrho(\vec{r})\,\vec{\nabla} W(\vec{r}) + \vec{\nabla}\cdot[\vec{\vec{T}}(\vec{r}) + \vec{\vec{\Pi}}(\vec{r})]. \quad (33)$$

Here $p(\vec{r})$ is the hydrostatic pressure accounting for all short range interactions not explicity accounted for otherwise. $\varrho(\vec{r})$ is the mass density assumed uniform in each bulk phase. $W(\vec{r})$ is the van der Waals potential derived from the long range part of the two body potential between molecules [34]. The Maxwell stress is given by

$$\vec{\vec{T}}(\vec{r}) = \frac{1}{4\pi}\{\varepsilon \vec{E}(\vec{r})\,\vec{E}(\vec{r}) - 1/2\,\vec{E}^2(\vec{r})\cdot\vec{\vec{I}}\}. \quad (34)$$

Here, ε is the dielectric constant of the medium under consideration, $\vec{E}(\vec{r})$ is the electric field, and $\vec{\vec{I}}$ represents the 3d unit tensor.

The steric stress, $\vec{\vec{\Pi}}(\vec{r})$, has been defined in equation (25).

Before evaluating equation (31) we note that the electric plus steric forces derive from a potential, i. e., [18]

$$\vec{\nabla}\cdot(\vec{\vec{T}}(\vec{r}) + \vec{\vec{\Pi}}(\vec{r})) = \vec{\nabla}\Phi(\vec{r}) \quad (35)$$

with

$$\Phi(\vec{r}) = \frac{\varepsilon-1}{8\pi}\vec{E}^2(\vec{r}) + \frac{1}{2}A[\eta_b]\,\beta^2(\vec{\eta})^2. \quad (36)$$

Equation (33) implies that for two points $\vec{r}_2$ and $\vec{r}_2$ in the film differing only by their respective z-coordinates, it holds that

$$p(\vec{r}_2) - T_{zz}(\vec{r}_2) - \Pi_{zz}(\vec{r}_2) + \varrho_2 W(\vec{r}_2) = p(\vec{r}_2) - T_{zz}(\vec{r}_2) - \Pi_{zz}(\vec{r}_2) + \varrho_2 W(\vec{r}_2) \quad (37)$$

or,

$$p(\vec{r}_2) - T_{zz}(\vec{r}_2) - \Pi_{zz}(\vec{r}_2) + \varrho_2 W(\vec{r}_2) = p_{N_2}. \quad (38)$$

Equation (37) is obtained upon integrating the z-component of equation (33) along z within the film, where T_{xz}, T_{yz}, Π_{xz} and Π_{yz} are identically zero. Thus, in a mechanically equilibrated film with constant mass density ϱ_2, no work is required to transfer an infinitesimal volume element parallel to the film normal, provided $p(\vec{r})$, $\vec{E}(\vec{r})$, $\vec{\eta}(\vec{r})$, and $W(\vec{r})$ vary according to their equilibrium profiles. The constant p_{N_2} in equation (38) represents the total normal stress at any point in the film.

Similar arguments for the external phase yield

$$p(\vec{r}_1) - T_{zz}(\vec{r}_1) + \varrho_1 W(\vec{r}_1) = p_1 \quad (39)$$

with $\vec{r}_1$ being an arbitrary point in the aqueous phase. Again the stress is constant; it has been chosen to equal the isotropic pressure p_1 far off the interface.

The stress states on either side of a surface relate to each other via the Laplace condition

$$[|p - (T_{zz} + \Pi_{zz})|] = 0 \quad (40)$$

with $[|\quad|]$ denoting the jump across the surface of the quantity between brackets. The van der Waals stress is continuous across the interface, for which reason it does not appear in equation (40).

Combining equation (38) with equation (40) leads to

$$p_1 - p_{N_2} = (\varrho_1 - \varrho_2)\,W^S = -\frac{H}{6\pi h^3}. \quad (41)$$

W^S indicates that the van der Waals potential has been evaluated at the film surface and H is the effective Hamaker constant [18]. Thus the lipid film exhibits a pressure increase due to the van der Waals interaction when passing from the external phase to the film interior.

In order to guarantee the mechanical equilibrium between the meniscus and the film phase we integrate the x-component of equation (33) from a point in the meniscus to a point $\vec{r}_2$ in the film phase along a path between and parallel to the (plane) film surfaces. While accounting for equation (35), this procedure yields

$$p_2 = p(\vec{r}_2) + \Phi(\vec{r}_2) + \varrho_2 W(\vec{r}_2), \quad (42)$$

with p_2 being the (isotropic) pressure in the meniscus. We note that $\Phi(\vec{r})$ vanishes identically in the hydrocarbon bulk phase.

A combination of equations (36) and (40) leaves us with

$$p_{N_2} - p_2 = -[T_{zz}(\vec{r}_2) + \Pi_{zz}(\vec{r}_2) + \Phi(\vec{r}_2)] = -\frac{\varepsilon}{8\pi h^2}\cdot\Delta^2 + \frac{1}{2}A[\eta_b]\,\beta^2\frac{\eta_s^2}{\cosh^2(\beta h/2)}. \quad (43)$$

Here we have made use of the electric field, E, and the electric cross film voltage, Δ, being related to each other via $E = \Delta/h$ [18]. Equation (43) demonstrates that two effects make up for the pressure variation between film and hydrocabon bulk: a) The electric cross film voltage tends to lower the pressure modulation; b) the increase of order from meniscus to film phase (manifested by the growing influence of the surface order with decreasing film thickness) correlates with a corresponding increase in normal pressure.

Finally, adding equations (41) and (43) together yields the *total disjoining pressure*

$$\Pi_D = p_1 - p_2 = -\frac{H}{6\pi h^3} - \frac{\varepsilon}{8\pi h^2}\Delta^2 + \frac{1}{2}A[\eta_b]\beta^2 \frac{\eta_s^2}{\cos h^2(\beta h/2)}. \quad (44)$$

Equations (41), (43), and (44) demonstrate how the total disjoining pressure samples pressure jumps across *both* the external phase/film interface and across the film/meniscus transition zone. This result stands in striking contrast to Toshev and Ivanov [31] as well as Eriksson and Toshev [32] on one side, and de Feijter et al. [33] on the other. The former postulate the disjoining pressure to be localized exclusively in the transition zone between film and meniscus; reference [33], however, has localized Π_D exclusively in the interface between film and external phase.

In order to evaluate their capacity measurements, Andrews et al. [19] assumed Π_D to vanish in their films resulting in the thickness dependence of the repulsive force as depicted in figure 1 (cf. Section 2). In fact, we fitted our formula for the steric disjoining pressure to the experimental results, which procedure yielded $\beta = 4.55\ \mathrm{nm}^{-1}$ and $\frac{1}{2}A\beta^2\eta_s^2 = 1.29 \times 10^8\ \mathrm{N\ mm}^{-2}$. A comparison between theory and experiment is shown in figure 5.

We note that the theory differs from experiment at "small" film thickness. One can think of a number of reasons for this discrepancy. For example does our theory strictly hold for small overlap only of the adjacent lipid surfactant layers; this restriction may not be satisfied any more at small film thickness. Yet, in our opinion, the deviation of theory from experiment is not worthy of being followed up any further. The experimental film thickness is certain to within some 10 % only, due to the uncertainty in magnitude of the film surface effective in the capacity measurements. Thus within the error set by experiment our theory agrees satisfactorily with experiment for all h.

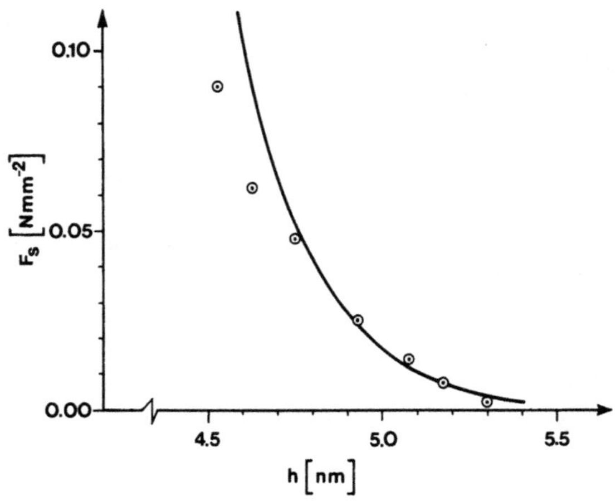

Fig. 5. Comparison of the thickness dependence of experimental (dots) and theoretical (solid line) steric disjoining pressure F_S. The theoretical curve was calculated from equation (44) with $\beta = 4.55\ \mathrm{nm}^{-1}$ and $\frac{1}{2}A\beta^2\eta_s^2 = 1.29 \times 10^8\ \mathrm{N\ mm}^{-2}$

5. Film tension and film elasticity [18]

In our previous work we have borne out [12] how the study of the stability of a symmetric liquid film against small amplitude shape fluctuations comes down to investigating the stability characteristics of bending (BE) and squeezing (SQ) deformations of the film.

The former couples velocity fields in which the normal displacements of the two interfaces are in phase while the tangential interface displacements are out of phase. In SQ, the phases are the other way around. We have shown [18] that at long wavelengths a thin film is stable against BE type fluctuations provided a quantity γ_F (which we called "film tension") is positive; the corresponding SQ stability depends on the sign of a quantity ε_F we termed "film elasticity". In this Section we want to discuss both γ_F and ε_F within the framework of our model.

We define [18] the film elasticity via

$$\varepsilon_F = -\left(\frac{\partial}{\partial h}\Pi_D\right)h^2 \quad (45)$$

i. e., ε_F is related to the *variation of the disjoining pressure with film thickness.*

We warn the reader not to confuse ε_F defined here with the surface elasticity ε or the elasticity $\varepsilon_{\mathrm{eff}}$, respectively, employed in previous publications [12]. ε there was defined as the change of surface tension with a relative change of film surface, A, i. e., $\varepsilon = d\gamma/d\ln A$.

ε thus relates to surface compression whereas ε_F, here, is linked to film thickness changes.

Employing equation (44) one derives for our model

$$\varepsilon_F = -\frac{H}{2\pi h^2} - \frac{\varepsilon}{4\pi h}\Delta^2 + \frac{1}{2} A[\eta_b](\eta_s)^2 (\beta h)^2 \frac{\tan h(\beta h/2)}{\cos h^2(\beta h/2)}. \quad (46)$$

In figure 6 we depict ε_F as a function of film thickness, h, for zero electric field (ε_1), and for a cross film voltage of $\Delta = 100$ mV (ε_2).

ε_F comprises the steric (ε_R), the van der Waals (ε_w), and the electrostatic (ε_E) part. ε_w is inversely proportional to the square of the film thickness and $\varepsilon_E \propto h^{-1}$. Therefore they contribute significantly for all film thicknesses. Moreover they contribute negatively. Thus for all h superior to a critical film thickness h_o (the exact value of which depends on the applied voltage), the film is unstable against SQ fluctuations. This instability has already been investigated in great detail for coloured films with $h \geq 1000$ Å [12]. At a film thickness of the order of 70 Å, the steric contribution, ε_R, starts to become important and to counteract van der Waals and electrostatics. At h_o, ε_R exactly compensates $\varepsilon_w + \varepsilon_E$. For film thicknesses inferior to h_o, ε_R makes ε_F become positive, i. e., it renders the film stable against SQ fluctuations.

In mechanical equilibrium, the film tension, γ_F, is obtained from the Bakker integral

$$\gamma_F = \int_{-\infty}^{\infty} (p_N - p_T)\, dz. \quad (47)$$

Here p_N and p_T are the normal and transverse components, respectively, of the *total* stress. That is, γ_F results by summing up the excess of the transverse over the normal stress in the film system. For the detailed derivation of γ_F from definition (47) for our model we refer the reader to reference [18]. Here we state only the result:

$$\gamma_F = 2\gamma_s - \frac{H}{4\pi h^2} - \frac{\varepsilon_2 \Delta^2}{4\pi h} + 2A[\eta_b](\eta_s)^2 \beta^2 h \exp(-\beta h). \quad (48)$$

γ_s is the (positive) interfacial tension which a single *non-interacting* interface between two semi-infinite phases of water and hydrocarbon, respectively, would have. We note especially that γ_s also contains the contribution of the steric interaction of the lipid surfactants at the single interface.

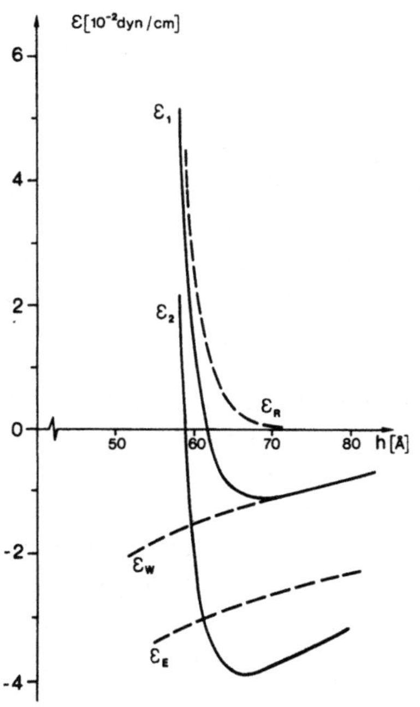

Fig. 6. Film elasticity ε_F as a function of film thickness h (eq. (46)) for two different cross film voltages Δ (ε_1: $\Delta = 0$ mV; ε_2: $\Delta = 100$ mV). Shown are also the contributions due to the van der Waals interaction, ε_w, due to electrostatics, ε_E, and due to sterics, ε_R

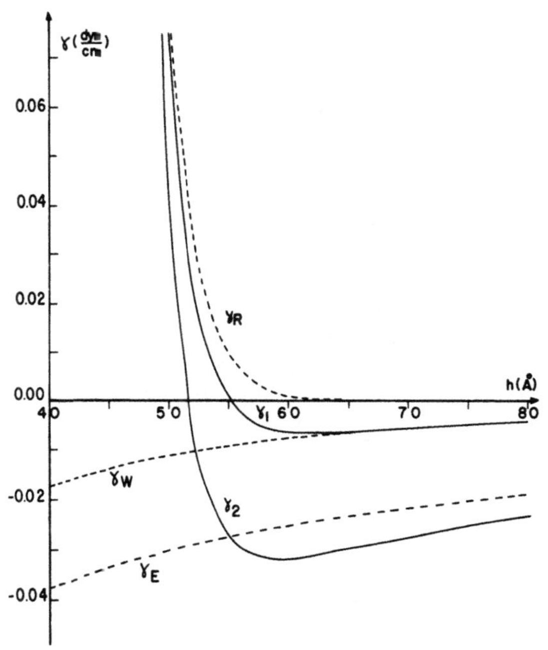

Fig. 7. $\gamma_F - 2\gamma_s$ (eq. (48)) as a function of film thickness h for two different cross film voltages Δ (γ_1: $\Delta = 0$ mV; γ_2: $\Delta = 100$ mV). Also shown are the individual contributions due to the van der Waals interaction, γ_w, due to electrostatics, ε_E, and due to sterics, ε_R

The remaining terms on r. h. s. of equation (48) arise upon the approach of two interfaces in forming the film, i. e., these terms are due to the interactions between the (film) interfaces. The van der Waals interactions cause the free energy of the film of thickness h to be lower by $-H/4\pi h^2$ than the free energy of an "infinitely thick" film. Electrostatics lowers the free energy by $-\varepsilon\Delta^2/4\pi h$ corresponding to an increase of the capacitance upon film thinning. The last term in equation (48) augments the free energy. It is caused by the increase of order in the center of the film as the two lipid surfactant layers are superposed. Thus the inner energy increases because of the enhanced short range van der Waals attraction while at the same time the entropy is decreasing.

$\gamma_F - 2\gamma_s$ is depicted in figure 7 as a function of h for zero voltage (γ_1)and for $\Delta = 100$ mV (γ_2). The figure demonstrates that for sufficiently large γ_s, the film can

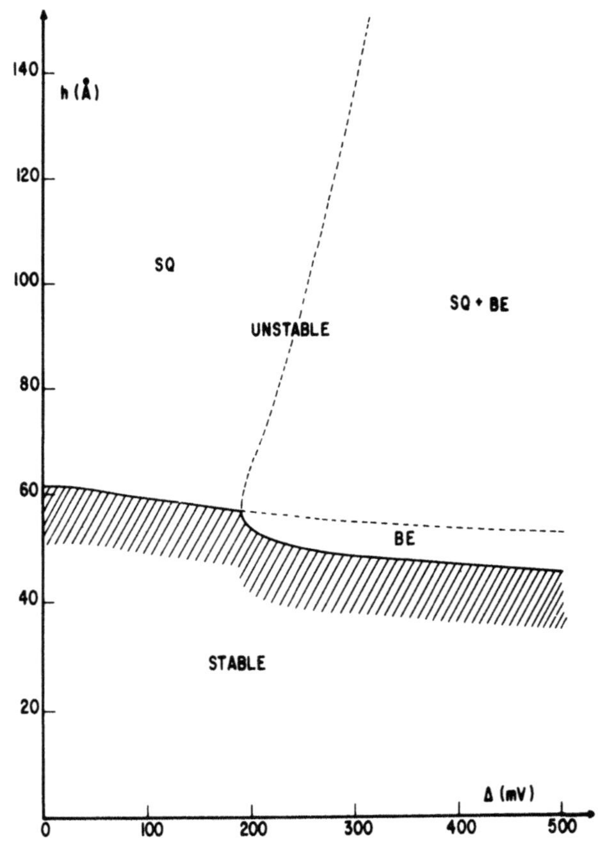

Fig. 8. Stability diagram for the film as a function both of film thickness h (in Å) and applied cross film voltage Δ (in mV). Depending on the choice of the parameter pair (h, Δ) in the unstable regime, the film is unstable against periodic thickness fluctuations (SQ), against bending (BE), or against both (SQ + BE)

be prevented from undergoing a BE instability at all thicknesses. For fixed γ_s, however, one can always switch an electric field which renders films of a certain thickness BE unstable. For example, from figure 7 one sees that low tension films with $\gamma_s = 1.5 \cdot 10^{-2}$ dyn/cm become BE unstable at thicknesses 56 Å $< h <$ 64 Å upon application of $\Delta = 100$ mV.

The aspect of a film being unstable with respect to BE type fluctuations gets even more interesting in view of figure 8. For a low tension film ($\gamma_s = 0.05$ dyn/cm), there we have plotted in a Δ-h-diagram the lines of marginal stability separating the stable from unstable regions. Marginal BE (SQ) stability was defined by $\gamma_F = 0$ ($\varepsilon_F = 0$). Figure 8 shows that there is an h-interval in which films can be BE de-stabilized by applying appropriate voltages. Those films will still be stable against thickness (SQ) fluctuations. Large amplitude bending deformation of lipid membranes has been observed already by Coakley et al. [35] after heating of erythrocytes.

We would like to conclude this Section by raising a few speculations. We hope to provoke thus "A Whack on th Side of the Head" [36] which will stimulate further thinking in the present context.

In view of the large value of the tension γ_s of an interface separating aqueous from hydrocarbon bulk, the steric effect on the film tension, γ_F, appears to be of minor importance. Yet, in contrast with the model system dealt with here, actual biomembranes exhibit an extremely low interfacial tension of the order of 10^{-2} dyn/cm [37], due to selective adsorption of proteins. Would not the steric contribution be vital then?

Another lesson of this Section is that there are three possibilities to trigger a BE instability of the film while having it stable against SQ deformations: i) to lower its interfacial tension, γ_s, spontaneously; ii) to enhance the cross film potential difference; iii) to change film thickness from a "stable" thickness to a thickness of potential BE instability.

The capability of biological system to change γ_s due to specific surface adsorption has already been mentioned. Variation of cross membrane potentials is common in the performance of biological systems [38]. To our knowledge there has been no systematic exploration of the possibility of film thickness changes prior to deformation processes of biological surfaces, e. g., as in the mode of fusion or cytotic engulfment [39]. A simple mechanism would be lipidic phase separation: this would transform part of the membrane into a "thin" area rich in short chain molecules and the complementary part into a "thick" area rich in long chain lipids.

Finally, let us mention that both γ_s and the film thickness depend on the lateral packing of the lipids. Lipid-protein interactions can induce local changes of this lipid packing.

6. Conclusion

In this article we have reviewed the gross features of a macroscopic model of steric repulsion in thin short chain hydrocarbon films, i. e., in black solvent-saturated lipid films. In a further step we have detailed its consequences for the film stability against bending and squeezing type deformations.

For this we first recalled two key experiments together with their interpretations in order to familiarize the reader with the specific problem: the macroscopic experiment quantified the term "repulsion" by measuring the repulsive force as a function of film thickness; the microscopic nmr-experiment and its theoretical interpretation linked the term "repulsion" to the order prevailing along individual lipid chains or, more exactly, to the order profile across the film, i. e., to the average orientation of the hydrocarbon segments as a function of the position in the film. We then analysed the problem from the point of view of interface physics where an interface represents a non-uniform region of transition between two different homogeneous bulk phases. In this picture, the overlap of the adjacent interfaces in the black film increased the order in the film center, thus causing the mutual repulsion of the two film surfaces.

For the mathematical formulation and exemplification of the theory it was of basic importance that the order parameter problem could be reduced to the problem of deriving the density profile between two parallel plane liquid-vapour interfaces of a monomolecular system. The mapping enabled us to employ the statistical mechanical formulation of the van der Waals theory of liquid-vapour interfaces and to derive a differential equation fixing asymptotically the order parameter profile in and close to the film center. The resulting order density profile depended i) on the ratio of the derivative of the chemical potential with respect to the order parameter density and the second moment of the two point Ornstein-Zernike correlation function, and ii) on the order at the film surface. We treated the two quantities as adjustable parameters and fixed them by means of the results of the macroscopic experiment in the large film thickness limit.

The order parameter met in this paper is supposed to coincide with the real order parameter in a more or less extended domain in the film center where there is need of a good description of the interfacial overlap; it deviates from the real order parameter in the surface region where the surface tension accounts for the molecular inhomogenities.

Consistently with the intuitive picture we found that a non-zero overlap of the interfacial layers was necessary, i. e., the order in the film center had to be raised over that in the homogeneous and isotropic bulk phase, in order to generate repulsion.

The theory also enabled us to calculate the anisotropic state of stress due to the steric repulsion, thus opening vistas for a stability analysis.

In carrying out this program we also showed how generally the disjoining pressure in a liquid film samples pressure jumps both across the transition zone between film and meniscus and across the interface between film and external bulk phase. Thus we put an end to a long-lasting controversy. In evaluating the disjoining pressure explicitly for our model we have demonstrated how a measurable microscopic order links quantitatively to the macroscopically measurable disjoining pressure.

Our discussion of the dynamic film stability started by stating that stability depends on the sign of the film tension and the first derivative of the disjoining pressure with respect to film thickness ("film elasticity"). We gave the explicit functional forms of these quantities within our model. Especially we demonstrated the way in which dispersion, electrostatic, and steric interactions enter as a function of film thickness: van der Waals and electrostatics lower tension and elasticity for all film thicknesses; at small film thicknesses, sterics counteracts this lowering and eventually raises tension and elasticity. Most important was our finding that periodic thickness fluctuations are stable from a critical film thickness down, due to steric action. These squeezing modes were unstable in thick (coloured) films. Instead, upon application of a critical electric field the film bends. We established a diagram depicting the regimes of stability as a function of film thickness and cross film potential. Finally we speculated on the implications of our results for biological interfaces.

In concluding this paper we express our believe that the fundamental principle of *phase stabilization due to surface interaction induced order* exemplified here must have a number of applications in many diverse fields. We would like to ask the reader to communicate such application in his respective field of specialization.

Acknowledgements

This work was made possible by the grant We 742/3 of the Deutsche Forschungsgemeinschaft, by financial support of the

Internationale Büro (KFA Germany) under contract 30.3. A. L. as well as by the CNPq (Brazil) under Proc. 1.11.10.011-81. HW also appreciates the award of a Heisenberg Fellowship (We 742/4-1).

References

1. Wendel H, Bisch PM, Gallez D (1982) Coll & Polym Sci 260:425
2. Tien HT (1974) Bilayer Lipid Membranes, Marcel Dekker, New York
3. Fettiplace R, et al (1974) (ed) Korn ED, in: Methods in Membrane Biology, Plenum Press, New York 4:1
4. Israelachvili JN, Marcelja S, Horn RG (1980) Quart Rev Biophys 13:121
5. Grabowski EF, Cowen JA (1977) Biophys J 18:23
6. Crilly JF, Earnshaw JC (1983) Biophys J 41:197
7. Vodyanoy V, Halverson P, Murphy RM (1982) J Coll Interf Sci 88:247, and references contained therein
8. Singer SJ, Nicolson GL (1972) Science 175:720; Raff MC (1976) Sci Amer 234:30; Schindler M, Osborn MJ, Koppel DE (1980) Nature, London 283:346
9. Albrecht O, Gruler H, Sackmann E (1978) J Phys, Paris 39:301
10. Vrij A, Joosten JGH, Fijnaut HM (1981) (eds) Prigogine I, Rice SA, Adv Chem Phys XLVIII, John Wiley and Sons, p 328, and references contained therein
11. Maldarelli C, Jain RK, Ivanov IB, Ruckenstein E (1981) J Coll Interf Sci 78:118, and references contained therein
12. Wendel H, Gallez D, Bisch PM (1981) J Coll Interf Sci 84:1; Bisch PM, Wendel H, Gallez D (1982) J Coll Interf Sci 92:105; Gallez D, Bisch PM, Wendel H (1982) J Coll Interf Sci 92:121
13. Derjaguin BV, Landau L (1941) Acta Physicochimica URSS 14:633; Verwey EJW, Overbeck JTG (1948) Theory of the Stability of Lyophobic Colloids, Esevier, Amsterdam
14. Steinchen A, Gallez D, Sanfeld A (1982) J Coll Interf Sci 85:5
15. Maldarelli C, Jain RK (1982) J Coll Interf Sci 90:233 and J Coll Interf Sci 90:263
16. Gallez D (1983) Biophysical Chemistry 18:165
17. Hladky SB, Gruen DWR (1982) Biophys J 38:251
18. Bisch PM, Wendel H, submitted; preprint, Notas de Fisica – CBPF, ns, 040/83 and 043/83
19. Andrews DM, Manev ED, Haydon DA (1970) Spec Disc Farad Soc 1:46
20. Seelig J, Seelig A (1980) Quart Rev Biophys 13:19
21. Gruen DWR (1980) BBA 595:161
22. Gruen DWR (1981) Biophys J 33:149; Gruen DWR, Haydon DA (1981) Biophys J 33:167
23. Yang AJM, Fleming III PD, Gibbs JH (1976) J Chem Phys 64:3732
24. Fleming III PD, Yang AJM, Gibbs JH (1976) J Chem Phys 65:7
25. Evans R (1979) Adv in Physics 28:143
26. Jähnig F (1979) J Chem Phys 70:3279
27. Mayer W, Saupe A (1958) Z Naturf A 13:564; (1959) A 14:882; (1960) A 15:287
28. Derjaguin BV, Kussakov MM (1939) Acta Physicochim URSS 10:25
29. Bisch PM, Wendel H (1983) J Coll Interf Sci 96:555
30. Wendel H, Bisch PM (1983) Chem Phys Lett 102:361
31. Toshev BV, Ivanov IB (1975) Coll & Polym Sci 253:558
32. Eriksson JC, Toshev BV (1982) Colloids and Surfaces 5:241
33. De Feijter JA, Rijnbout JB, Vrij AJ (1978) J Coll Interf Sci 64:258
34. Felderhof BU (1968) J Chem Phys 49:44
35. Coakley WT, Deeley JOT (1980) Biochim Biophys Acta 602:355
36. Van Oech R (1983) A Whack on the Side of the Head, Creative Think, Menlo Park
37. Harvey EN, Danielli JF (1938) Biol Rev 13:319
38. See e g: (1977) (ed) Roux E, Proceedings of the 29th International Meeting of the Société de Chimie Physique, Elsevier Scientific Publ Comp, Amsterdam
39. Van Oss CJ (1978) Ann Rev Microbiol 32:19–39

Authors' addresses:

H. Wendel
Abteilung für Biophysik
Universität Ulm
Oberer Eselsberg
D-7900 Ulm/Donau

P. M. Bisch
Centro Brasileiro de Pesquisas Fisicas
Rua Xavier Sigaued, 150
22.290 Rio de Janeiro
Brazil

Progress in Colloid & Polymer Science Progr Colloid & Polymer Sci 69:127–133 (1984)

Kernspinlabel-Untersuchungen zur Struktur und Dynamik von flüssigkristallinen Hauptkettenpolymeren*

K. Müller, C. Eisenbach**), A. Schneller***), H. Ringsdorf***) und G. Kothe

Institut für Physikalische Chemie der Universität Stuttgart, F.R.G.

Zusammenfassung: Thermotrope nematische Polyester wurden an verschiedenen Positionen der Polymerkette selektiv deuteriert und durch Magnetfelder und Festphasenextrusion makroskopisch ausgerichtet. Zur Analyse der beobachteten ^{2}H NMR-Spektren diente ein dynamisches NMR-Modell, das auf der stochastischen Liouville Gleichung basiert. Das Modell beschreibt die NMR-Spektren für alle gängigen Pulsfolgen der FT-NMR-Spektroskopie.

Aus den Simulationen der experimentellen ^{2}H-NMR-Spektren ergeben sich die Orientierungsverteilungen und Konformationen der Polymerkette und die Korrelationszeiten für die verschiedenen Bewegungen. In der anisotropen Schmelze betragen die Korrelationszeiten für Kettenfluktuation bzw. Kettenisomerisierung $10^{-10}\,s \leq \tau_{R\perp}, \tau_J \leq 10^{-8}\,s$. Abkühlen des festen Polymeren führt zum allmählichen Einfrieren der intermolekularen Bewegungen. Unterhalb der Glastemperatur lassen sich nur noch intramolekulare Bewegungen wie Trans-Gauche-Isomerisierung oder Ringflips detektieren.

Der Kettenordnungsparameter in der Schmelze ist überraschend groß und beträgt $S_{ZZ} = 0.85$. Eine Transpopulation von $n_t \sim 0.8$ für sämtliche Spacersegmente zeigt eine weitgehende Ausrichtung der Polymerkette an. Diese hohe Mikroordnung bleibt erhalten, wenn man den Polyester unter den Schmelz- und Glaspunkt abkühlt. Durch Variation der Spacerlänge läßt sich die Ordnungsstruktur der Polymeren signifikant verändern (odd-even effect).

Abstract: Thermotropic nematic polyesters, specifically deuterated at different positions of the polymer chain, were macroscopically aligned by strong magnetic fields and solid state extrusion. Analysis of the observed ^{2}H NMR spectra was achieved, employing a dynamic NMR model, based on the stochastic Liouville equation. The model considers various double and multiple pulse sequences, recently employed in FT-NMR spectroscopy.

Computer simulations provide the orientational distributions and conformations of the polymer chains and the correlation times of the various motions. In the anisotropic melt the correlation times for chain reorientation and trans-gauche isomerization are in the range $10^{-10}\,s \leq \tau_{R\perp}, \tau_J \leq 10^{-8}\,s$. Decreasing the temperature of the solid polymer first freezes the intermolecular motions. Thus, below the glass transition only intramolecular motions such as trans-gauche isomerization or ring flips can be detected.

The chain order parameter of the nematic melt is $S_{ZZ} = 0.85$, a value considerably higher than those observed in low molecular weight nematogens. In addition, the chains adopt a highly extended conformation, evidenced by a trans population of $n_t \sim 0.8$ throughout the entire spacer. This micro-order is essentially retained when the polymer is cooled below the melting point and glass transition, respectively. Variation of the spacer length causes significant changes of the molecular order of the polyesters (odd-even effect).

Key words: Nuclear spinlabelling, liquid crystalline polymers.

*) Vortrag, gehalten auf der 31. Hauptversammlung der Kolloid-Gesellschaft, Bayreuth 11. bis 14. Oktober 1983.

**) Institut für Makromolekulare Chemie der Universität Freiburg, D-7800 Freiburg

***) Institut für Organische Chemie der Universität Mainz, D-6500 Mainz

KT-W 791

Einleitung

Kernspinlabel sind magnetische Reportergruppen, die Struktur und Dynamik der molekularen Umgebung im Kernresonanz(NMR)– Spektrum anzeigen. Im folgenden berichten wir über die Grundlage der Kernspinlabel-Methode und neue Anwendungen auf flüssigkristalline Hauptkettenpolymere.

Zuerst wird das zugrundeliegende NMR-Modell vorgestellt. Es basiert auf der stochastischen Liouville Gleichung und ist im gesamten Dynamikbereich gültig. Inter- und intramolekulare Dynamik werden explizit berücksichtigt. Für die molekulare Ordnung ist eine umfassende Verteilungsfunktion vorgesehen. Schließlich lassen sich NMR-Linienprofile für alle gängigen Pulsfolgen der Fourier-Transformations-NMR-Spektroskopie berechnen [1–3].

Das diskutierte NMR-Modell wird zur Analyse von temperatur- und winkelabhängigen ^{2}H NMR-Spektren flüssigkristalliner Polyester herangezogen. Sie waren an verschiedenen Positionen der mesogenen Gruppe und des aliphatischen Spacers spezifisch deuteriert. Aus den Computersimulationen ergeben sich die Orientierungsverteilungen und Konformationen der Polymerketten und die Korrelationszeiten für die verschiedenen Bewegungen. Sie stehen in direktem Zusammenhang mit den makroskopischen Eigenschaften der flüssigkristallinen Polymeren.

Theorie

Kernspinlabel-Experimente werden im allgemeinen in der Zeitdomäne durchgeführt. Das kernspinmarkierte System wird einer Folge starker HF-Pulse ausgesetzt und das Signal anschließend Fourier-transformiert, um ein Frequenzspektrum zu erhalten. Je nach Pulsfolge ergeben sich verschiedene NMR-Spektren, die empfindlich vom Pulsabstand abhängen. Entsprechendes gilt für die Spektren von makroskopisch ausgerichteten Proben bei unterschiedlicher Magnetfeldorientierung. Durch Variation typischer NMR-Parameter wie Pulsfolge, Pulsabstand oder Magnetfeldorientierung lassen sich also verschiedene experimentelle NMR-Spektren vom gleichen System erzeugen.

Zur Analyse dieser Spektren in bezug auf molekulare Ordnung und Dynamik benötigt man eine umfassende Theorie. Wir haben ein entsprechendes Modell entwickelt, das auf dem Dichtematrixformalismus basiert [1]. Vor Anwendung eines HF-Pulses befindet sich das Spinsystem im thermischen Gleichgewicht. Der erste HF-Puls erzeugt einen definierten Nichtgleichgewichtszustand der Spindichtematrix $\underset{\sim}{\varrho}$. Nach dem Puls entwickelt sich $\underset{\sim}{\varrho}$ unter dem Einfluß der magnetischen Wechselwirkungen des Spinsystems. Dann wird ein zweiter HF-Puls appliziert, gefolgt von einer zweiten Evolutionsphase, usw. Nach dem letzten Puls wird die Dichtematrix Fourier-transformiert und damit ein Frequenzspektrum berechnet.

Die Wirkung der verschiedenen Pulse auf $\underset{\sim}{\varrho}$ läßt sich mit Hilfe von Wignerschen Rotationsmatrizen beschreiben [1, 4]. Zwischen den Pulsen evolviert die Dichtematrix entsprechend der stochastischen Liouville Gleichung [1]:

$$\dot{\underset{\sim}{\varrho}}_{ABK} = (i/\hbar)\,[\underset{\sim}{\varrho}_{ABK}, \underset{\sim}{H}_{ABK}] + (\dot{\underset{\sim}{\varrho}}_{ABK})_{rot} + (\dot{\underset{\sim}{\varrho}}_{ABK})_{isom} \,. \qquad (1)$$

Dabei ist $\underset{\sim}{H}_{ABK}$ der Hamiltonoperator des Spinsystems und $(\dot{\underset{\sim}{\varrho}}_{ABK})_{rot}$, $(\dot{\underset{\sim}{\varrho}}_{ABK})_{isom}$ beschreiben die inter- und intramolekulare Dynamik. Wir nehmen an, daß diese Bewegungsprozesse über eine endliche Zahl diskreter Winkelpositionen, charakterisiert durch die Indizes ABK, verlaufen.

Der Spinhamiltonoperator für Deuteronen in C-D-Bindungen läßt sich in der Form

$$H = -\gamma \underset{\sim}{B} \cdot \underset{\sim}{I} + \underset{\sim}{I} \cdot \underset{\sim}{Q} \cdot \underset{\sim}{I} \qquad (2)$$

darstellen, wobei Q der Quadrupolkopplungstensor ist. Die Gleichgewichtsverteilung der C-D-Bindungen wird mit einer Orientierungsverteilungsfunktion beschrieben, die von inneren und äußeren Koordinaten abhängt [1]. Der innere Teil beschreibt verschiedene Konformationen und der äußere Teil

$$\begin{aligned} f(\theta,\Psi) &= N_1 \exp[A(\cos\xi\cos\theta - \sin\xi\sin\theta\cos\Psi)^2] \\ \cos\xi &= \cos\delta\cos\varrho - \sin\delta\sin\varrho\cos\varepsilon \qquad (3) \\ f(\delta,\varepsilon) &= N_2[\exp B\cos^2\delta] \end{aligned}$$

verschiedene Orientierungen. θ, Ψ, δ, ε und ϱ sind Eulersche Winkel, die verschiedene Molekül- und Laborkoordinatensysteme miteinander verbinden [1]. Der Koeffizient A charakterisiert die Orientierung der Hauptachse des Ordnungstensors relativ zu einem lokalen Direktor (Mikroordnung). Der Parameter B beschreibt die Verteilung der Direktorachsen in einem Laborsystem (Makroordnung). Die Ordnungsparameter S_{ZZ} und $S_{z'z'}$ berechnen sich über Mittelwertintegrale

$$S_{ZZ} = \frac{1}{2} N_1 \int_0^{\pi} (3\cos^2\beta - 1)\exp(A\cos^2\beta)\sin\beta d\beta$$

$$S_{z'z'} = \frac{1}{2} N_2 \int_0^{\pi} (3\cos^2\delta - 1)\exp(B\cos^2\delta)\sin\delta d\delta \quad (4)$$

aus den Koeffizienten A und B.

Die Form der dynamischen Terme in Gleichung 1 hängt vom zugrundeliegenden Bewegungsmodell ab. Die kinetischen Annahmen für die intermolekulare Bewegung entsprechen einem diffusiven Prozess, charakterisiert durch die Rotationskorrelationszeiten $\tau_{R\perp}$ (Umorientierung der Symmetrieachse des Rotationsdiffusionstensors) und $\tau_{R\parallel}$ (Drehung um die Symmetrieachse des Rotationsdiffusionstensors). Für die intramolekulare Bewegung haben wir einen Sprungprozess angenommen. Die Isomerisierung in unserem Modell erfolgt also durch Sprünge zwischen verschiedenen Konformationen mit der mittleren Lebensdauer τ_J.

Experimentelles

Der untersuchte flüssigkristalline Polyester hat folgende molekulare Struktur [5, 6]:

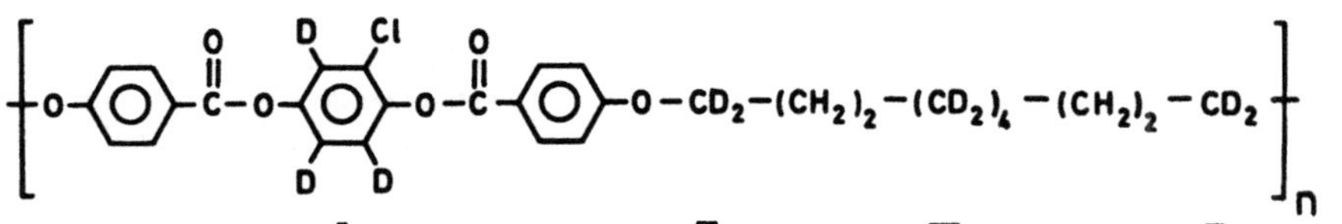

I II III II

Er zeigt eine Glasstufe bei 303 K, einen Schmelzpunkt bei 433 K und einen Klärpunkt bei 553 K (DSC-Messungen). Nach polarisationsmikroskopischen Untersuchungen liegt eine nematische Phase vor. Das mittlere Molekulargewicht M_n der Proben variierte zwischen $3000 \leq M_n \leq 10000$ (Dampfdruckosmose). Deuteronen-Kernspinlabel wurden in den zentralen Phenylring der mesogenen Gruppe (I) und verschiedene Positionen (II, III) des aliphatischen Spacers eingeführt. Einzelheiten dieser Synthese sind an anderer Stelle beschrieben [7]. Die makroskopische Ausrichtung der Proben erfolgte mit einem Magnetfeld von 7.0 T [8].

Für die NMR-Experimente stand ein Hochleistungs-Puls-Spektrometer CXP 300 der Firma BRUKER zur Verfügung. Die Meßfrequenz für Deuteronen betrug 46.1 MHz. Zur Registrierung winkelabhängiger Frequenzspektren wurde ein neuer Probenkopf mit Goniometer entwickelt. Die damit erzielten 90°-Pulslängen für Deuteronen betrugen 3.5 µs. Bei allen Experimenten wurden die Sequenz $180^\circ_x, -\tau_1 - 90^\circ_x, -\tau_2 - 90^\circ_y$, (Inversion Recovery) bzw. $90^\circ_x, -\tau - 90^\circ_y$, (Quadrupol-Echo-Sequenz) und Quadratur-Detektion verwendet. Für ein ausreichendes Signal-Rauschverhältnis waren zwischen 500 und 2000 Akkumulationen notwendig.

Das FORTRAN-Programm DEUROTJUMP diente zur Analyse der experimentellen Spektren. Es berechnet dynamische NMR-Linienprofile für I = 1 Spinsysteme mit inter- und intramolekularer Dynamik in anisotropem Medium [1]. Die verwendete Quadrupolkopplungskonstante betrug $e^2qQ/h = 165$ kHz.

Die Interpretation der ^{2}H NMR Daten setzt eine Kenntnis der molekularen Symmetrie und Konformation voraus. Wir nehmen an, daß sich jede molekulare Konformation durch einen Zylinder approximieren läßt, dessen Längsachse mit der langen Molekülachse zusammenfällt (Ordnungstensorachse). Weiterhin nehmen wir an, daß diese lange Molekülachse auch die Symmetrieachse des Rotationsdiffusionstensor ist. Neben der intermolekularen Bewegung (Rotationsdiffusion der langen Molekülachse) sind intramolekulare Konformationsänderungen möglich. Aus Symmetriegründen wurde angenommen, daß die beiden Gauche-Konformationen eines jeden Segments gleich besetzt sind.

Ergebnisse und Diskussion

In den Abbildungen 1–3 sind ^{2}H NMR-Spektren des Polymeren II wiedergegeben, das an beiden Enden des aliphatischen Spacers spezifisch deuteriert war. Die Spektren beziehen sich auf drei verschiedene Temperaturen und charakterisieren die anisotrope Schmelze (Abb. 1) und das feste Polymere oberhalb (Abb. 2) und unterhalb der Glastemperatur (Abb. 3). Abbildung 1 zeigt ^{2}H NMR-Linienprofile einer $180^\circ_x, -\tau_1 - 90^\circ_x, -\tau_2 - 90^\circ_y$, –Sequenz (Inversion Recovery) für verschiedene Pulsabstände τ_1. Man erkennt, daß mit zunehmendem Pulsabstand drastische spektrale Veränderungen auftreten, die wertvolle Information über die molekulare Dynamik enthalten. Entsprechendes gilt für die Linienprofile einer $90^\circ_x, -\tau - 90^\circ_y$, – Sequenz (Quadrupol-Echo-Sequenz). Abbildung 3 zeigt ^{2}H NMR-Spektren für verschiedene Orientierungen ϱ zwischen Ausrichtungsachse und Magnetfeld. Wiederum beobachtet man signifikante Veränderungen, die die Mikro- und Makroordnung des Systems charakterisieren. Eine iterative Anpassung verschiedener winkel- und pulsabhängiger ^{2}H NMR-Linienprofile für jeden Temperaturpunkt liefert zuverlässige Werte für die Simulationsparameter, d. h. die Orientierung der langen Molekülachse, die Mikro- und Makroordnungsparameter, die Rotationskorrelationszeiten und die Lebensdauern und Populationen bestimmter Konformationen. Gestrichelte Kurven in der Abbildungen 1–3 geben entsprechende Simulationen wieder. Sie stimmen mit den experimentellen Spektren gut überein.

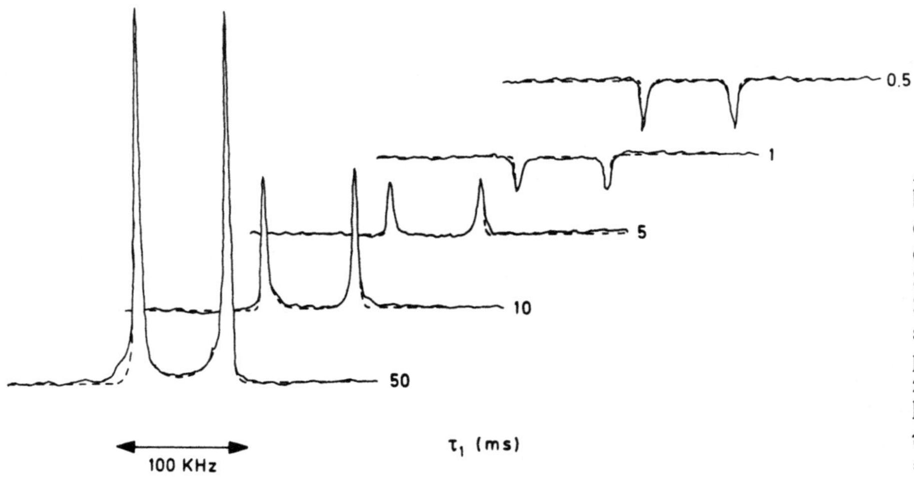

Abb. 1. Experimentelle (——) und berechnete (– – –) ^{2}H NMR-Spektren des flüssigkristallinen Polyesters II in einer $180^\circ_x - \tau_1 - 90^\circ_x - \tau_2 - 90^\circ_y$-Sequenz be $T = 443$ K (anisotrope Schmelze) und verschiedenen Pulsabständen τ_1. Die Probe war makroskopisch ausgerichtet (Direktor parallel zum Magnetfeld). Den Simulationen liegen folgende Parameter zugrunde: $\tau_2 = 20\mu s$, $\tau_{R\perp} = 21\ ns$, $\tau_J = 2.5\ ns$, $n_t = 0.8$, $S_{ZZ} = 0.86$ und $S_{z'z'} = 1.0$

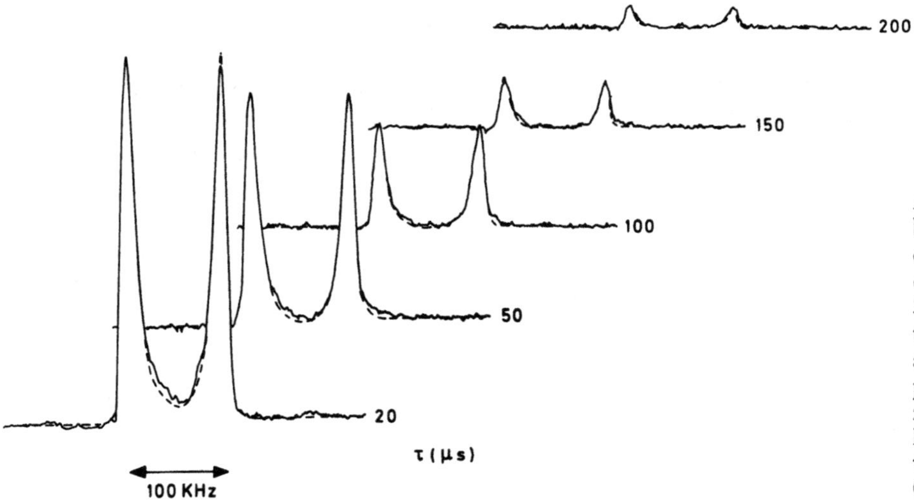

Abb. 2. Experimentelle (——) und berechnete (– – –) ^{2}H NMR-Spektren des flüssigkristallinen Polyesters II in einer $90^\circ_x - \tau - 90^\circ_y$-Sequenz bei $T = 313$ K (Festzustand oberhalb der Glastemperatur) und verschiedenen Pulsabständen τ. Die Probe war makroskopisch ausgerichtet (Direktor parallel zum Magnetfeld). Den Simulationen liegen folgende Parameter zugrunde: $\tau_{R\perp} = 7\ \mu s$, $\tau_J = 5\ \mu s$, $n_t = 0.85$, $S_{ZZ} = 0.89$ und $S_{z'z'} = 1.0$

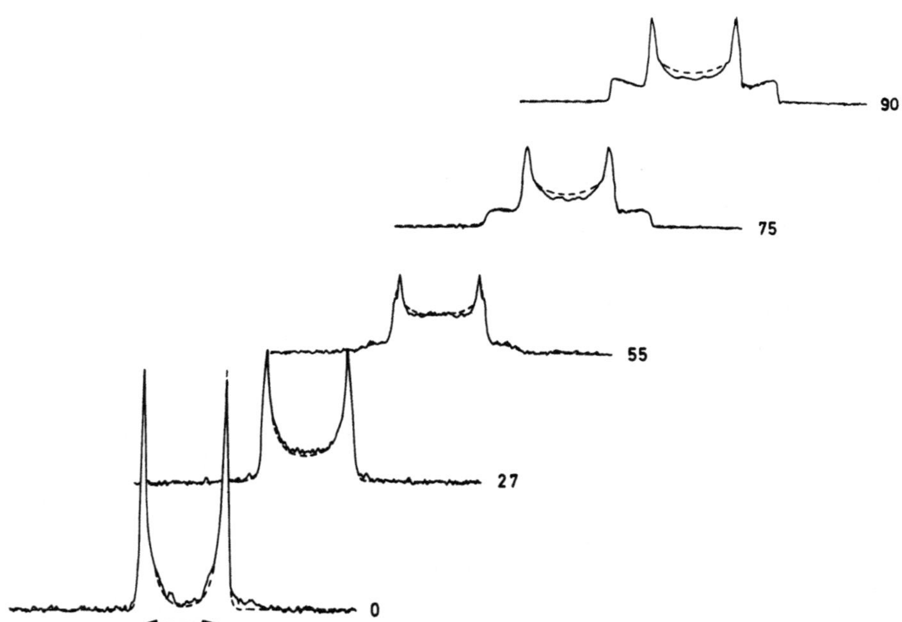

Abb. 3. Experimentelle (——) und berechnete (– – –) ^{2}H NMR-Spektren des flüssigkristallinen Polyesters II in einer $90^\circ_x - \tau - 90^\circ_y$-Sequenz bei 130 K (Glaszustand) und verschiedenen Winkeln ϱ zwischen Orientierungsachse und Magnetfeld. Verzerrungen des experimentellen Linienprofils aufgrund der endlichen Pulsbreite wurden korrigiert [24, 25]. Den Simulationen liegen folgende Parameter zugrunde: $\tau = 20\mu s$, $\tau_{R\parallel} > 1\ ms$, $\tau_{R\perp} > 1\ ms$, $\tau_J = 80\ \mu s$, $n_t = 0.88$, $S_{ZZ} = 0.89$ und $S_{z'z'} = 1.0$

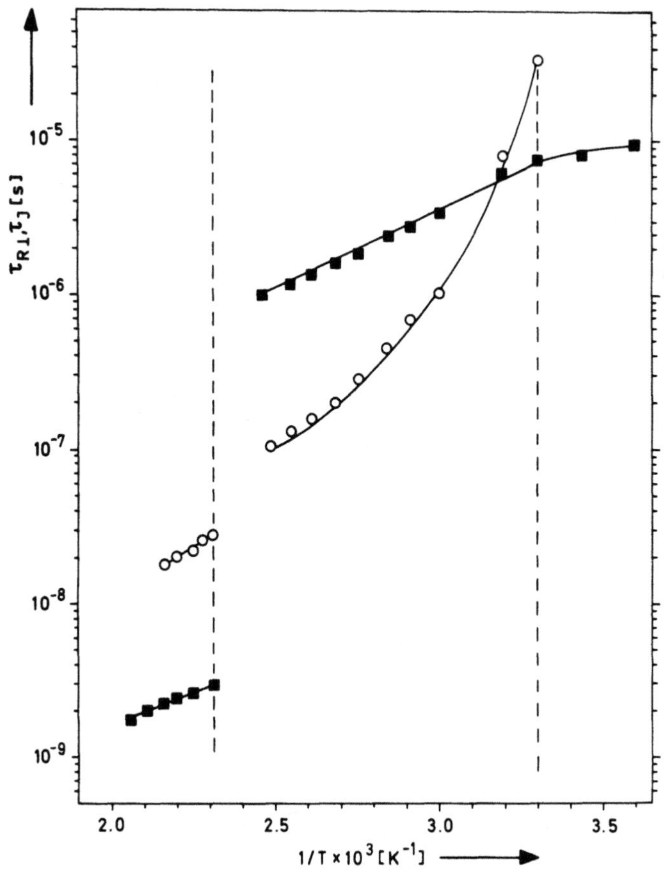

Abb. 4. Debye-Auftragung der Korrelationszeiten für die verschiedenen Bewegungen des flüssigkristallinen Polyesters II. Offene Kreise beziehen sich auf die intermolekulare Bewegung (Umorientierung der langen Molekülachse), ausgefüllte Quadrate kennzeichnen die intramolekulare Bewegung (Trans-Gauche-Isomerisierung des ersten Spacersegmentes). Schmelzpunkt ($T = 433$ K) und Glasstufe ($T = 303$ K) sind durch gestrichelte Kurven eingezeichnet

In Abbildung 4 sind die Korrelationszeiten für die verschiedenen Bewegungen des Polymeren II über der reziproken Temperatur $1/T$ aufgetragen. Sie beschreiben die Umorientierung der langen Molekülachse (offene Kreise) und die Trans-Gauche-Isomerisierung des äußeren Spacersegments (ausgefüllte Quadrate). In der Schmelze liegt die Korrelationszeit für die Umorientierung der langen Molekülachse bei $\tau_{R\perp} = 10^{-8}$ s, während die Trans-Gauche-Isomerisierung mit $\tau_J \sim 10^{-9}$ s erfolgt. Diese überraschend schnelle Molekularbewegung bedingt offensichtlich die ungewöhnlichen rheologischen Eigenschaften der flüssigkristallinen Polymeren [9]. Abkühlen unter den Schmelzpunkt führt innerhalb von 20 K zu einer drastischen Verlangsamung der Moleküldynamik. Ein weiteres Absenken der Temperatur im festen Zustand läßt zuerst die intermolekularen Bewegungen „einfrieren". Unterhalb der Glastemperatur können nur noch intramolekulare Bewegungen detektiert werden. Selbst bei $T = 130$ K haben wir noch Trans-Gauche-Isomerisierung des Spacers mit einer Korrelationszeit von $\tau_J \sim 10^{-4}$ s nachgewiesen. Die niedrige Aktivierungsenergie von $E_J = 6$ kJ/mol entspricht dem lokalen Charakter dieses Prozesses in Übereinstimmung mit T_1-Dispersionsmessungen an Paraffinen [10]. Damit konnte erstmals die komplexe Dynamik von flüssigkristallinen Polyestern in drei verschiedenen Phasen, einschließlich des Glaszustands, aufgeklärt werden.

Außerdem haben wir wertvolle Information über die Mikroordnung der Polymeren erhalten. Sie wird mit Hilfe der Parameter S_{ZZ} und n_t beschrieben. S_{ZZ} gibt die mittlere Orientierung der langen Molekülachsen in bezug auf den Direktor an, während n_t die Trans-Population des markierten Segments charakterisiert. Aus den winkelabhängigen Linienprofilen des Polymeren I wurde der Winkel α zwischen der langen Molekülachse und der p-Achse des zentralen Phenylrings zu $\alpha = 14°$ bestimmt. In Abbildung 5 sind die Mikroordnungsparameter S_{ZZ} (ausgefüllte Kreise) und die Trans-Populationen n_t der end- und mittelständigen Spacersegmente (offene Kreise) über der Temperatur aufgetragen. Man erkennt, daß die Polymerketten in der Schmelze weitgehend parallel orientiert sind. Der Ordnungsparameter ist mit $S_{ZZ} = 0.85$ bedeutend größer als bei niedermolekularen Nematogenen. Diese hohe Mikroordnung wurde auch mit Röntgenbeugungs- [11, 12], ESR- [13] und ^{1}H NMR-Untersuchungen [14, 15] ermittelt. Darüber hinaus zeigt diese Arbeit, daß die Ketten nahezu ausgerichtet sind, entsprechend einer Trans-Population von $n_t \sim 0.8$ für endständige und mittlere Spacersegmente.

Die Mikroordnung bleibt erhalten, wenn man das Polymer unter den Schmelz- bzw. Glaspunkt abkühlt. Damit kann die nematische Ordnungsstruktur von Hauptkettenpolymeren dauerhaft fixiert werden. In allen untersuchten Systemen war der Ordnungsparameter S_{ZZ} im Bereich $3000 \leq M_n \leq 10000$ unabhängig vom Molekulargewicht. Dieser Plateau-Effekt wurde auch bei anderen flüssigkristallinen Polymeren beobachtet [16–18].

Die bisherige Diskussion bezog sich auf Polyester mit zehn Methylengruppen im aliphatischen Spacer. Eine Reduktion der Spacerlänge auf neun Segmente führt zu einer signifikanten Veränderung der Mikroordnung. Erste Untersuchungen zeigen deutlich kleinere Werte des Ordnungsparameters S_{ZZ} in der anisotropen Schmelze und eine weitere Abnahme beim Abkühlen in den festen Zustand. Dieser Befund

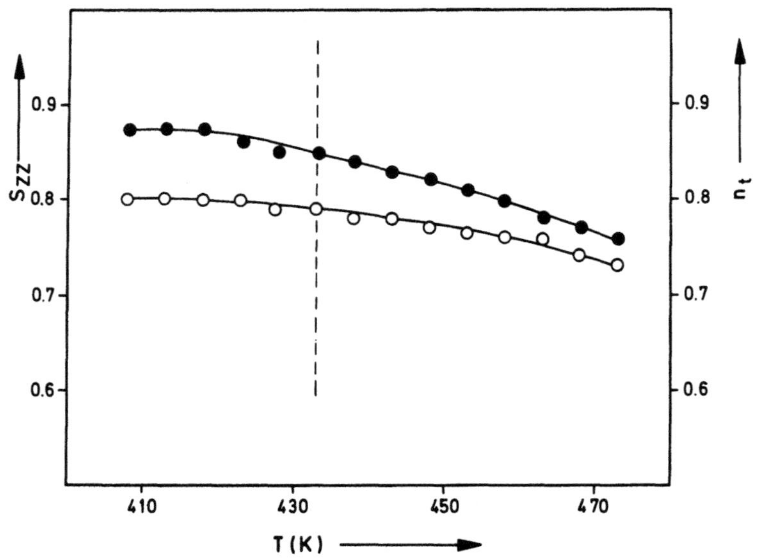

Abb. 5. Temperaturabhängigkeit des Mikroordnungsparameters S_{ZZ} und der Trans-Population n_t der flüssigkristallinen Polyester I-III. Ausgefüllte Kreise kennzeichnen S_{ZZ} (linke Ordinate), offene Kreise charakterisieren n_t (rechte Ordinate). Der Schmelzpunkt bei $T = 433$ K ist durch eine gestrichelte Linie angezeigt

entspricht dem „Gerade-Ungerade-Effekt", der auch bei anderen thermotropen Polymeren beobachtet wird [19–22]. Eine theoretische Deutung dieses Effekts steht noch aus.

Der makroskopische Ausrichtungsgrad $S_{z'z'}$ der flüssigkristallinen Polymeren hängt von der Orientierungsmethode ab. Da die dielektrische Anisotropie des Polyesters negativ ist, kann mit starken elektrischen Feldern nur eine zweidimensionale Verteilung der Direktorachse erzielt werden. Demgegenüber gelingt mit einem Magnetfeld von 7.0 T eine einheitliche Ausrichtung der Domänen. Eine detaillierte Analyse der winkelabhängigen Linienprofile lieferte einen Makroordnungsparameter von $S_{z'z'} = 1.0$. Durch Festphasen-Extrusion lassen sich Fasern mit $S_{z'z'} = 0.9$ [13] und potentiell hohem Elastizitätsmodul [23] erzeugen. Versuche zur Herstellung von Fasern durch Verspinnen der anisotropen Schmelze sind gegenwärtig im Gange.

Danksagung

Wir danken Herrn Dr. B. Hisgen (Universität Mainz) für die Hilfe bei den synthetischen Arbeiten und den Herren Dr. P. Meier und Dr. E. Ohmes (Universität Stuttgart) für wertvolle Diskussionen. Der Deutschen Forschungsgemeinschaft und dem Fonds der Chemischen Industrie sind wir für die finanzielle Unterstützung dieser Untersuchungen sehr zu Dank verpflichtet.

Literatur

1. Meier P, Ohmes E, Kothe G, Blume A, Weidner J, Eibl H-J (1983) J Phys Chem, im Druck
2. Lausch M, Spiess HW (1983) J Magn Reson 54:466
3. Schwartz LJ, Meirovitch E, Ripmeester JA, Freed JH (1983) J Phys Chem 87:4453
4. Wigner EP (1959) Group Theory and its Application to Quantum Mechanics of Atomic Spectra, Academic Press, New York
5. Jin J-I, Antoun S, Ober C, Lenz RW (1980) Br Polym J 12:132
6. Antoun S, Lenz RW, Jin J-I (1981) J Polym Sci Polym Chem Ed 19:1901
7. Mueller K, Eisenbach C, Hisgen B, Ringsdorf H, Schneller A, Lenz RW, Kothe G, in Vorbereitung
8. Noel C, Monnerie L, Achard MF, Hardouin F, Sigaud G, Gasparoux H (1981) Polymer 22:578
9. Jerman RE, Baird DG (1981) J Rheol 25:275
10. Stohrer M, Noack F (1977) J Chem Phys 67:3729
11. Liebert L, Strzelecki L, Van Luyen D, Levelut AM (1981) Eur Polym J 17:71
12. Blumstein A, Vilasagar S, Ponrathnam S, Clough SB, Blumstein RB (1982) J Polym Sci Polym Phys Ed 20:877
13. Mueller K, Wassmer K-H, Lenz RW, Kothe G (1983) J Polym Sci Polym Lett Ed 21:785
14. Volino F, Martins AF, Blumstein RB, Blumstein A (1981) J Phys (Lett) 42:L305
15. Martins AF, Ferreira JB, Volino F, Blumstein A, Blumstein RB (1983) Macromolecules 16:279
16. Finkelmann H (1982) (ed) Ciferri A, Krigbaum WR, Meyer RB, Polymer Liquid Crystals, Academic Press, New York
17. Wassmer K-H, Ohmes E, Kothe G, Portugall M, Ringsdorf H (1982) Macromol Chem Rapid Commun 3:281
18. Blumstein RB, Stickless EM, Blumstein A (1982) Mol Cryst Liq Cryst (Lett) 82:205
19. Blumstein A, Thomas O (1982) Macromolecules 15:1264
20. Roviello A, Sirigu A (1982) Macromol Chem 183:895

21. Blumstein A, Blumstein RB, Gauthier MM, Thomas O, Asrar J (1983) Mol Cryst Liq Cryst (Lett) 92:87
22. Griffin AC, Havens SJ (1981) J Polym Sci Polym Phys Ed 19:951
23. Schaefgen JR, Plechter TC, Kleinschuster JJ (1975) Belg Patent Nr 828.935
24. Hentschel R, Spiess HW (1979) J Magn Reson 35:157
25. Bloom M, Davis JH, Valic MI (1980) Can J Phys 58:1510

Anschrift der Verfasser:

K. Müller
Institut für Physikalische Chemie
der Universität Stuttgart
Pfaffenwaldring 55
D-7000 Stuttgart 80

Progress in Colloid & Polymer Science Progr Colloid & Polymer Sci 69:134–138 (1984)

Aggregationseffekte von Polymeren mit mesogenen Seitengruppen in Lösung*

H. Cackovic, J. Springer und F. W. Weigelt

Institut für Technische Chemie der Technischen Universität Berlin, Fachgebiet Makromolekulare Chemie, Berlin, F. R. G.

Zusammenfassung: Aus Streulichtmessungen und Messungen des osmotischen Druckes, aus Untersuchungen der Lösungsviskosität und aus DSC-Messungen folgt, daß flüssigkristalline Polymere auf Methacrylatbasis in Lösung Aggregate bilden. Der Aggregationsgrad hängt von der Temperatur und der Art des Lösungsmittels ab. Mit steigender Temperatur werden die Aggregate bis 348 K langsam abgebaut, um dann schnell zu zerfallen. Damit ist eine sprunghafte Erniedrigung der Grenzviskositätszahl und eine deutliche Abnahme der scheinbaren Molmasse verbunden. Werden die Lösungen von Temperaturen oberhalb 348 K ohne äußere Krafteinwirkung abgekühlt, dann bilden sich keine Aggregate, bzw. dann ist die Aggregation ein sehr langsam ablaufender Prozess. Werden dagegen die Lösungen beim Abkühlen einer Scherbeanspruchung ausgesetzt, dann wird die Aggregation der Kettenmoleküle induziert. Aus dem Exponenten der Viskosität/Molmassenbeziehung und aus der Streufunktion geht hervor, daß die Moleküle oberhalb 348 K eine längliche Gestalt annehmen. Es wird ein aus Röntgendaten ermitteltes Strukturmodell der Aggregate vorgestellt.

Abstract: We have found from light scattering and osmotic measurements, from the viscosity in solution and from DSC-measurements, that liquid crystalline polymers on a methacrylate basis form aggregates in solution. The degree of aggregation depends on temperature and on the choosen solvent. With a temperature rise up to 348 K the aggregates get steadily smaller, and then dissolve quickly. At this point a sudden decrease in intrinsic viscosity and apparent molecular weight occurs. Cooling the solutions down from above 348 K without applying mechanical stress results in no aggregation, or rather the aggregation is then a very slow process. On the other hand, shearing the solution during cooling induces aggregation. The exponent of the viscosity versus molecular weight function and the light scattering function show that above 348 K molecules have an oblong shape. From X-ray data a structural model of the aggregates is deduced.

Key words: LC-poly(methacryl acid) derivate, aggregation, viscosity, DSC, X-ray.

1. Einleitung

Untersuchungen über die thermotropen Eigenschaften von flüssigkristallinen Polymeren sowie deren Abhängigkeit von der chemischen Struktur der Grundbausteine sind mehrfach durchgeführt worden [1–3]. Dabei handelt es sich meistens um Polymere, bei denen die mesogenen Gruppen über bewegliche Abstandshalter (Spacer) an die Polymerhauptkette gebunden sind [4]. Die thermotropen Eigenschaften hängen aber nicht nur von der Art der Ankopplung der mesogenen Gruppen an die Hauptkette [5] und von der Beweglichkeit der Kettenglieder ab, sondern auch von Parametern wie Molmasse und Molmassenverteilung, Taktizität und Wechselwirkungen der Polymermoleküle untereinander [6–8]. In der vorliegenden Arbeit wird der Einfluß dieser Größen untersucht, und es werden einige Ergebnisse vorgestellt, die an einem Polymethacrylsäurederivat des Typs[1])

* Vortrag, gehalten auf der 31. Hauptversammlung der Kolloid-Gesellschaft, Bayreuth 11. bis 14. Oktober 1983.

[1]) Synthetischer Name: Poly{1-methyl-1-[4-(4'-methoxyphenoxycarbonyl)-phenoxyhexoxycarbonyl]ethylen}

KT-W 789

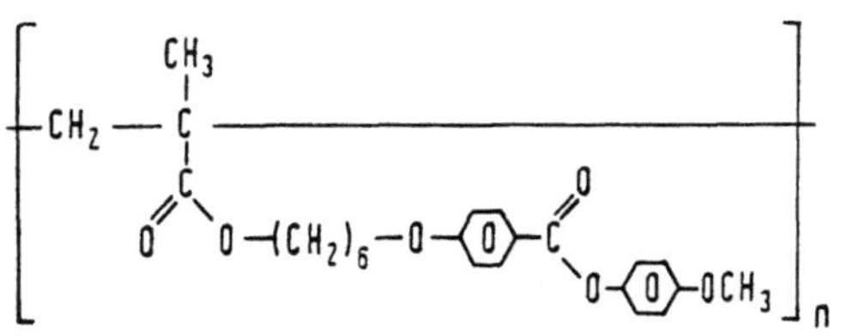

erhalten worden sind.

2. Molmassenbestimmung

In einer zeitweise parallel laufenden Arbeit [9] ist festgestellt worden, daß die mittlere Molmasse einer unfraktionierten Probe wider Erwarten mit der Anzahl der Fällungen abnimmt.

Bis zur vierten Fällung nimmt die Molmasse ab, um dann konstant zu bleiben. Zur Erklärung dieses Effektes wurde angenommen, daß die Kettenmoleküle als Aggregate vorliegen (zwischen „Aggregation" und „Assoziation" wird in den folgenden Ausführungen nicht unterschieden). Die Aggregate werden möglicherweise bei der mit dem Umfällen verbundenen mechanischen Beanspruchung bzw. durch die Änderung der thermodynamischen Bedingungen zerstört, was durch die Beobachtung gestützt wird, daß nach dem Aufschmelzen einer fünffach gefällten Probe wieder eine wesentlich höhere Molmasse bestimmt wird. Solche Effekte sind von Polymethylmethacrylaten (PMMA) her bekannt und die Aggregation wird als

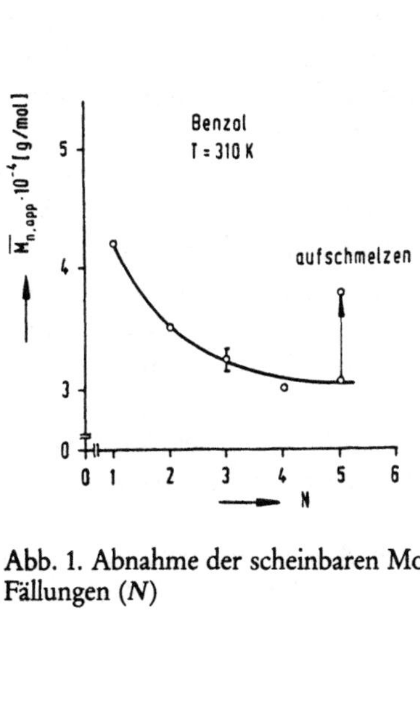

Abb. 1. Abnahme der scheinbaren Molmasse von der Anzahl der Fällungen (N)

Tab. 1. Temperaturabhängigkeit des Massenmittels der scheinbaren Molmasse ($M_{w,\,app.}$), des zweiten osmotischen Virialkoeffizienten (A_2) und des mittleren quadratischen Trägheitsradius $(\overline{r^2})^{1/2}$

Lösungsmittel	T K	$M_{w,\,app.} \cdot 10^{-6}$ g/mol	$A_2 \cdot 10^5$ cm³ mol/g²	$\sqrt{\overline{r^2}}$ nm
	298	0,92	16,1	53
	308	0,88	16,2	51
Chloroform	318	0,87	16,4	50
	328	0,87	16,3	48
	298	4,7	2,8	84
	308	4,1	3,3	84
Benzol	318	3,8	4,2	83
	328	3,5	4,6	82
	348	3,2	3,3	93
THF[a])	298	24,1	0,9	115

[a]) Wir danken Herrn Jr. W. Mächtle (BASF) für die Durchführung der Messungen.

Stereoassoziation bezeichnet. Da der Aggregationsgrad von der Temperatur und vom Lösungsmittel abhängt, sind in Anlehnung an Untersuchungen am PMMA [10] die Lösungsmittel nach ihrer die Aggregation hemmenden oder fördernden Wirkung ausgewählt worden. Chloroform wirkt als entassoziierendes, Benzol als schwach und Tetrahydrofuran (THF) als stark assoziierendes Lösungsmittel. Die Ergebnisse sind in Tabelle 1 zusammengestellt. Aufgrund der Temperatur und Lösungsmittelabhängigkeit wird die Molmasse als „scheinbare (apparent) Molmasse" bezeichnet.

Es zeigt sich, daß das Lösungsmittel einen beachtlichen Einfluß auf die scheinbare Molmasse hat. In dem entassoziierenden Lösungsmittel ist die kleinste Molmasse erhalten worden. In dem schwach assoziierend wirkenden Benzol liegt sie deutlich höher und in THF, das stark assoziierend wirkt, ist der größte Wert gemessen worden. Die Trägheitsradien sind im Vergleich zu den Massen der Aggregate verhältnismäßig klein. Erstaunlich ist, daß selbst in dem entassoziierend wirkenden Chloroform die Molmasse und der Trägheitsradius mit steigender Temperatur abnehmen, was bedeutet, daß auch in diesem Lösungsmittel noch Aggregate vorzuliegen scheinen.

3. Viskosimetrische Untersuchungen

Da mit der Aggregation Veränderungen der hydrodynamischen Eigenschaften einhergehen, sind viskosimetrische Untersuchungen an verdünnten Lösungen durchgeführt worden.

Als Lösungsmittel wurde 1,2-Dichlorbenzol ausgewählt, das als schwach aggregierend einzuordnen ist und dessen Siedetemperatur Messungen innerhalb eines größeren Temperaturbereiches zuläßt. Abbildung 2 gibt die Temperaturabhängikeit der Grenzviskositätszahl $[\eta]$ wieder.

Nach einem relativ schwachen Abfall bis etwa 348 K nimmt $[\eta]$ sprunghaft ab, um ab 358 K nahezu konstant zu bleiben. Dieser Befund ist unabhänig vom Geschwindigkeitsgefälle. Der schwache Abfall kann als ein langsamer Zerfall von Aggregaten interpretiert werden. Der stufenförmige Abfall der Grenzviskositätszahl kann zwei Ursachen haben. Einerseits ist eine Konformationsänderung - wie z. B. eine Helix-Knäuel Umwandlung - vorstellbar, und andererseits kann er auf den Zerfall der Aggregate zurückzuführen sein. Auch ist nicht auszuschließen, daß beide Vorgänge gemeinsam ablaufen.

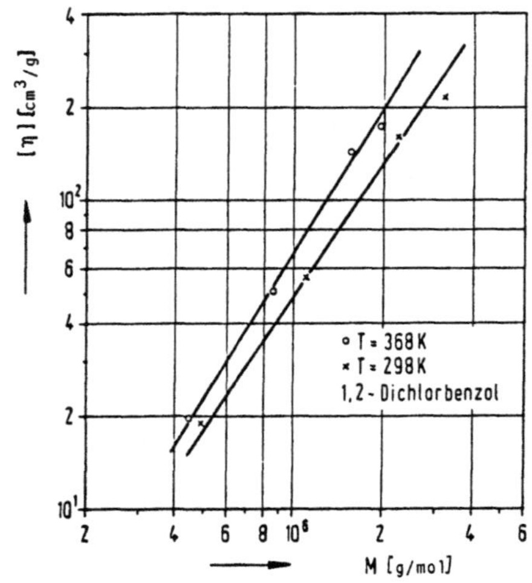

Abb. 3. Viskositäts-Molmassen-Beziehung

4. Struktur und hydrodynamische Eigenschaften

Eine Möglichkeit, temperaturabhängige konformationsbedingte Veränderungen der hydrodynamischen Eigenschaften von Polymeren in Lösung zu erkennen, besteht in der Bestimmung der Koeffizienten K und a der Staudinger'schen Gleichung

$$[\eta] = K\,M^a\,.$$

Dazu wurden von vier Fraktionen des Polymeren in 1,2-Dichlorbenzol bei 298 und 368 K die Grenzviskositätzahlen bestimmt, um mit den scheinbaren Molmassen die Konstanten ermitteln zu können.

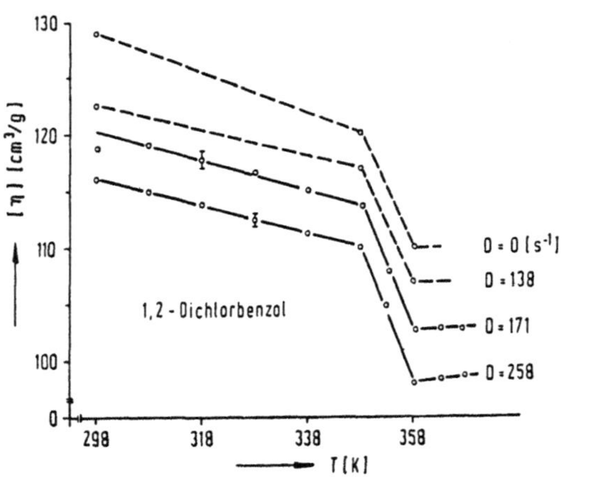

Abb. 2. Temperaturabhängigkeit der Grenzviskositätszahl

Sowohl bei 298 K als auch bei 368 K, also unter- und oberhalb der sprunghaften Änderung von $[\eta]$, haben die beiden Geraden die gleiche Steigung von 1,5, was bedeutet, daß die Aggregate in Lösung als verhältnismäßig langgestreckte Teilchen vorliegen.

Zur Untermauerung dieser Annahme ist an vier Fraktionen bei diesen Temperaturen im gleichen Lösungsmittel aus der Intensitätsverteilung des seitlich gestreuten Lichtes die Streufunktion $P(\vartheta)$ bestimmt worden. In Abbildung 4 sind die Messwerte $1/P\,(90°)$ gegen die Unsymmetrie $Z = P\,(45°)/P\,(135°)$ aufgetragen. Zum Vergleich wurden die theoretischen Streukurven für stäbchenförmige Partikel und für polymolekulare Knäuel eingezeichnet [11].

Wider Erwarten deckt sich nur die Streukurve bei 368 K mit der von Stäbchen, während die bei 298 K gemessene der für polymolekulare Knäuel berechneten folgt. Möglicherweise liegt die Diskrepanz an den verschiedenen Bestimmungsmethoden. Während bei den Streulichtmessungen die Lösung ruht und die gelösten Moleküle nur Brownsche Molekularbewegungen ausführen, wirkt auf sie bei den Viskositätsmessungen durch das Geschwindigkeitsgefälle ein Drehmoment.

Zur Überprüfung der Annahme, daß es zu der sprunghaften Abnahme von $[\eta]$ durch den Zerfall von

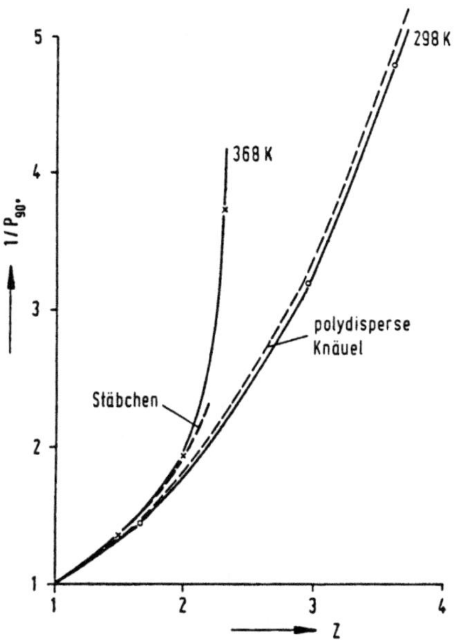

Abb. 4. Reziproke Streufunktion in Abhängigkeit von der Unsymmetrie ($\lambda = 436$ nm)

Aggregaten kommt, sind Streulichtmessungen bei verschiedenen Temperaturen durchgeführt worden. Beim Erwärmen der Lösungen von 298 auf 368 K ergibt sich für die scheinbare Molmasse eine dem Verlauf von $[\eta]$ entsprechende Abhängigkeit (Abb. 5).

Werden die Lösungen im Anschluß daran wieder abgekühlt, dann behalten die streuenden Partikel die Masse, die sie bei 368 K angenommen haben. Der Trägheitsradius fällt in einem Maße ab, wie es nach der Größenordnung für den Temperaturkoeffizienten dieses Parameters [12] zu erwarten ist. Im scheinbaren Widerspruch dazu steht der vorgestellte Befund (vgl. 3.), daß entsprechende Messungen zum Temperaturverhalten von $[\eta]$ eine reversible Abhängigkeit ergeben. Um zu prüfen, ob die Ursache dafür die Scherbeanspruchung der Lösungen bei der Bestimmung von $[\eta]$ ist, sind diese nach dem Abkühlen vor den Streulichtmessungen durch die Kapillare eines Viskosimeters geschickt worden. Wie aus Abbildung 5 ersichtlich, hat dies eine sprunghafte Zunahme der scheinbaren Molmasse zur Folge. Sie ist beispielhaft für die Temperatur 298 K gezeigt und kann folgendermaßen erklärt werden:

Da die gelösten Teilchen länglich sind, werden sie in der Strömung durch die auf sie wirkenden Scherkräfte eine Drehbewegung ausführen und in der Strömung orientiert bzw. im Mittel eine Vorzugsrichtung erhalten. Dadurch kommen verschiedene Teilchen in Positionen zueinander, die eine Aggregatbildung ermöglichen, von der angenommen wird, daß sie durch partielle Kristallisation erfolgt.

4.1 Kalorische Untersuchungen

Die Hypothese der Aggregation durch partielle Kristallisation kann durch kalorische Messungen an Lösungen gestützt werden. Abbildung 6 zeigt die Diagramme in 1,2-Dichlorbenzol ($c = 2{,}5 \cdot 10^{-3}$ g/cm^3)

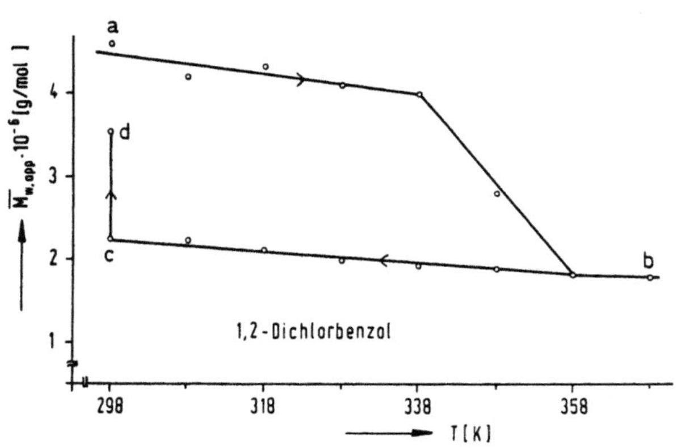

Abb. 5. Einfluß der Temperatur auf die scheinbare Molmasse

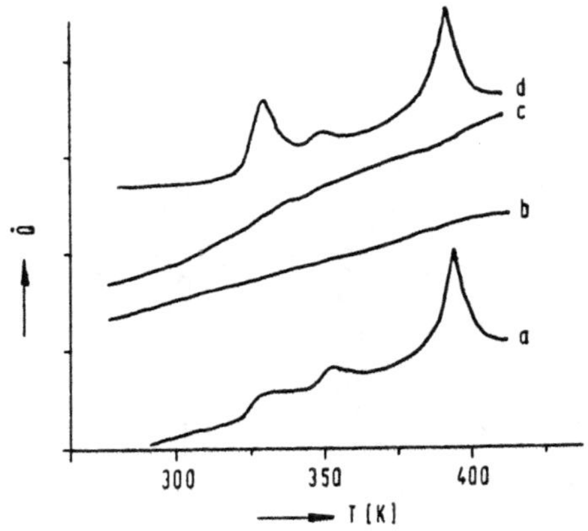

Abb. 6. DTA-Thermogramme von Lösungen nach verschiedener Vorbehandlung (vgl. Text)

nach verschiedener Vorbehandlung. Kurve a wurde an einer Lösung des Polymeren, das nach der Synthese längere Zeit gelagert worden ist, gemessen. Der Kurvenverlauf ist dem DTA-Digramm des Feststoffes ähnlich. Kurve b wurde nach Erwärmung der Lösung auf 368 K und anschließend schnellem Abkühlen erhalten. Die Kristallite sind dabei geschmolzen bzw. die Aggregate zerstört worden. Kurve c ist nach langsamen Abkühlen der Lösung auf 298 K gemessen worden. Die gelösten Teilchen können dabei zwar alle Lagen im Raum gleich wahrscheinlich einnehmen, jedoch zu einer Aggregation kann es nur dann kommen, wenn sich zwei oder mehrere Teilchen zufällig in der richtigen Lage zueinander befinden. Die Aggregation wird entsprechend langsam vonstatten gehen und konnte zur Zeit der Messung noch nicht beobachtet werden. Vor Messung der Kurve d schließlich ist die Lösung einer Scherbeanspruchung durch Kapillarströmung ausgesetzt worden. Hierbei sind die Teilchen orientiert worden und es ist zur Aggregation durch Kristallisation gekommen. Der Verlauf der DTA-Kurve gleicht deshalb wieder dem von Kurve a.

5. Struktur der Aggregate im Feststoff

Da sich die Struktur dieser durch Kristallisation gebildeten Aggregate im Feststoff einfacher bestimmen läßt, wurden Röntgenuntersuchungen an einer im nematischen Zustand versteckten Probe durchgeführt. Die aus dieser Röntgenstrukturanalyse gewonnenen Daten (Gitterabstände und deren Ausdehnung der geordneten Bereiche [13]) führten zu einem Strukturvorschlag, der in Abbildung 7 dargestellt ist.

Nach der Lage der Reflexe müssen die Seitengruppen orthogonal zur Hauptkette und aufgrund ihrer Raumerfüllung entlang der Hauptkette versetzt angeordnet sein, da zur Rotation eines Ringsystems der Seitengruppe ein größerer Abstand als der gefundene erforderlich wäre. Von der Breite der Reflexe her gesehen sind die geordneten Bereiche sehr klein bzw. liegt ein stark gestörtes Gitter vor. Die Dimensionen der Mikrokristallite lassen sich mit Hilfe der mittleren Reflexausdehnung abschätzen. Demnach sind sie in z-Richtung (Längsrichtung der mesogenen Gruppen) aus etwa 14 und in x- und y-Richtung aus etwa fünf mesogenen Seitengruppen aufgebaut. Sie sind in eine amorphe Matrix eingebettet.

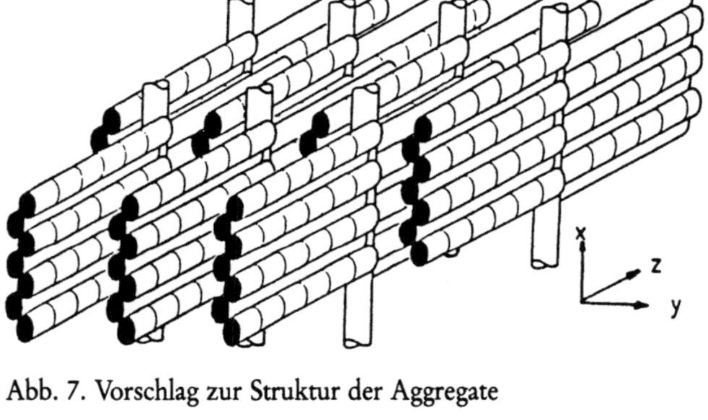

Abb. 7. Vorschlag zur Struktur der Aggregate

Die Röntgenuntersuchungen haben weiter gezeigt, daß die Monomerproben bis zu 100 % kristallin sind und die gefällten Polymerproben keine nachweisbare Kristallinität besitzen.

Breitlinienkernresonanzuntersuchungen an Monomer und Polymerproben [14] haben gezeigt, daß das Polymere nur 7 % mehr bewegliche Gruppen als das Monomere beinhaltet, und daß bei beiden Proben die Phenylgruppen beweglich sind. Das deutet auf eine verhältnismäßig dichte Packung hin, die beim Polymeren im festen Zustand eine Aggregat- bzw. Kristallitbildung fördert.

Literatur

1. Shibaev VP, Freidzon YS, Plate NA (1978) Polym Sci, USSR 20:94
2. Perplies E, Ringsdorf H, Wendorff JH (1976) J Polym Sci, Polym Lett Ed 13:243
3. Finkelmann H, Day D (1979) Makromol Chem 180:2269
4. Finkelmann H, Happ M, Portugall M, Ringsdorf H (1978) Makromol Chem 179:2541
5. Ringsdorf H, Schneller A (1981) Br Polym J 13:43
6. Springer J, Weigelt FW (1983) Makromol Chem 184:1489
7. Springer J, Weigelt FW (1983) Makromol Chem 184:2635
8. Cackovic H, Springer J, Weigelt FW, in Vorbereitung
9. Siebke W (1980) Diplomarbeit, TU Berlin
10. Vorenkamp J, Bosscher F, Challa G (1979) Polymer 20:59
11. Huglin MB (1972) Light Scattering from Polymer Solutions, Academic Press, London, New York, Chapt 7, p 333
12. Brandrup J, Immergut EH (1975) Polymer Handbook, J Wiley & Sons, New York, p IV-34
13. Klug HP, Alexander LE (1962) X-Ray Diffraction Procedures, J Wiley & Sons, New York, p 509
14. Cackovic H, Springer J, in Vorbereitung

Anschrift der Verfasser:

J. Springer
Institut für Technische Chemie
der Technischen Universität Berlin
Fachgebiet Makromolekulare Chemie
Straße des 17. Juni 135
D-1000 Berlin 12

Properties of liquid crystalline polymers in the electric field*)

W. Haase and H. Pranoto

Institut für physikalische Chemie der Technischen Hochschule Darmstadt, Darmstadt, FRG

Abstract: Some liquid crystalline side chain polymers are described with respect to some of their properties in the electric field. Compounds investigated were polyacryl and polysiloxanes with mesogenic p-substituted phenylbenzoates. The compounds show a static dielectric anisotropy $\Delta\varepsilon > 0$. This allows the two-frequency technique to be applied. Some response times are in the 100 ms range at temperatures not too much below the clearing point.

The relaxation frequencies of $\varepsilon_{\parallel}$ are $\sim 10^3$ smaller compared to the low molecular liquid crystal data. This depends on the remarkable higher viscosity of the polymers.

Key words: liquid crystalline polymers, electric field.

Introduction

The properties of low molecular liquid crystals in the electric field are well characterized. The orientation of the liquid crystals by the electric field is the basis of the applicability of liquid crystals in LCD's[1]), utilizing the field effect. The realization of the homogeneous or homeotropic orientation depends on the molecular properties of the liquid crystalline materials.

According to their structures one distinguishes two different kinds of polymeric liquid crystals [1, 2]. The liquid crystalline main chain polymers have mesogenic fragments as parts of the polymeric chain. In opposition to this, the mesogenic groups in the liquid crystalline side chain polymers are coupled via a flexible spacer to a non-mesogenic chain, e. g. polyacryl or polysiloxane.

The aim of this work is to describe some properties of liquid crystalline side chain polymers especially in the electric field and to compare this with the properties of low molecular liquid crystals.

Some properties of low molecular liquid crystals in the electric field [3]

The orientation of the liquid crystals in the electric field depends on the sign of the dielectric anisotropy $\Delta\varepsilon$; $\Delta\varepsilon = \varepsilon_{\parallel} - \varepsilon_{\perp}$. $\varepsilon_{\parallel}$ and $\varepsilon_{\perp}$ are the components parallel and perpendicular to the long molecular axis. In the case of $\Delta\varepsilon > 0$ the molecules are oriented with their long axis parallel to the electric field, whereas in the case of $\Delta\varepsilon < 0$ the orientation is perpendicular to the electric field.

The orientation by the electric field takes place if the applied voltage is higher than the threshold-voltage, which depends on the properties of the liquid crystalline material. For a characteristic threshold-voltage e. g. ~ 5 V, a field strength of ~ 0.5 MV/m results, if a thickness of 10 µm is assumed.

The orientation (preorientation) of the molecules on the glas surface acts on surface effects. Under the influence of the electric field, the performed orientation will be changed in some cases. This is demonstrated in figure 1. The definition of the response time for the twisted-nematic (TN) cell, which is the time necessary for the orientation or reorientation processes, measured e. g. by transmission experiments, is given in figure 2 [3]. For the Freedericksz cell [3] we used [3] a definition of the response time which is comparable with the time for the analogous process in the TN cell.

*) Lecture given at the 31th Conference of the Kolloid Gesellschaft, Bayreuth October 11–14, 1983.

[1]) Liquid Crystal Display

KT-W 803

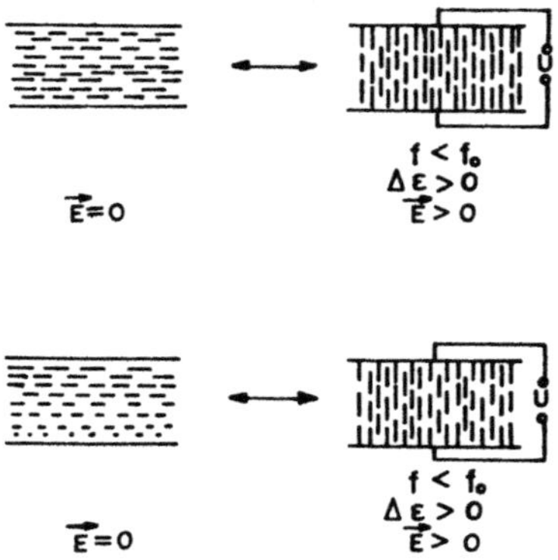

Fig. 1. Influence of the electric field. top: Freedericksz-cell, botton: TN-cell

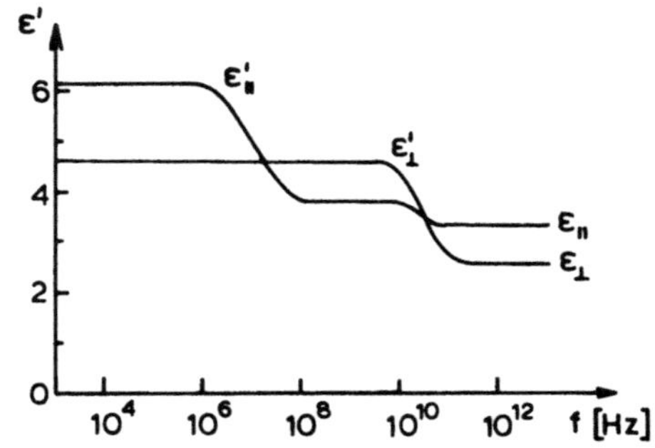

Fig. 3. Relaxation processes dependent upon the frequency of the electric field

The rise times of liquid crystalline materials used in the LCD's lie in the region of ~ 50 ms comparable with the turn-on-delay, whereas the fall time is longer, ~ 100 to 200 ms.

Maier and Meier [6] described the dielectric properties depending namely on the molecular polarizability anisotropy $\Delta\alpha$ $(=\alpha_{\|} - \alpha_{\perp})$ and the value and angular position of the permanent electric dipole moment μ, in extension of the Onsager theory. For $\Delta\varepsilon$ the formula (1)

$$\Delta\varepsilon = 4\pi\ NhF\left[\Delta\alpha - F\frac{m^2}{3kT}\frac{3}{2}\left(1-3\cos^2\beta\right)\right]S \tag{1}$$

is valid. h is the cavaty factor, F the Onsager reaction field and β the angle between the mean position of the single molecule and the director. N is the particle density and S is the order parameter.

Compounds with a p-cyano group exhibit normally $\Delta\varepsilon > 0$ because the dipole moment of the cyano group is nearly parallel to the long molecular axis.

Contrary to the isotropic liquids, in the liquid crystalline state a relaxation process of the dipole around the short molecular axis takes place in a frequency range which is lower than the normal Debye relaxation frequency, which lies in the microwave region. This is demonstrated in figure 3. For 4-pentyl-4'-cyano-biphenyl the frequency of the dielectric relaxation f_R is at room temperature ~ 6.5 MHz [7]. In comparison to this, compounds or mixtures are known with relaxation frequencis in the kHz region, for example the Merck mixture ZLI 518 with ~ 100 kHz at room temperature [8] or the Hoffmann-La Roche mixtures M1, M2 and M3 with ~ 1 kHz [9]. The different dispersion regions depend on the molecular properties and the viscosity coefficients.

The dispersion of $\varepsilon_{\|}$ in the lower frequency region can be used to shorten the response times of the compounds. In this method, the so called two-frequency technique, one does not use the field-on and field-off state for the orientation and the reorientation of the molecules, but a change in the frequenncy of the applied voltage. Thus the turn off time resulting after a change of the frequency from $f < f_o$ to $f > f_o$ is remarkably shorter than the decay time in the passive state. The effect is demonstrated in figure 4. f_o is the frequency of the dielectric isotropy.

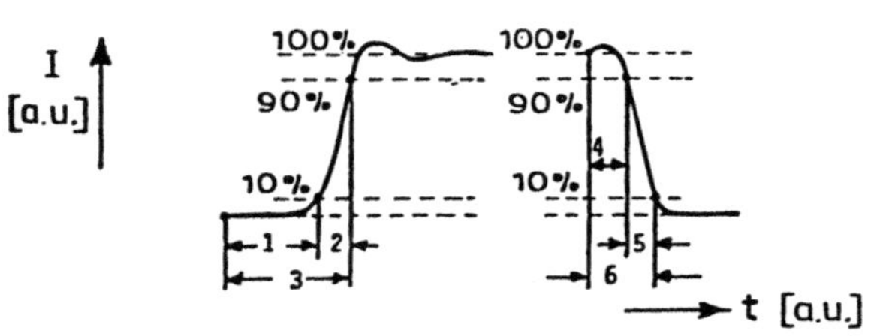

Fig. 2. Response properties of the TN-cell. 1 turn-on delay, 2 rise time, 3 turn-on time, 4 turn-off delay, 5 fall time, 6 turn-off time

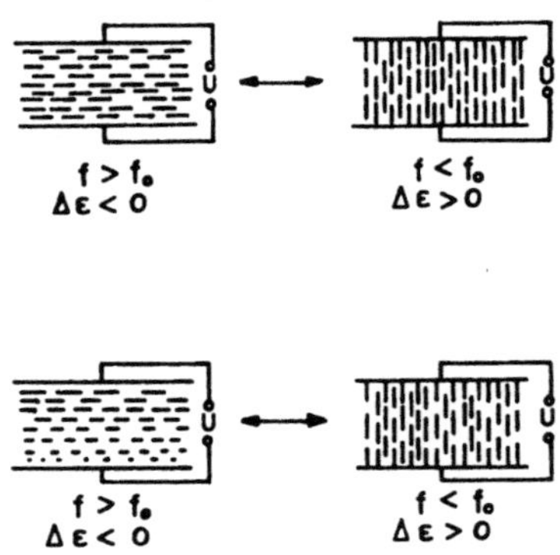

Fig. 4. Orientation change in the two-frequency technique. top: Freedericksz-cell, botton: TN-cell

Properties of some liquid crystalline side chain polymers in the electric field

Compounds investigated

The compounds investigated are presented in figure 5. All the compounds show a broad liquid crystalline range. The polyacryl compound was first described by [10, 11], and the polysiloxanes by [12, 13]. Our preparation of the polyacryl compound follows the method described by [10, 11]. Another part of the polyacryl compound was kindly placed at our disposal by Professor Ringsdorf. The polysiloxane compounds were kindly prepared by Dr. Finkelmann.

Orientation through glass surfaces and cell preparation techniques

In comparison to the low molecular liquid crystals the orientation of the liquid crystalline side chain polymers is more complicated. A comparably good orientation is achieved if the glass surface is coated with a thin film of polyimide. The quality of the orientation is better if it is attained by slow cooling of the liquid crystalline material rather than by a rapid temperature change. The latter mode of treatment leads to a destruction of the orientated layer.

In comparing the compounds investigated one can say that the orientation of the polysiloxanes is much better than the orientation of the polyacryls.

The construction of the polysiloxane cells follows the method described previously [5]. The filling of a cell of $d \sim 12$ μm thickness is effected through capillary forces and at a temperature slightly below the clearing point. It was also possible to obtain well orientated layers of ~ 1 cm^2 for the polyacryls by capillary forces, but for the filling procedure we required ~ 3 days at a temperature in the vicinity of the clearing point.

Summing up, the orientation of liquid crystalline side chain polymers (of the sort investigated) and the cell preparation techniques are both comparable to the methods used for low molecular liquid crystals if we consider some particularities. The principal fact that the liquid crystalline side chain polymers orientate through the electric field confirms the orientation through the magnetic field [14].

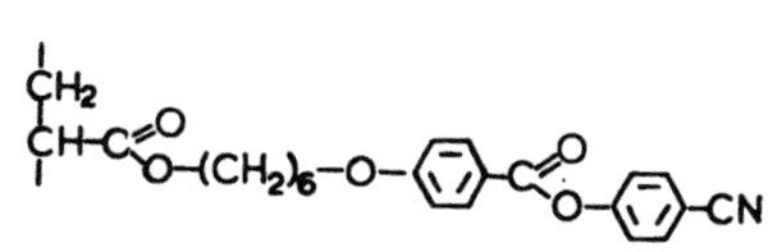

g 309 n 406 i

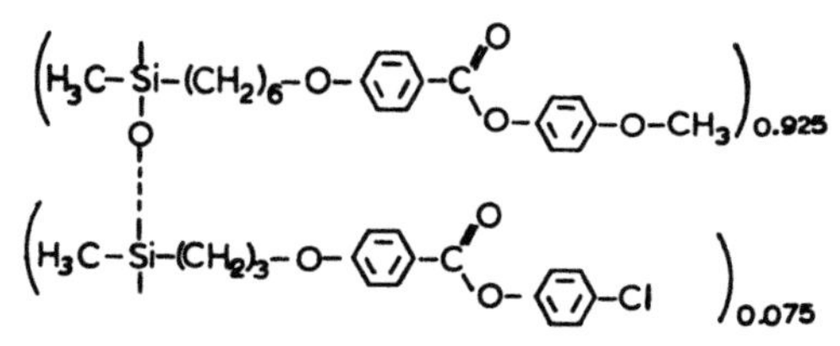

g 280 s 329 n 381 i

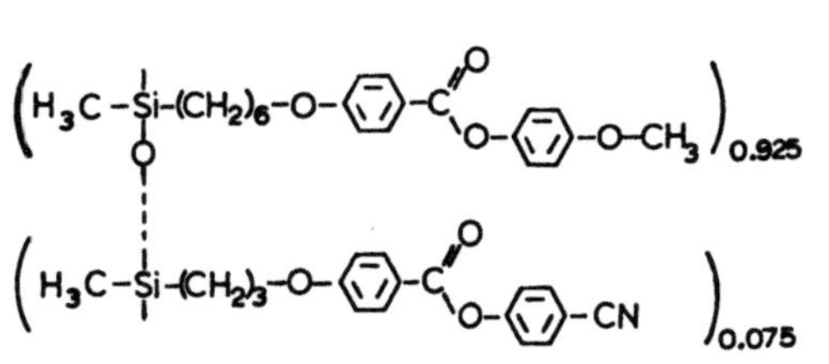

g 280 s 343 n 386 i

Fig. 5. Compounds investigated and their transition temperatures. The following abbreviations are used: top: Ac–CN, centre: Si–Cl, bottom: Si–CN

Switching properties

The measurements were performed with a Freedericksz cell as well as a TN-cell. The experimental setup that was used for our measurements especially with the Freedericksz cell was described elsewhere [5]. This arrangement is also applicable for the TN-cell.

Some typical response times for two temperatures, one slightly below the clearing point, the other 15 degrees lower than the clearing point, are presented in table 1. For the determination of the active decay time we used the two-frequency technique. It is remarkable that some response times at temperatures slightly below the clearing point lie in the 100 ms-region.

The compounds under investigation showed frequencies of the dielectric isotropy in the kHz-region. These values are also included in table 1. In comparison to the polysiloxanes the f_0-frequency of the polyacryl compounds is higher at comparable reduced temperatures.

Si–Cl and Si–CN are copolymers. Si–Cl contains mesogenic parts 92.5 percent p-methoxy group and 7.5 percent p-chloro group whereas Si–CN contains 92.5 percent p-methoxy group and 7.5 percent p-cyano group (see fig. 5). In the low molecular liquid crystalline p-methoxy benzoates $\Delta\varepsilon$ is negative over the whole frequency range [15]. This means that the relatively small chloro-part or cyano-part contributed to the entire positive value.

Table 1. For TN-cell

Polymer	U_o/V	U_o^x/V	t_d^o/s	t_d^{15}/s	t_r^{30}/s	$d/\mu m$	T/K	T_{ni}/K
Ac-CN	0.86	–	99	–	0.08	18.4	403	406 $f_o > 600$ kHz
	1.3	3.7	630	40	1.1	18.4	373	406 0.2 kHz $< f_o$ 200 kHz $> f_o$
Si-Cl	6.0	13.6	4.9	2.3	0.19	8.6	378	380 $f_o = 66$ kHz
	10.0	12.6	8.9	4.1	1.0	8.6	369	380 $f_o = 15$ kHz
Si-CN	5.4	12.5	20.7	8.3	0.7	19.1	379	386 $f_o = 82$ kHz
	4.1	4.0	64.2	4.5	1.2	19.1	368	386 $f_o = 20$ kHz

U_o is threshold-voltage for $f < f_o$; U_o^x is threshold-voltage for $f > f_o$; t_d^o is passive decay time; t_d^{15} is active decay time for 15 Volt applied AC-voltage, 200 kHz; t_r^{30} is rise time for 30 Volt applied AC-voltage, 0.2 kHz; d is cell thickness; T is experimental temperature; T_{ni} is clearing point

The frequencies of the dielectric relaxation are strongly temperature dependent. In figure 6 a plot $\ln f_R$ versus $1/T$ for Si-Cl is given. f_R is the frequency of the dielectric relaxation, obtained at dielectric measurements. The activation energy E_A of liquid crystalline side chain polymers is according to equation (2)

$$\ln f_R = \ln A - \frac{E_A}{RT} \tag{2}$$

(with A as factor) in the order of 200 kJ/mole and so 2 to 3 times higher than that of comparable low molecular liquid crystals. A detailed discussion of the activation processes will be given later [16].

The absolute values of the response times are by a factor 10^2 to 10^3 higher for the liquid crystalline side chain polymers, than the comparable values for the low molecular liquid crystals. This depends on the viscosity, which is remarkably higher in polymeric liquid crystals. The viscosity of the polyacryl is also higher than that of the polysiloxanes. This is in agreement with the higher passive decay times of the former compounds. The experimental determination of the response processes are accordingly easier to observe.

A plot of the reciprocal response time as a function of the square of the voltage is given in figure 7. This indicates that the equation (3)

$$\frac{1}{t_d} = \left(\frac{\pi}{d}\right)^2 \cdot \frac{k_{11}}{\eta_i} \left[\left(\frac{U}{U_o}\right)^2 + 1\right] \tag{3}$$

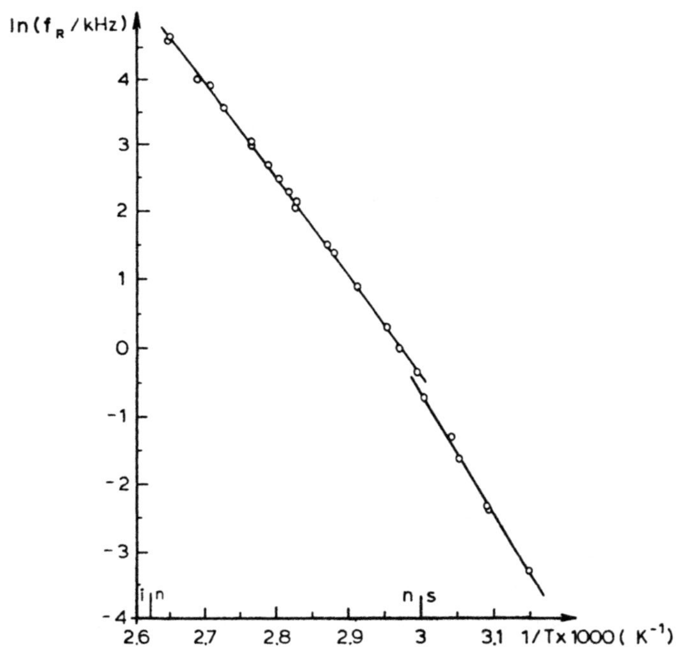

Fig. 6. $\ln f_R$ versus $1/T$ for Si-Cl

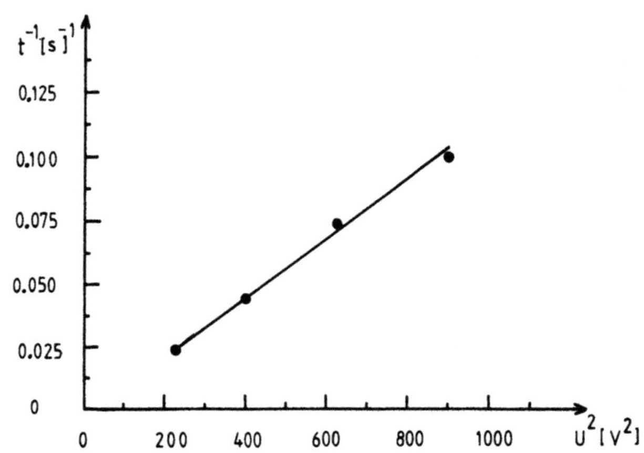

Fig. 7. The reciprocal decay time t_d versus U^2 for the TN-cell of Ac-CN, 373 K, 200 kHz

is also valid for the liquid crystalline side chain polymers. From equation (3) and using the response time we can calculate the viscosity coefficient η_i. The elastic constant k_{11} can be estimated from equation (4) if the threshold-voltage U_o is known:

$$U_o = \pi \sqrt{\frac{k_{11}}{\varepsilon_o \cdot \Delta\varepsilon}} \quad (4)$$

ε_o is the dielectric constant in vacuum. $\Delta\varepsilon$, the dielectric anisotropy, is to be obtained from dielectric measurements.

Whereas the elastic constants for the liquid crystalline side chain polymers are in the same order as the elastic constants of the low molecular liquid crystals, the viscosity is 2 to 3 times greater than the latter. The viscosity can also be estimated based on the knowledge of the glass temperature under application of the well-known WLF-equation. After this procedure we receive for the polyacryl compound Ac–CN at 376 K $\eta \sim 1.6 \times 10^5$ cP and for the polysiloxanes Si–Cl and Si–CN at 372 K $\eta \sim 8 \times 10^3$ cP.

Dielectric properties

The dielectric properties as function of the temperature and the frequency are measured with the help of a general radio bridge in the range 10 Hz to 100 kHz. Above 100 kHz a Wayne Kerr bridge B 602 and a source/detector SR 268 were used. The employed cell for the measurements was prepared with a mask constructed from a gold film. Experimental details will be reported later [16], together with a detailed discussion of the rather complicated relaxation processes. The frequency dependence of $\varepsilon_\parallel$ is presented in figure 8. The relaxation frequency of the liquid crystalline side chain polymers is lower by approximately the factor thousand than the comparable data of the low molecular liquid crystals. This also depends on the molecular properties.

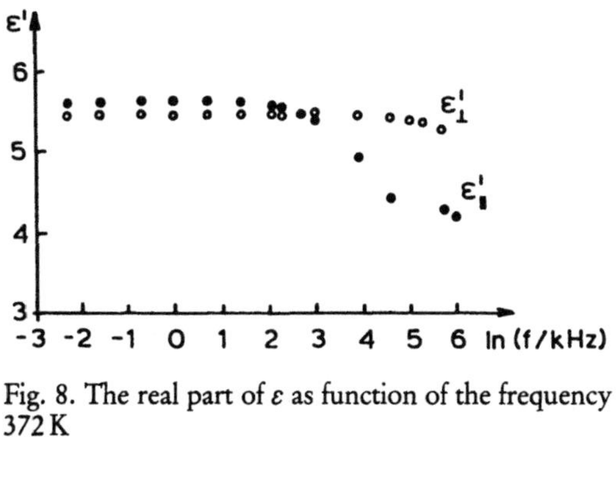

Fig. 8. The real part of ε as function of the frequency for Si–Cl at 372 K

From the Meier-Saupe equation [17], equation (5)

$$\tau_R = g \cdot \tau_D \quad (5)$$

with τ_R (= $1/2\,\pi\, f_R$) relaxation time for the $\varepsilon_\parallel$ relaxation and τ_D relaxation time for the Debye relaxation we can conclude that τ_R is higher if

g, the retardation factor, is higher and/or
τ_D is higher.

g depends on the strength of the nematic potential and τ_D depends on the viscosity. Since the viscosity of the polymeric liquid crystals is higher, the normal Debye relaxation shall likewise be changed from the GHz-region to the MHz-region.

Acknowledgement

The authors thank Professor Ringsdorf and Dr. Finkelmann for providing the substances. The work was supported by the Deutsche Forschungsgemeinschaft.

References

1. Shibaev VP (1977) Vysokomol Soedin A9:923
2. Finkelmann H, Ringsdorf H, Wendorff JH (1978) Makromol Chem 179:273
3. Meier G, Sackmann E, Grabmeier JG (1975) Applications of liquid crystals, Springer-Verlag, Berlin
4. Schadt M, Helfrich W (1971) Appl Phys Lett 18:127
5. Haase W, Pranoto H (1983) Mol Cryst Liq Cryst 98:299
6. Maier W, Meier G (1961) Z Elektrochem 65:556, (1961) Z Naturforsch 16a:262, 470
7. Cummins PG, Dunmur DA, Laidler DA (1975) Mol Cryst Liq Cryst 30:109
8. Haase W, Pötzsch D (1977) Mol Cryst Liq Cryst 38:77
9. Schadt M (1982) Mol Cryst Liq Cryst 89:77
10. Portugall M, Ringsdorf H, Zentel R (1982) Makromol Chem 183:2311
11. Finkelmann H, Happ M, Portugall M, Ringsdorf H (1978) Makromol Chem 179:2541
12. Finkelmann H, Rehage G (1980) Makromol Chem, Rapid Commun 1:31
13. Finkelmann H, Kiechle U, Rehage G (1983) Mol Cryst Liq Cryst 94:343
14. Achard MF, Sigaud G, Hardouin F, Weill C, Finkelmann H (1983) Mol Cryst Liq Cryst Lett 92:111

15. de Jeu WH (1978) Dielectric permittivity of liquid crystals, Solid State Physics Suppl 14:109
16. Pranoto H, Haase W, to be published
17. Meier G, Saupe A (1966) Mol Cryst Liq Cryst 1:515

Authors' address:

W. Haase
Institut für Physikalische Chemie
der Technischen Hochschule Darmstadt
Petersenstraße 20
6100 Darmstadt

Tensideinfluß auf den Wasserablauf an harten Oberflächen*)

C.-P. Kurzendörfer, Th. Altenschöpfer und H.-J. Völkel

Henkel KGaA, Düsseldorf

Zusammenfassung: Das Ablaufverhalten (Restflüssigkeitsmenge, Ablaufgeschwindigkeit, Ablaufcharakteristik) wäßriger Lösungen von Na-Dodecylsulfat und Nonylphenoldecaglykolether an vertikalen Glas- und Kunststoffoberflächen wurde mit einem kontinuierlich gravimetrischen Zweistufenverfahren untersucht.

Die Glasoberfläche benetzendes Wasser mit „hydrophiler" Ablaufcharakteristik ergibt Restflüssigkeitsmengen, die entsprechend der Ablauftheorie durch die Flüssigkeitsparameter Viskosität und Dichte bedingt sind. Benetzende Tensidlösungen führen in einem begrenzten Bereich niedriger Konzentration entgegen der Theorie zu erhöhten Restflüssigkeitsmengen. Das unterhalb der kritischen Mizellbildungskonzentration auftretende Restflüssigkeitsmaximum ist auf eine verringerte Ablaufgeschwindigkeit des Wassers durch die Tenside zurückzuführen. Erklärt wird dieser Befund mit dem Gibbs-Marangoni-Effekt. An Kunststoffen zeigen verdünnte Tensidlösungen eine „hydrophobe" Ablaufcharakteristik mit geringeren Restflüssigkeitsmengen als an Glas. Die „hydrophile" Ablaufcharakteristik von Tensidlösungen an Kunststoffen ist erst oberhalb der kritischen Mizellbildungskonzentration gewährleistet, was zu Restflüssigkeitsmengen führt, die unabhängig von der Hydrophobie der harten Oberfläche sind.

Abstract: The drainage behaviour (residual quantity of liquid, rate of flow, drainage characteristics) of aqueous solutions of Na-dodecylsulfate and nonylphenoldecaglycolether on vertical surfaces of glass and plastics has been investigated by means of a continuous gravimetric two-step process.

Water with „hydrophilic" drainage characteristics, wetting the surface of glass, gives residual quantities of liquid that depend on viscosity and density according to the theory of drainage. Against the theory, wetting solutions of surfactants in a limited range of low concentration results in higher residual quantities of liquid. The maximum of the residual quantity of liquid occurring below the critical micelle concentration is to be attributed to a reduced rate of flow of the water by the surfactants. This result is explained by the Gibbs-Marangoni Effect.

On plastics, diluted solution of surfactants show "hydrophobic" drainage characteristics with lower residual quantities of liquid than on glass. The "hydrophilic" drainage characteristics of solutions of surfactants on plastics is only sure to occur above the critical micelle concentration. In this case, it results in residual quantities of liquid that are independent of the hydrophobicity of the hard surface.

Schlüsselwörter: Wasserablauf, Tenside, Glas, Kunststoff.

Einleitung

Das Ablaufverhalten tensidhaltiger wäßriger Lösungen an harten Oberflächen spielt in der Anwendung eine wichtige Rolle. Beim maschinellen und manuellen Geschirrspülen beispielsweise übt das Ablaufverhalten an Gläsern einen entscheidenden Einfluß auf das Aussehen der Oberfläche nach dem Trocknen aus. Hierbei ist für eine glänzende Oberfläche ohne Eintrocknungsrückstände unter anderem von Bedeutung, wie schnell die Lösung abläuft und welche Menge auf dem Glas beim Einsetzen der Trocknung vorliegt.

Zur Zeit wird der Flüssigkeitsablauf vornehmlich visuell beurteilt. Zur Objektivierung der Bestimmung

*) Vortrag, gehalten auf der 31. Hauptversammlung der Kolloid-Gesellschaft, Bayreuth 11. bis 14. Oktober 1983.

KT-W 799

des Ablaufverhaltens von tensidhaltigen Lösungen an harten Oberflächen wurde ein meßtechnisches Verfahren entwickelt. Mit ihm sollten die beim Flüssigkeitsablauf an vertikalen Oberflächen zu bestimmten Zeiten auftretenden Restflüssigkeitsmengen, Ablaufgeschwindigkeiten und Ablaufcharakteristiken bestimmt werden können. Die Ablaufcharakteristiken betreffend wird zwischen dem „hydrophilen" Ablauf, d. h. Auslaufen des die Oberfläche benetzenden Flüssigkeitsfilms unter Verringerung der Filmdicke, und dem „hydrophoben" Ablauf, d. h. Zurückziehen des die Oberfläche nicht benetzenden Flüssigkeitsfilms, unterschieden.

Die vorliegende Arbeit beschreibt das Ablaufverhalten verschiedener reiner Flüssigkeiten an Glas und wäßriger Lösungen der Tenside Natriumdodecylsulfat und Nonylphenoldecaglykolether an Glas und Kunststoffen.

2. Grundlagen des kontinuierlich gravimetrischen Zweistufenverfahrens

Die Anforderungen an das zu entwickelnde Meßverfahren und die meßtechnischen Möglichkeiten führten zum Meßprinzip eines Zweistufenverfahrens. Hierbei wird ein hohlzylindrischer Meßkörper in die Flüssigkeit eingetaucht und anschließend mit konstanter Geschwindigkeit soweit herausgezogen, daß sein unterer Rand mit der Flüssigkeitsoberfläche in Kontakt bleibt. Unmittelbar danach wird der Gewichtsverlust des Meßkörpers, der durch den Flüssigkeitsablauf an der Meßkörperoberfläche bei konstanter Herausziehhöhe bedingt ist, kontinuierlich gravimetrisch gemessen. Für die theoretische Beschreibung des Flüssigkeitsablaufes beim Zweistufenverfahren wurde auf eine Funktion von Lang und Tallmadge [1] zurückgegriffen, die auf der Ablauftheorie von Flüssigkeiten an vertikalen Oberflächen nach der Kontinuitätsgleichung für Flüssigkeitsströmungen beruht [2, 3, 4]. Hierin wird zwischen dem Flüssigkeitsablauf während der Herausziehzeit t_w (withdrawal time) einer eingetauchten Platte und dem Flüssigkeitsablauf bei konstanter Plattenhöhe in der Ablaufzeit t_d (drainage time) unterschieden. Für das konzipierte Zweistufenverfahren kann diese Funktion herangezogen werden, wenn t_w konstant gehalten wird und t_d die Variable ist. Ergänzt werden mußte die Funktion durch den Ausdruck über die Restflüssigkeitsmenge an der inneren und äußeren Oberfläche des holzylindrischen Meßkörpers.

Beim Zweistufenverfahren wird die Restflüssigkeitsmenge G auf der Oberfläche des zylindrischen Meßkörpers in Abhängigkeit von der Ablaufzeit t_d nach Gleichung 1 beschrieben.

$$G = \frac{8}{3} \cdot \bar{r} \cdot \pi \cdot h^{3/2} \cdot \frac{\eta^{1/2} \cdot \delta^{1/2}}{g^{1/2}} \cdot (t_w + t_d)^{-1/2} \quad (1)$$

mit

G	=	Restflüssigkeitsmenge	(g)
$\bar{r}$	=	mittlerer Zylinderradius	(cm)
h	=	Herausziehhöhe des Zylinders	(cm)
η	=	dynamische Viskosität	(g/cm · s)
δ	=	spez. Gewicht	(g/cm³)
t_w	=	Herausziehzeit des Zylinders	(s)
t_d	=	Ablaufzeit der Flüssigkeit	(s)
g	=	Gravitationskonstante	(cm/s²)

Gleichung 1 besagt, daß die Restflüssigkeitsmenge einer die Oberfläche benetzenden Flüssigkeit abhängig ist von den geometrischen Abmessungen des Meßkörpers, den physikalischen Parametern Viskosität und Dichte der Flüssigkeit, der Herausziehzeit des Meßkörpers und der Ablaufzeit der Flüssigkeit bei konstanter Herausziehhöhe des Meßkörpers. Die Temperaturabhängigkeit der Restflüssigkeitsmenge wird mit den temperaturabhängigen Flüssigkeitsparametern berücksichtigt. Dagegen liegt nach der Ablauftheorie von Flüssigkeiten [2, 3, 4] keine Beeinflussung durch die Oberflächenspannung der Flüssigkeit vor.

Gleichung 1 als $G = f\,((t_w + t_d)^{-1/2})$ graphisch aufgetragen, stellt eine lineare Funktion durch den Nullpunkt mit der Steigung S (Gleichung 2) dar.

$$S = \frac{8}{3} \cdot \bar{r} \cdot \pi \cdot h^{3/2} \cdot \frac{\eta^{1/2} \cdot \delta^{1/2}}{g^{1/2}} \qquad (g \cdot s^{1/2}) \quad (2)$$

Nach dem ersten Teilschritt des Meßverfahrens, d. h. unmittelbar nach dem Herausziehen des Meßkörpers zur Zeit t_w mit $t_d = 0$ liegt eine Restflüssigkeitsmenge G_{t_w} vor, die die Ausgangsflüssigkeitsmenge für den anschließenden Ablaufvorgang im t_d-Bereich darstellt.

Die Ablaufgeschwindigkeit u der Flüssigkeit im t_d-Bereich erhält man durch Bildung von $-dG/dt_d$ bei konstantem t_w (Gleichung 3).

$$u = -\left(\frac{dG}{dt_d}\right)_{t_w} = \frac{G_{t_w}}{2} \cdot \frac{t_w^{1/2}}{(t_w + t_d)^{3/2}} \qquad (g/s) \quad (3)$$

Die Ablaufgeschwindigkeit nimmt demnach mit zunehmender Zeit t_d ab. Je größer die Ausgangsflüssigkeitsmenge G_{t_w} ist, um so höher wird die Ablaufgeschwindigkeit im t_d-Bereich.

3. Experimenteller Teil

3.1 Untersuchte Substanzen

Natriumdodecylsulfat lag als kettenlängenreine Substanz vor, die frei von oberflächenaktiven Verunreinigungen war (kritische Mizellbildungskonzentration $c_{M,\,25\,°C} = 8 \cdot 10^{-3}$ mol/1). Nonylphenoldecaglykolether, techn. lag als kettenlängenreine Substanz vor mit statistischer Verteilung der Zahl der Glykolethergruppen ($c_{M,\,25\,°C} = 7{,}5 \cdot 10^{-5}$ mol/1, Trübungspunkt = 60 – 65 °C).

3.2 Meßkörper

Die Glaszylinder mit den Abmessungen h = 80 mm, d_i = 34 mm, d_a = 38 mm wurden aus DURAN-Glas gefertigt. Die Kunststoffzylinder (Polyethylen PE, Polypropylen PP, Polyvinylchlorid PVC) gleicher Abmessungen wurden aus Rohlingen hergestellt, die sorgfältig und weitgehend rückstandsfrei poliert wurden. Die Charakterisierung der polierten Kunststoffoberflächen erfolgte durch Messung der Benetzungsspannung j_R (mN/m) von Wasser bei 25 °C (PE: 27.1, PP: 25,1, PVC: 30.6 mN/m). Die geometrische Oberfläche der hohlzylindrischen Meßkörper bei der Herausziehhöhe h = 60 mm betrug 136 cm².

Für den Vergleich von Restflüssigkeitsmengen und Ablaufcharakteristiken an harten Oberflächen unterschiedlicher Materialien muß gleiche Oberflächenrauheit vorausgesetzt werden. Mit einer mechanischen Methode wurden die Oberflächenprofildiagramme der zylindrischen Meßkörper bestimmt und daraus als Maß für die Oberflächenrauheit der normierte Mittenrauhwert Ra(μ) berechnet. Die Meßkörper aus Glas, PE, PP und PVC zeigen in der Größenordnung übereinstimmende Mittenrauhwerte (Glas: 0.012, PE: 0.05, PP: 0.05, PVC: 0.03 μ), was durch rasterelektronenmikroskopische Aufnahmen bestätigt werden konnte.

3.3 Meßapparatur und Messung mit dem Zweistufenverfahren

Die kontinuierlich gravimetrische Messung des Flüssigkeitsablaufes am Meßkörper (Abb. 1) erfolgt mit einem induktiven Kraftaufnehmer der Firma Hottinger Baldwin Meßtechnik, Q 11 (Fehler: < ± 0.5 %). Mittels einer pneumatischen Hebebühne wird das mit der zu untersuchenden Flüssigkeit gefüllte temperierbare Gefäß gehoben, so daß der an dem Kraftaufnehmer hängende Meßzylinder fast vollständig in die Flüssigkeit eintaucht. Nach einer Kontaktzeit von 60 s wird das Gefäß innerhalb t_w = 2 s so weit gesenkt, daß der hierbei herausgezogene Meßzylinder mit seinem Rand noch 5 mm tief in die Flüssigkeit eintaucht. Unmittelbar danach erfolgt eine Schreiberaufzeichnung des Gewichts der Restflüssigkeit auf der Meßkörperoberfläche in Abhängigkeit von der Zeit t_d. Ein Gewichtsverlust durch Verdunstung des Flüssigkeitsfilms auf der Zylinderoberfläche wird durch eine den

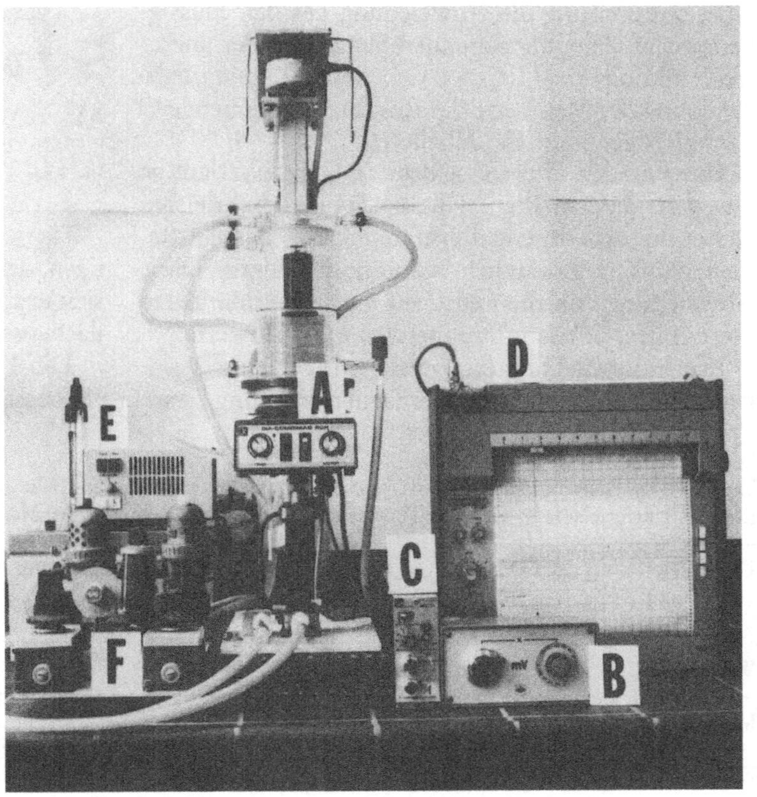

Abb. 1. Apparatur des gravimetrisch-kontinuierlichen Zweistufenverfahrens zur Messung des Flüssigkeitsablaufs an harten Oberflächen (A: pneumatische Hebebühne, temperierbares Meßgefäß mit Meßkörper, induktiver Kraftaufnehmer; B: Gegenspannungsgeber; C: Verstärker; D: Kompensationsschreiber; E: Thermostat; F: pneumatische Steuerung)

Meßzylinder umgebende wasserdampfgesättigte Atmosphäre verhindert.

Die beim Eintauchen des unteren Randes des Meßkörpers auftretenden Kräfte werden bei der Bestimmung des Gewichts der Restflüssigkeit auf der Meßkörperoberfläche berücksichtigt. Zu dem Gewicht des Meßkörpers kommt noch die Benetzungsspannung $j_R = \sigma \cdot \cos \theta_R$ (Oberflächenspannung σ und Rückzugsrandwinkle θ_R der Flüssigkeit) und der Auftrieb. Bei positivem j_R (an einem benetzbaren Körper) wird die Kraft erhöht, bei negativem j_R (an einem nicht benetzbaren Körper) verringert.

4. Ablauf reiner Flüssigkeiten an Glas

Durch Messung des Flüssigkeitsablaufs an Glas bei 25 °C mit den benetzenden reinen Flüssigkeiten Wasser, Methanol und Ethylenglykol, die unterschiedliche Viskosität, Dichte und Oberflächenspannung aufweisen (Tabelle 1), kann das Zweistufenverfahren überprüft werden. Hierfür wurde die den Flüssigkeitsablauf beschreibende theoretische Beziehung für die Restflüssigkeitsmenge (Gleichung 1) mit den experimentell ermittelten Restflüssigkeitsmengen in Abhängigkeit von $(t_w + t_d)^{-1/2}$ verglichen (Abb. 2).

Es liegt eine aureichende Übereinstimmung von Theorie und Experiment vor, wenn als Maß für die Übereinstimmung die Abweichung der den Flüssigkeitsablauf charakterisierenden Steigung S der lineraren Funktion $G = f((t_w + t_d)^{-1/2})$ und die Standardabweichung der Meßwerte der Restflüssigkeitsmenge G herangezogen werden (Tabelle 1).

Der von der Theorie geforderte lineare Verlauf ist bei den untersuchten Flüssigkeiten gewährleistet, wobei die experimentell ermittelten Restflüssigkeitsmengen G grundsätzlich etwas höher liegen. Diese Abweichung von theoretischen und experimentellen Werten liegt etwa im Streubereich der Meßwerte.

Die Untersuchungsergebnisse mit reinen Flüssigkeiten unterschiedlicher Oberflächenspannung bestä-

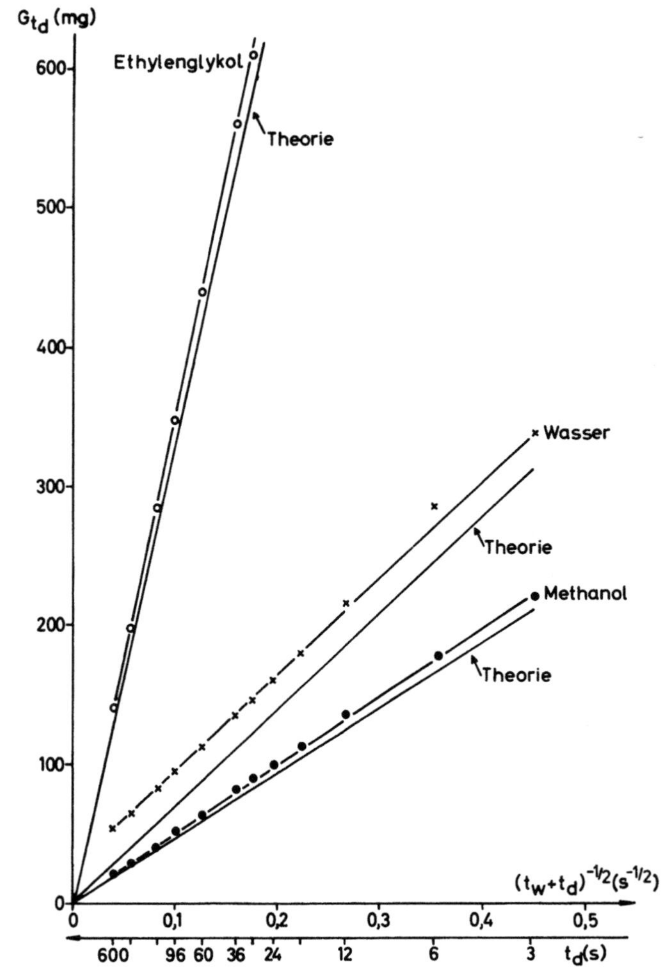

Abb. 2. Vergleich gemessener und berechneter Restflüssigkeitsmengen reiner Flüssigkeiten an Glas in Abhängigkeit von $(t_w + t_d)^{-1/2}$ bei 25 °C

tigen, daß von den Flüssigkeitsparametern nur Viskosität und Dichte den Ablauf bestimmen und die Oberflächenspannung keinen Einfluß ausübt.

Die Theorie geht von der formal-mathematischen Voraussetzung aus, daß bei $t_d \to \infty$ die Flüssigkeit voll-

Tabelle 1. Physikalische Daten der Flüssigkeiten, Abweichung der Steigung S der experimentellen linearen Beziehung $G_{t_d} = f((t_w + t_d)^{-1/2})$ des Flüssigkeitsablaufs an Glas zur Theorie und Standardabweichung s der Meßwerte G_{t_d} bei 25 °C

Flüssigkeit	Dichte (g/cm^3)	Viskosität (g/cm · s)	Oberflächenspannung (mN/m)	Abweichung von S zur Theorie (%)	s (± mg/Zylinderoberfläche)
Wasser	0,99705	0,0089	72,4	0	15
Methanol (99,9 %)	0,7867	0,00544	22	6	3
Ethylenglykol (99,7 %)	1,1096	0,173	47,3	5	5

ständig ablaufen kann. Physikalisch ist jedoch zu erwarten, daß eine benetzende Flüssigkeit bei Ausschaltung der Verdunstung nicht vollständig ablaufen kann und ein sehr dünner, die Oberfläche benetzender Flüssigkeitsfilm zurückbleibt. Beispielsweise besitzt Wasser eine besonders starke Adsorptionsaffinität an Glas, die zu einer immobilen Wasserschicht an der Glasoberfläche führt.

5. Ablauf wäßriger Tensidlösungen an Glas

Zum Verständnis des Ablaufverhaltens von Tensidlösungen war eingangs zu überprüfen, ob die Gesetzmäßigkeit über den Wasserablauf an Glas durch die gelösten Tenside Dodecylsulfat und Nonylphenoldecaglykolether noch ihre Gültigkeit besitzt.

Nach der Ablauftheorie, wobei von den Flüssigkeitsparametern nur Viskosität und Dichte den Ablauf bestimmen, ergeben Wasser und verdünnte wäßrige Tensidlösungen (10^{-5} bis 10^{-3} mol/1, [5]) Restflüssigkeitsmengen, die sich nur um den Faktor 1,015 unterscheiden. Die theoretischen Ablaufkurven $G = f((t_w + t_d)^{-1/2})$ von Wasser und den Tensidlösungen an Glas stimmen demnach überein, wenn man die Meßgenauigkeit der Meßmethode (Tabelle 1) berücksichtigt.

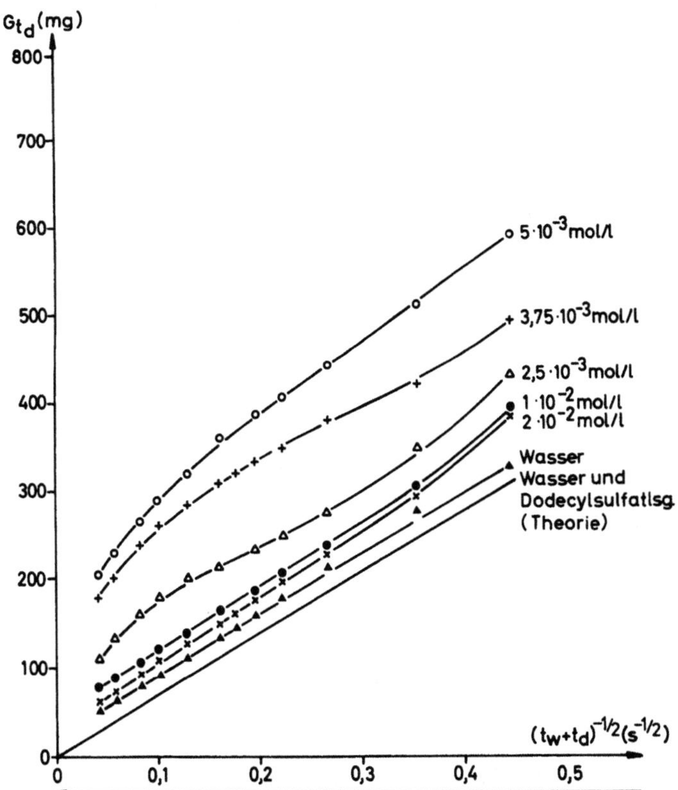

Abb. 3. Restflüssigkeitsmenge von Dodecylsulfatlösungen unterschiedlicher Konzentration an Glas in Abhängigkeit von $(t_w + t_d)^{-1/2}$ bei 25 °C

Die entsprechenden experimentell bestimmten Ablaufkurven der Tensidlösungen, wie am Beispiel von Dodecylsulfat gezeigt wird (Abb. 3), führen jedoch entgegen der Theorie in einem begrenzten Konzentrationsbereich zu stark erhöhten Restflüssigkeitsmengen. Bei Konzentrationen oberhalb der jeweiligen kritischen Mizellbildungskonzentration c_M eines Tensides nähern sich die Restflüssigkeitsmengen im gesamten t_d-Bereich denen der Theorie und damit denen von Wasser. Hinzu kommt, daß die von der Theorie geforderte Linearität der Funktion $G = f((t_w + t_d)^{-1/2})$ nur bei konzentrierteren mizellaren Lösungen annähernd erfüllt wird. Die „hydrophile" Ablaufcharakteristik an Glas ist sowohl bei den molekulardispersen als auch bei den mizellaren Tensidlösungen gewährleistet.

Es wurde somit gefunden, daß beim „hydrophilen" Ablauf der Anion- und Niotensidlösungen an Glas entgegen der Ablauftheorie die Restflüssigkeitsmengen im t_d-Bereich unterhalb der jeweiligen c_M eines Tensides durch ein Maximum verlaufen und sich darüber denen von reinem Wasser nähern (Abb. 4 und 5).

Zur Aufkärung des Tensideinflusses auf den Wasserablauf wurde für die experimentell ermittelte Zeitabhängigkeit der Restflüsssigkeitsmenge G_{t_d} von Wasser und den Tensidlösungen eine Näherungsgleichung (Gleichung 4) entwickelt, die die Meßwerte mit Korrelationskoeffizienten von etwa 0,99 beschreibt (Tab. 2). Die Konstanten A und B sind im Unterschied zu C von der Tensidkonzentration abhängig. Mit

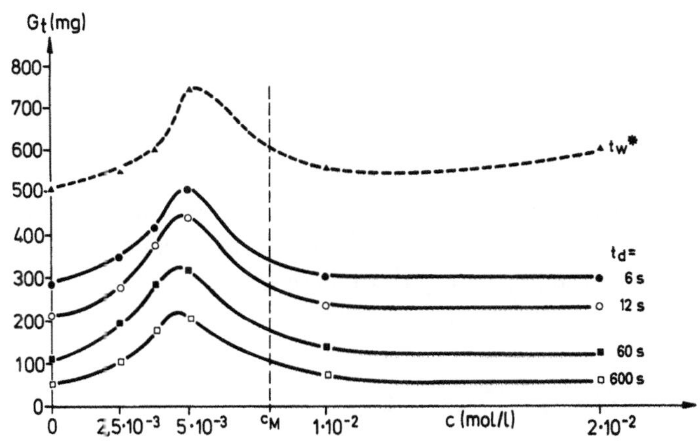

Abb. 4. Restflüssigkeitsmengen G_{t_w} und G_{t_d} von Dodecylsulfatlösungen an Glas in Abhängigkeit von der Konzentration bei 25 °C (* G_{t_w} wurde aus der Näherungsgleichung der gemessenen Ablaufkurve berechnet)

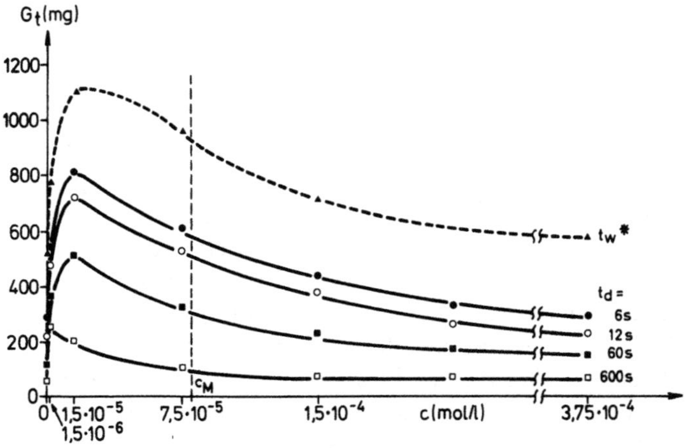

Abb. 5. Restflüssigkeitsmengen G_{t_w} und G_{t_d} von Nonylphenoldecaglykoletherlösungen an Glas in Abhängigkeit von der Konzentration bei 25 °C (* G_{t_w} wurde aus der Näherungsgleichung der gemessenen Ablaufkurve berechnet)

Tabelle 2: Die Konstanten A und B mit C = 23 der Näherungsgleichung für die Zeitabhängigkeit der Restflüssigkeitsmenge G_{t_d} von Dodecylsulfat- und Nonylphenoldecaglykoletherlösungen unterschiedlicher Konzentration c an Glas bei 25 °C und deren Korrelations-Koeffizienten r (Gleichung 4)

Tensid	c (mol/1)	A	B	r
Dodecylsulfat	0	687	0,4869	0,9982
	$2{,}5 \cdot 10^{-3}$	718	0,3500	0,9910
	$3{,}75 \cdot 10^{-3}$	683	0,2339	0,9968
	$4{,}38 \cdot 10^{-3}$	744	0,2646	0,9892
	$5 \cdot 10^{-3}$	891	0,2576	0,9984
	$1 \cdot 10^{-2}$	753	0,4450	0,9963
	$2 \cdot 10^{-2}$	846	0,4980	0,9989
Nonylphenol-decaglykolether	0	687	0,4869	0,9982
	$1{,}5 \cdot 10^{-6}$	863	0,2189	0,9941
	$1{,}5 \cdot 10^{-5}$	1267	0,2344	0,9882
	$7{,}5 \cdot 10^{-5}$	1198	0,3706	0,9825
	$1{,}5 \cdot 10^{-4}$	889	0,3859	0,9836
	$3{,}75 \cdot 10^{-4}$	708	0,3755	0,9950
	$7{,}5 \cdot 10^{-4}$	745	0,4196	0,9884
	$1{,}5 \cdot 10^{-3}$	802	0,4754	0,9924

Gleichung 4 kann durch Bildung von $-dG/dt_d$ bei konstantem t_w die Ablaufgeschwindigkeit zu bestimmten Zeiten t_d des Flüssigkeitsablaufs berechnet werden (Gleichung 5).

$$G = \frac{A}{(t_w + t_d)^B} + C \tag{4}$$

mit $C = 23$ für wäßrige Systeme

$$u = A \cdot B \cdot (t_w + t_d)^{-(B+1)} \tag{5}$$

Aus der Näherungsgleichung kann durch Einsetzen von $t_d = 0$ die unmittelbar nach dem Herausziehen des Meßkörpers erhaltene Restflüssigkeitsmenge G_{t_w} berechnet werden, die aus meßtechnischen Gründen experimentell nicht bestimmbar ist. Aus der Beziehung für die Ablaufgeschwindigkeit ist ebenso durch Einsetzen von $t_d = 0$ die Ablaufgeschwindigkeit u_{t_w} unmittelbar nach dem Herausziehen des Meßkörpers zu bestimmen.

Dei Ablaufgeschwindigkeiten der Tensidlösungen, wie am Beispiel von Dodecylsulfat gezeigt wird (Abb. 6), verlaufen bei t_w- und im t_d-Bereich in einem begrenzten Konzentrationsbereich unterhalb der jeweiligen kritischen Mizellbildungskonzentration des Tensides durch ein Minimum. Vergleichsweise stark ist hierbei die Verringerung der Ablaufgeschwindigkeit bei t_w, d. h. unmittelbar nach dem Herausziehen des Meßkörpers. Im anschließenden t_d-Bereich klingt die Verringerung der Ablaufgeschwindigkeit ab und die Tensidlösungen laufen dann annähernd so schnell ab wie reines Wasser.

Das Minimum der Ablaufgeschwindigkeit u_{t_w} und u_{t_d} liegt im gleichen Konzentrationsbereich unterhalb c_M wie das Maximum der Restflüssigkeitsmengen G_{t_w} und G_{t_d} (Abb. 4). Das bedeutet, daß Tenside in verdünnter Lösung die Ablaufgeschwindigkeit von Wasser verringern können und hierdurch die Restflüssigkeitsmengen erhöhen. Da die relative Verringerung der Ablaufgeschwindigkeit bei t_w besonders stark ist (Abb. 6), muß der Tensideinfluß durch einen Vorgang ausgelöst werden, der beim Herausziehen des Meßkörpers stärker ist als im anschließenden t_d-Bereich des Flüssigkeitsablaufs.

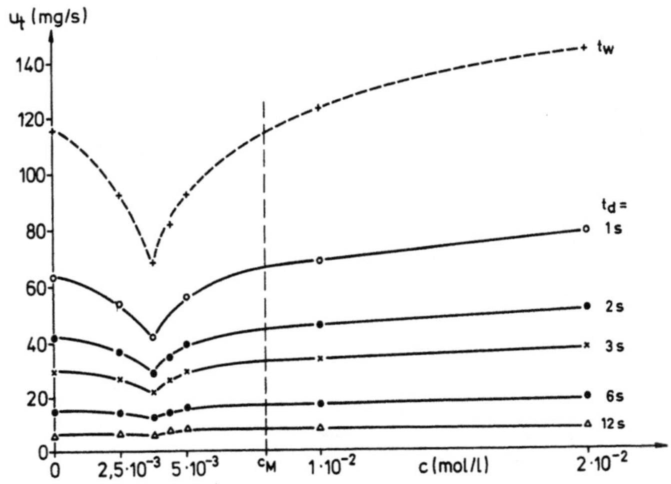

Abb. 6. Ablaufgeschwindigkeiten u_{t_w} und u_{t_d} von Dodecylsulfatlösungen an Glas in Abhängigkeit von der Konzentration bei 25 °C

Ein solcher beim Zweistufenverfahren auftretender Vorgang ist die Dehnung der monomolekularen Tensidschicht auf der Oberfläche des den Meßkörper benetzenden Flüssigkeitsfilms. Beim Herausziehen des Meßkörpers aus der Flüssigkeit, d. h. im t_w-Bereich wird die geometrische Oberfläche des Flüssigkeitsfilms und damit auch die Monoschicht gedehnt. Durch die Abwärtsströmung der Flüssigkeit im t_w-Bereich erfährt die Monoschicht nach dem Marangoni-Effekt [6] eine zusätzliche Dehnung. Durch die Abwärtsströmung bei konstanter Herausziehhöhe des Meßkörpers wird die Monoschicht auch im t_d-Bereich gedehnt. Folglich wird die Monoschicht im t_w-Bereich stärker gedehnt als im t_d-Bereich.

Nach der Theorie von Gibbbs [7] erzeugt eine örtlich begrenzte Dehnung der Monoschicht eines Flüssigkeitsfilms geringer Tensidkonzentration ein Oberflächenspannungsgefälle. Ein Oberflächenspannungsgefälle hat eine Rückstellkraft zur Wiederherstellung des ursprünglichen Zustandes von Monoschicht und Flüssigkeitsfilms zur Folge. Beim Flüssigkeitsablauf an vertikalen Oberflächen wirkt die Rückstellkraft entgegen der Abwärtsströmung der Flüssigkeit. Daraus resultiert eine Verringerung der Ablaufgeschwindigkeit von verdünnten Tensidlösungen, die im t_w-Bereich vergleichsweise stärker ist als im anschließenden t_d-Bereich.

In Übereinstimmung mit den Bedingungen der Gibbs'schen Theorie wird das Minimum der Ablaufgeschwindigkeit bei geringer Volumenkonzentration (Na-Dodecylsulfat: $5 \cdot 10^{-3}$ mol/1, Nonylphenoldecaglykolether: $1{,}5 \cdot 10^{-5}$ mol/1) und geringer, auf die geometrische Substratoberfläche bezogene Flüssigkeitsfilmdicke (Na-Dodecylsulfat: ≈ 33 μ, Nonylphenoldecaglykolether: ≈ 53 μ) beobachtet.

In Arbeiten von Stanley et. all [8] und Lange [9] wird bereits darauf hingewiesen, daß die Strömungsgeschwindigkeit von Dodecylsulfatlösungen beim Auslauf aus vertikalen Röhren und beim Durchlauf in geneigten Kapillaren unterhalb c_M verringert wird. Eine dynamische Erhöhung der Oberflächenspannung, verursacht durch eine vorübergehende Vergrößerung der Lösungsoberfläche, wird hierfür verantwortlich gemacht; eine Erklärung, die der Theorie von Gibbs entspricht.

6. Ablauf wäßriger Tensidlösungen an Kunststoffen

Das Ablaufverhalten von Tensidlösungen an Kunststoffen wurde mit Polyethylen, Polypropylen und Polyvinylchlorid untersucht, wobei durch Messung der Benetzungsspannung von Wasser an diesen Materialien eine geringere Hydrophobie von PVC gegenüber PE und PP festgestellt wurde (Abschnitt 3.2).

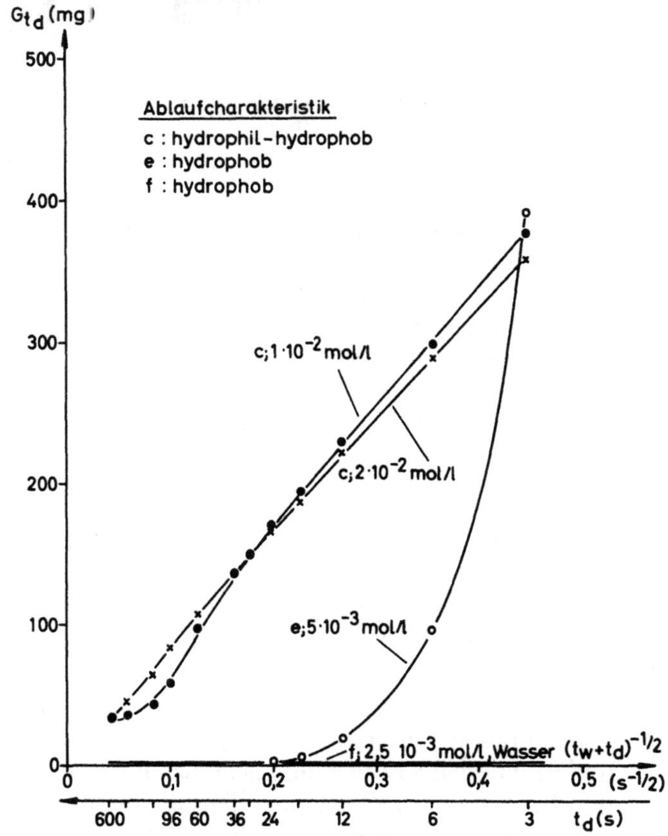

Abb. 7. Restflüssigkeitsmenge von Dodecylsulfatlösungen unterschiedlicher Konzentration an Polyethylen in Abhängigkeit von $(t_w + t_d)^{-1/2}$ bei 25 °C (Bezeichnung der Ablaufcharakteristiken c, e, f siehe Tabelle 3)

Am Beispiel von Dodecylsulfat werden die Ablaufkurven $G = f((t_w + t_d)^{-1/2})$ der wäßrigen Lösungen an PE diskutiert (Abb. 7). Wasser zeigt an PE und den anderen Kunststoffen eine „hydrophobe" Ablaufcharakteristik, d. h. Wasser zieht sich von der Kunststoffoberfläche so schnell zurück, daß bereits zu Beginn des t_d-Bereiches eine Restflüssigkeitsmenge von Null vorliegt. Mit steigender Tensidkonzentration werden Änderungen der Ablaufcharakteristik beobachtet, worauf die Bezeichnungen der Ablaufcharakteristiken f, e und c an den Kurven hinweisen (Tabelle 3). Ein allmählicher Übergang vom „hydrophoben" zum weitgehend „hydrophilen" Ablauf findet statt, zunehmende Benetzung der Oberfläche sowie wachsende Restflüssigkeitsmengen im gesamten t_d-Bereich sind die Folge.

Tabelle 3: Bezeichnungen der Ablaufcharakteristiken an harten Oberflächen

a: „hydrophil", Auslaufen des die Oberfläche benetzenden Films unter Verringerung der Filmdicke.

b: „hydrophil", schwaches Aufreißen des Films, ca. 80% der Oberfläche bleibt benetzt.

c: „hydrophil-hydrophob", starkes Aufreißen des Films, ca. 50% der Oberfläche bleibt benetzt.

d: „hydrophob", langsames unvollständiges Zurückziehen des Films von der Oberfläche unter Hinterlassen von die Oberfläche benetzenden Tropfen.

e: „hydrophob", langsames vollständiges Zurückziehen des Films von der Oberfläche.

f: „hydrophob", schnelles vollständiges Zurückziehen des Films von der Oberfläche.

Die experimentell ermittelten Restflüssigkeitsmengen der Dodecylsulfat- und Nonylphenoldecaglykolether-Lösungen an den ausgewählten Kunststoffen und an Glas wurden in Abhängigkeit von der Tensidkonzentration zusammengefaßt dargestellt (Abb. 8 und 9). Hiernach ist in einem begrenzten Konzentrationsbereich unterhalb c_M eines Tensides die Restflüssigkeitsmenge von der Tensidkonzentration und der Hydrophobie des Substrates abhängig. Die Hydrophobie der untersuchten Substrate von vergleichbarer Oberflächenrauheit nimmt ab in der Reihenfolge PE, PP > PVC > Glas (Abschnitt 3.2).

Im Konzentrationsbereich unterhalb von c_M eines Tensides werden mit steigender Konzentration bei stark hydrophoben Kunststoffen zunehmende Restflüssigkeitsmengen bis zu einem oberen Grenzwert und bei schwächer hydrophoben Kunststoffen sowie bei Glas Restflüssigkeitsmaxima beobachtet. Die Restflüssigkeitsmaxima verstärken sich dabei mit abnehmender Hydrophobie der Substrate.

Bei schwach hydrophobem PVC wird mit steigender Tensidkonzentration bereits unterhalb c_M die Substratoberfläche so stark mit einem benetzenden Flüssigkeitsfilm belegt, daß ein weitgehend „hydrophiler" Ablauf vorliegt. Wie beim „hydrophilen" Ablauf an Glas kann sich somit bei geringer Volumenkonzentration eines Tensides durch den Gibbs-Marangoni-Effekt ein Restflüssigkeitsmaximum ausbilden. Das Restflüssigkeitsmaximum an PVC ist gegenüber dem an Glas abgeschwächt, da der Umfang der Belegung der Substratoberfläche mit einem benetzenden Film beim PVC geringer ist.

Bei stark hydrophobem PE wird mit steigender Tensidkonzentration unterhalb c_M nur eine schwach zunehmende Belegung der Substratoberfläche mit einem benetzenden Film unter Erhöhung der Restflüssigkeitsmenge beobachtet, ohne einen weitgehend „hydrophilen" Ablauf zu erreichen.

Bei Konzentrationen oberhalb c_M eines Tensides, wenn die Tensidlösungen durch verstärkte Benetzung auch an stark hydrophoben Oberflächen weitgehend „hydrophil" ablaufen, liegen an Kunststoffen und Glas übereinstimmende Restflüssigkeitsmengen vor. Die Restflüssigkeitsmengen mizellarer Tensidlösungen an Kunststoffen nähern sich bei Konzentrationserhöhung denen an Glas und schließlich, wenn der Gibbs-Marangoni-Effekt bei höheren Konzentrationen unwirksam wird, den nach der Ablauftheorie gefor-

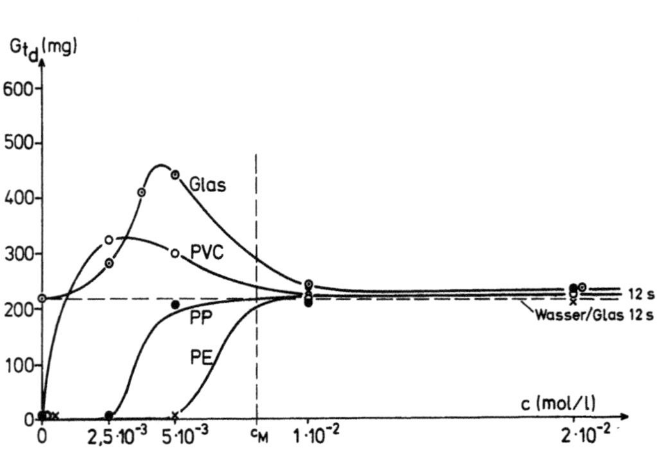

Abb. 8. Restflüssigkeitsmenge G_{t_d} von Dodecylsufatlösungen an unterschiedlichen Substraten in Abhängigkeit von der Konzentration bei 25 °C

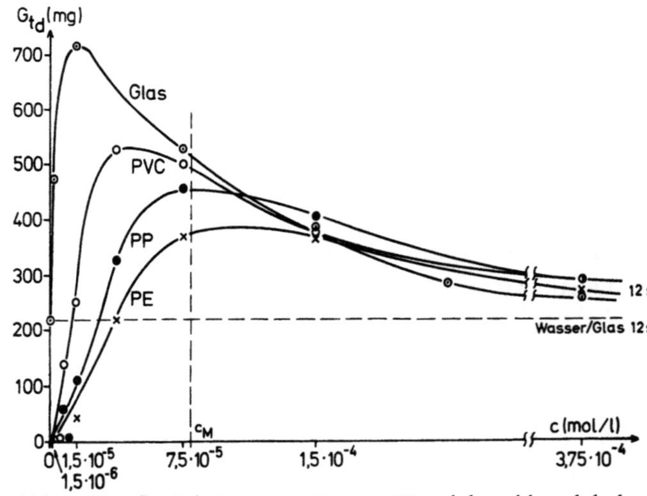

Abb. 9. Restflüssigkeitsmenge G_{t_d} von Nonylphenoldecaglykoletherlösungen an unterschiedlichen Substraten in Abhängigkeit von der Konzentration bei 25 °C

derten Restflüssigkeitsmengen. Hierbei können die Restflüssigkeitsmengen von Wasser an Glas annähernd erreicht werden, wenn sich Dichte und Viskosität von Wasser und Tensidlösung nur geringfügig unterscheiden.

Daraus ergibt sich, daß bei weitgehend „hydrophiler" Ablaufcharakteristik mizellarer Tensidlösungen an harten Oberflächen die Restflüssigkeitsmengen unabhängig von der Hydrophobie des Substrates sind.

7. Schlußbetrachtung

Zur Bestimmung des Ablaufverhaltens von Tensidlösungen an vertikalen harten Oberflächen wurde ein Zweistufenverfahren nach dem Tauchprinzip entwikkelt. Bei diesem Verfahren erfolgt wegen der geometrischen Vergrößerung der Flüssigkeitsfilmoberfläche beim Herausziehen des Meßkörpers und der Abwärtsströmung des Flüssigkeitsfilms an vertikaler Oberfläche eine Dehnung der monomolekularen Tensidschicht auf der Flüssigkeitsfilmoberfläche. Dadurch entsteht auf der Flüssigkeitsfilmoberfläche ein Oberflächenspannungsgradient. Das führt nach Gibbs-Marangoni in einem begrenzten Konzentrationsbereich unterhalb der kritischen Mizellbildungskonzentration c_M zur Verringerung der Ablaufgeschwindigkeit und damit im Vergleich zur Ablauftheorie zu erhöhten Restflüssigkeitsmengen auf dem Substrat.

Demnach kann die Ablaufgeschwindigkeit von Tensidlösungen durch den Oberflächenspannungsgradienten, der verfahrens- und tensidspezifisch ist, gesteuert werden. Die nach dem Zweistufenverfahren an harten Oberflächen sich einstellenden Restflüssigkeitsmengen werden durch eine Vielzahl von Parametern bestimmt.

Bei vollständiger Benetzung („hydrophiler" Ablauf) ist die Restflüssigkeitsmenge in Übereinstimmung mit der Ablauftheorie direkt proportional zur Viskosität und Dichte der Flüssigkeit und umgekehrt proportional zur Ablaufzeit. In einem begrenzten Konzentrationsbereich unterhalb c_M wird die Restflüssigkeitsmenge durch einen sich ausbildenden Oberflächenspannungsgradienten erhöht.

Bei unvollständiger Benetzung („hydrophober" Ablauf) wird die Restflüssigkeitsmenge durch den Umfang der Belegung der Substratoberfläche mit einem benetzenden Flüssigkeitsfilm bestimmt. Hierfür sind die sich wechselseitig bedingenden Parameter Benetzungsvermögen, Tensidkonzentration und Hydrophobie der Substratoberfläche verantwortlich. Die Restflüssigkeitsmenge wird an einer stark hydrophoben Oberfläche niedrig sein, wenn der Umfang der Belegung der Oberfläche mit einem benetzenden Film selbst bei höherer Tensidkonzentration nur gering ist. Eine schwach hydrophobe Oberfläche kann bereits bei niedriger Tensidkonzentration stark mit einem benetzenden Film belegt sein, so daß durch Oberflächenspannungsgradienten erhöhte Restflüssigkeitsmengen auftreten können.

Literatur

1. Lang IC, Tallmadge JA (1971) Ind Eng Chem Fundam 10:648
2. Nusselt W (1916) Zeitschrift des Vereins Deutscher Ingenieure, 27:541
3. Jeffreys H (1930) Proceedings of the Cambridge Philosophical Soc 26:204
4. Bikerman JJ (1956) J of Colloid Sci 11:299
5. Poskanzer AM, Goodrich FC (1975) J Phys Chem 79:20 S. 2122
6. Ewers WE, Sutherland KL (1952) Austral J Sci Res, A, 5:697 Sriven LE, Sternling CV (1960) Nature 187:186
7. Gibbs JW (1972) Collected Works, 1:269, 300 (Langmanns, Green, New York 1931), Lange H, VDI-Berichte 182:71
8. Stanley JS, Radley JA (1960) III. Internationaler Kongreß für Grenzflächenaktive Stoffe, Köln B/III/2, 68:464
9. Lange H (1954) Kolloid-Zeitschrift 136:136

Anschrift der Verfasser:

C.-P. Kurzendörfer
Henkel KGaA
4000 Düsseldorf

Progress in Colloid & Polymer Science Progr Colloid & Polymer Sci 69:154–158 (1984)

Röntgenkleinwinkeluntersuchungen an tensidbehandelten Faserkeratinen*)

M. Spei

Lehrstuhl für Makromolekulare Chemie an der RWTH Aachen und Deutsches Wollforschungsinstitut an der Technischen Hochschule Aachen

Zusammenfassung: Röntgenkleinwinkeluntersuchungen an tensidbehandelten Faserkeratinen haben zu folgenden Ergebnissen geführt:

1. Alle ionogenen Aminosäure-Seitenketten der Mikrofibrillen weisen eine axiale Periodizität von 39 Å auf. Entlang der Faserachse existieren Zonen mit vorwiegend ionogenen Aminosäure-Seitenketten gefolgt von Zonen mit vorwiegend hydrophoben Aminosäure-Seitenketten.
2. Die Einlagerung von anionischen Tensiden (n-Alkylsulfaten) erfolgt in zwei Stufen: Zuerst wird die Matrix durchdrungen (Verstärkung des axialen 28 Å Reflexes) und dann erst – bei höheren Tensidkonzentrationen und geringerem pH-Wert – die mikrofibrillären Bereiche.
3. In einigen Röntgenkleinwinkeldiagrammen wird der axiale 198 Å Reflex (Interferenz 1. Ordnung) beobachtet.

Abstract: Low-angle X-ray investigations of detergent-treated fibre keratins have yielded the following results:

1. All ionic amino acid residues of the microfibrils display a common axial repeat of 39 Å. It therefore seems that along the fibre axis of α-keratins there exist bands with preferably ionic amino acid residues followed by bands with hydrophobic amino acid residues.
2. The deposition of anionic detergents (n-alkylsulfates) occurs in two stages: First the matrix is penetrated and then -at higher detergent concentrations and lower pH-values – the microfibrillar regions.
3. In some low-angle X-ray patterns the axial 198 Å repeat (first order interference) is observed.

Schlüsselwörter: Röntgenkleinwinkeluntersuchungen, Faserkeratine, Tensideinlagerungsmechanismen, 28 Å Matrixreflex, 198 Å Reflex (Interferenz 1. Ordnung).

1. Einleitung

Keratine sind in der Natur weit verbreitet und bilden den Hauptbestandteil der Hornschicht der Epidermis und von Epidermisausstülpungen wie Wolle, Haare, Nägel, Klauen, Schuppen und Federn. Lichtmikroskopische, elektronenmikroskopische und Röntgenkleinwinkeluntersuchungen haben ergeben, daß die cystinarmen (teilhelikalen) Mikrofibrillen in einer cystinreichen (nichthelikalen) Matrix eingelagert sind. Zur weiteren Charakterisierung des Mikrofibrillen-Matrix-Komplexes haben wir die Röntgenkleinwinkeluntersuchungen auf chemisch modifizierte Faserkeratine ausgedehnt. Die meisten Untersuchungen wurden an (südafrikanischem) Mohair vorgenommen, das im unbehandelten – bzw. schonend modifizierten – Zustand neben Stachelschweinkielen das schärfste Röntgenkleinwinkeldiagramm aller Faserkeratine liefert. Die zahlreichen meridionalen und nahmeridionalen Röntgenkleinwinkelreflexe lassen sich in erster Näherung als höhere Ordnungen (Untereinheiten) einer axialen Periodizität indizieren [1, 2] (vgl. Abb. 1).

*) Vortrag auf der 31. Hauptversammlung der Kolloidgesellschaft e. V. in Bayreuth, 11. bis 14. Oktober 1983

KT-W 826

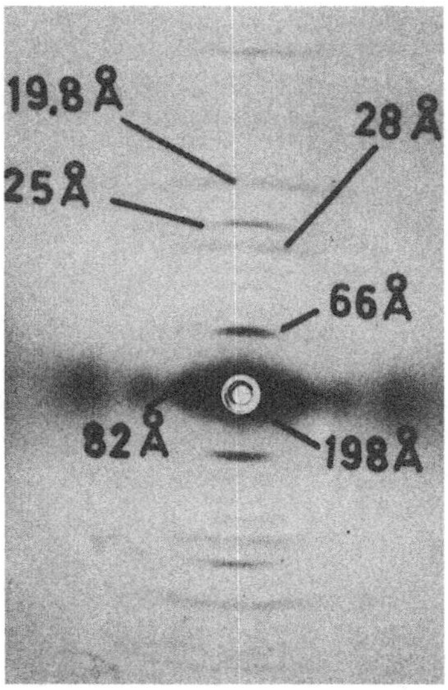

Abb. 1. Röntgenkleinwinkeldiagramm einer Mohairprobe mit schwachem 198 Å Reflex (Interferenz 1. Ordnung): Die 3., 7., 8. und 10. Ordnung bei 66 Å, 28 Å, 25 Å und 19,8 Å sowie der innerste Äquatorreflex bei 82 Å sind markiert

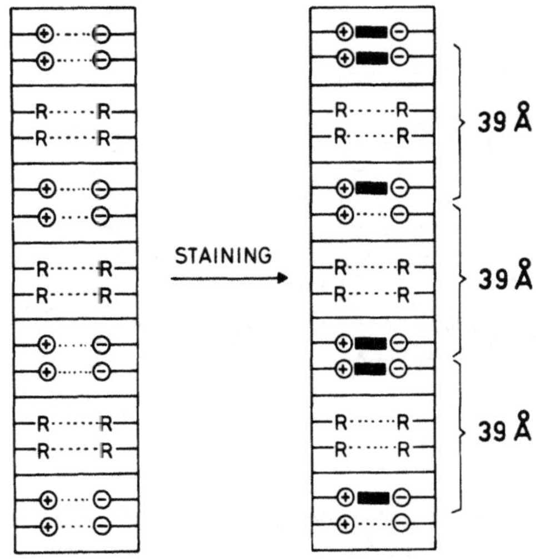

Abb. 3. Erklärung für die Verstärkung des meridionalen 39 Å Reflexes

Einige dieser meridionalen Kleinwinkelreflexe werden nach gezielten chemischen Modifizierungen beträchtlich verstärkt und geben somit zusätzliche Auskunft über den Feinbau der Mikrofibrillen-Matrix-Verbundstruktur. Als besonders geeignet erwiesen sich tensidbehandelte Faserkeratine. Hierüber soll im Folgenden kurz berichtet werden.

2. Die Verteilung der ionogenen Aminosäurereste entlang der Hauptkette

Bei der Einlagerung von Natrium-Alkylsulfaten ($C_6 - C_{12}$) in Faserkeratine – im vorliegenden Fall Natrium-Decylsulfat in Mohairfasern – werden unabhängig von der Kettenlänge des Tensids zwei Meridianreflexe bei 39 Å und bei 28 Å beträchtlich verstärkt [3] (vgl. Abb. 2). Die Verstärkung des meridionalen 39 Å Reflexes wurde von uns bereits früher nach der Schweratommarkierung der Lysinreste beobachtet und als eine regelmäßige Verteilung der Lysinreste entlang der Hauptkette gedeutet [4, 5]. Später zeigte sich dann, daß auch nach der Carboxylgruppenmarkierung eine Verstärkung des 39 Å Reflexes auftritt [6]. Bei der Umsetzung von Faserkeratinen mit Alkylsulfaten werden die Tensidmoleküle im ersten Schritt von den ionogenen Lysin- und Argininresten salzartig gebunden. Um zu sehen ob die enorme Verstärkung des 39 Å Reflexes allein von der Lysinmarkierung herrührte oder auch die Argininreste daran beteiligt waren, wurde auch Dnp – Mohair mit blockierten Lysinresten mit Natrium-Decylsufat umgesetzt. Auch in diesem Fall wurde der 39 Å Reflex verstärkt [3]. Alle ionogenen Seitenketten weisen also eine gemeinsame axiale Periodizität von 39 Å auf. Vermutlich existieren in Faserkeratinen ähnlich wie bei Kollagen – wenn auch bei weitem nicht so ausgeprägt – Zonen mit vorwiegend ionogenen Aminosäureresten gefolgt von Zonen mit hydrophoben Aminosäureresten [7]. Auf diese Weise ist im nassen Zustand eine optimale Stabilisierung der nativen Faserproteinstruktur durch Salzbrücken und hydrophobe Wechselwirkungen gewährleistet. Ein entsprechender Modellvorschlag ist in Abbildung 3 wiedergegeben [8]. Hierbei ist es nun völlig irrelevant, welcher ionogene Rest reagiert hat. Entscheidend ist lediglich die Tatsache, daß die Elek-

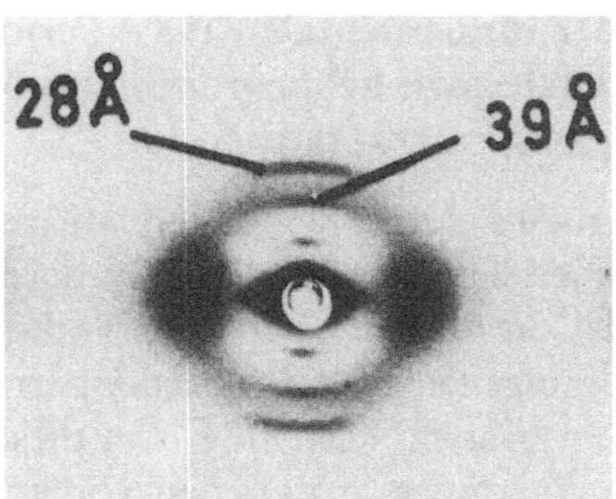

Abb. 2. Röntgenkleinwinkeldiagramm einer mit Natrium-Decylsulfat behandelten Mohairprobe (0,03n Lösung von pH 2)

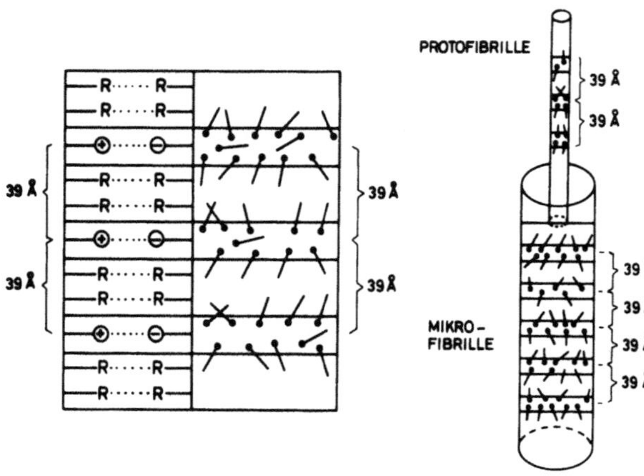

Abb. 4. Schematische Darstellung der Wechselwirkung von anionischen Tensidmolekülen mit den mikrofibrillären Bereichen von Faserkeratinen

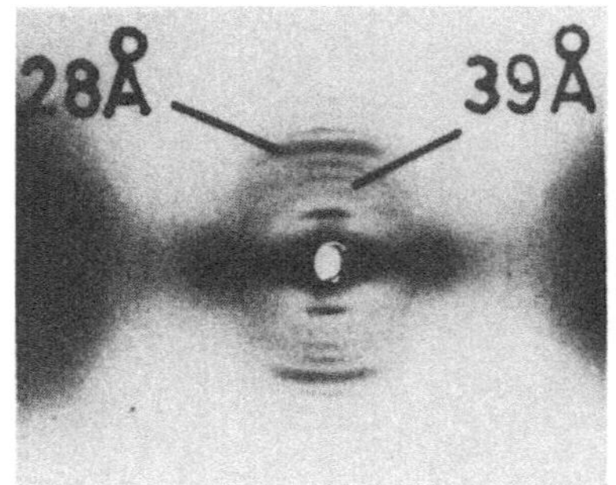

Abb. 5. Röntgenkleinwinkeldiagramm einer mit Natrium-Decylsulfat behandelten Mohairprobe (0,01 n Lösung von pH 4)

tronendichte in den hydrophilen Zonen zugenommen hat. Ein erweitertes und verfeinertes Zonenmodell, das die Wechselwirkung der Natrium-Alkylsulfatmoleküle mit den mikrofibrillären und protofibrillären Proteinen berücksichtigt, ist in Abbildung 4 wiedergegeben. Die negativen Sulfatreste der Tensidmoleküle reagieren mit den positiven Lysin- und Argininseitenketten in den ionogenen Zonen, während es zwischen den hydrophoben Alkylresten und den hydrophoben Zonen der Mikrofibrillen zur Ausbildung von hydrophoben Wechselwirkungen kommt [9]. In der Zwischenzeit ist die 39 Å Periodizität als Wiederholungseinheit für die ionogenen Aminosäurereste auf völlig anderem Weg bestätigt worden Parry et al. [10] berechneten die Wechselwirkungsmöglichkeiten von zwei schwefelarmen (helikalen) mikrofibrillären Wollproteinen, fanden ebenfalls bei 40 Å eine Häufung von ionogenen Aminosäureresten und bestätigten in ihrer Arbeit ausdrücklich die Ergebnisse der röntgenographischen Untersuchungen.

3. Nachweis von zwei geordneten Bereichen entlang der Faserachse

Weitere Untersuchungen der Einlagerung von Natrium-Decylsulfat in Mohairfasern zeigten, daß die Einlagerung in zwei Stufen verläuft. Unter milden Versuchsbedingungen (geringe Tensidkonzentration und relativ hoher pH-Wert) wird zuerst der 28 Å Reflex verstärkt (Abb. 5); mit steigender Tensidkonzentration und Erniedrigung des pH-Wertes erfolgt dann erst die enorme Verstärkung des 39 Å Reflexes (vgl. noch einmal Abb. 2). Unter Berücksichtigung der Ausführungen des vorangehenden Kapitels und der Tatsache, daß sich die Lysinreste des Mikrofibrillen-Matrix-Komplexes fast ausschließlich in den Mikrofibrillenproteinen befinden, kann man davon ausgehen, daß der 39 Å Reflex eine mikrofibrilläre Wiederholungseinheit darstellt. Tensideinlagerungstudien in partiell reduzierte bzw. oxidierte Faserkeratine, bei denen ein Teil der Cystinbrücken in der Matrix zerstört worden war, ergaben eine Abschwächung des 28 Å Reflexes sowie eine weitere Zunahme der Intensität des 39 Å Reflexes [3, 11]. Hieraus und aus weiteren (Dehnungs-) Untersuchungen [12] wurde geschlossen, daß der 28 Å Reflex ein „Matrixreflex" ist. Die Tensidmoleküle durchdringen also zuerst die Matrix (Abb. 6a) und dann erst die mikrofibrillären Bereiche (Abb. 6b). Entlang der Faserachse müssen also zwei geordnete Bereiche existieren, und nicht mehr alle axialen Kleinwinkelreflexe können zwanglos als mikrofibrilläre Wiederholungseinheiten angesehen werden.

4. Nachweis der fundamentalen 198 Å Periodizität in den Röntgenkleinwinkeldiagrammen von tensidbehandelten Faserkeratinen

Röntgenkleinwinkeluntersuchungen an verschiedenen α-Keratinen hatten bereits 1943 zu der Auffassung geführt, daß die meridionalen Kleinwinkelreflexe als höhere Ordnungen einer 198 Å Pseudoperiodizität aufgefaßt werden können [1, 2]. Ein Reflex dieser Größenordnung ist von uns erstmals in den Röntgenkleinwinkeldiagrammen von einigen lösungsmittelbehandelten Mohairproben beobachtet worden [13] (vgl. Abb. 1). In Zusammenhang mit Untersuchungen über die Einlagerung von Tensiden in unbehandelte und

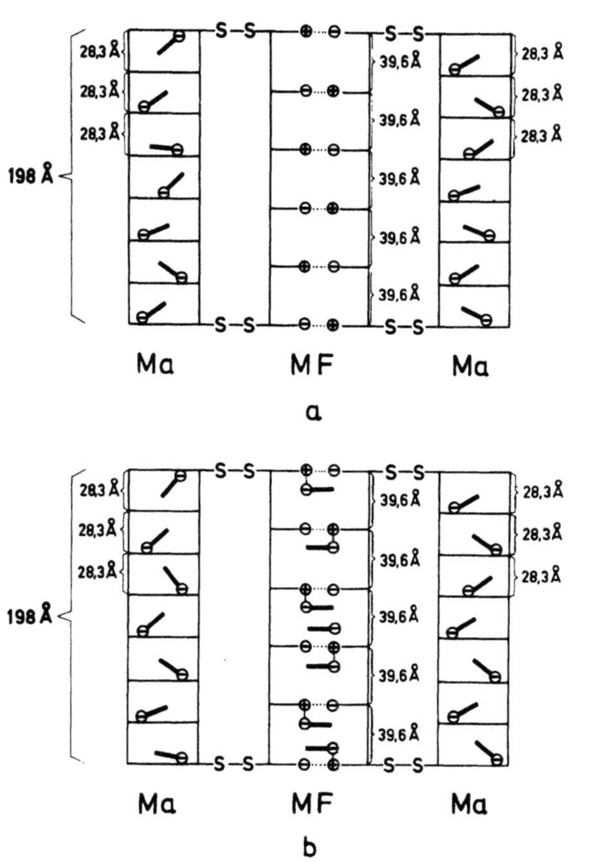

Abb. 6. Schematische Darstellung des Faserkeratinaufbaus aus zwei geordneten Bereichen

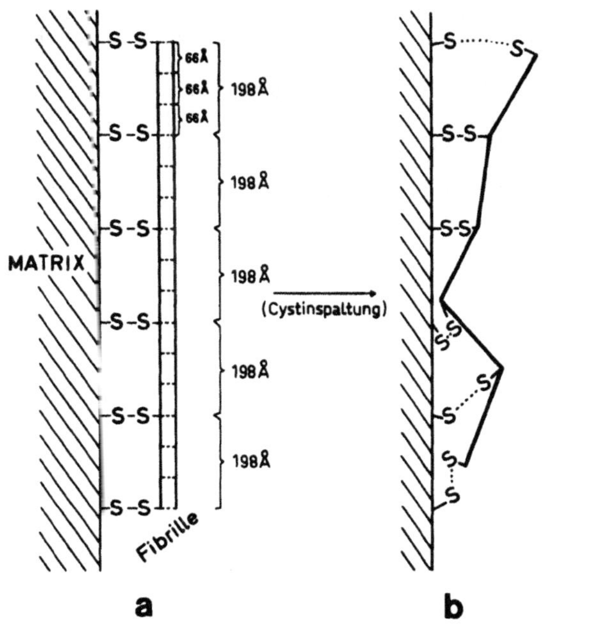

Abb. 8. Chemische Begründung für eine statistische Knickperiode von 198 Å durch eine teilweise, unregelmäßige Spaltung der Cystinbrücken zwischen den Mikrofibrillen und der Matrix

chemisch modifizierte α-Keratine beobachteten wir in einigen Röntgenkleinwinkeldiagrammen den 198 Å Reflex und seine 2. Ordnung [14, 15]; (vgl. Abb. 7) In Zusammenarbeit mit Bonart [16] konnte das äußerst überraschende und sporadische Auftreten dieses lang gesuchten Reflexes durch eine teilweise und unregelmäßig verteilte Spaltung der Cystinbrücken zwischen den Mikrofibrillen und der Matrix erklärt werden (Abb. 8).

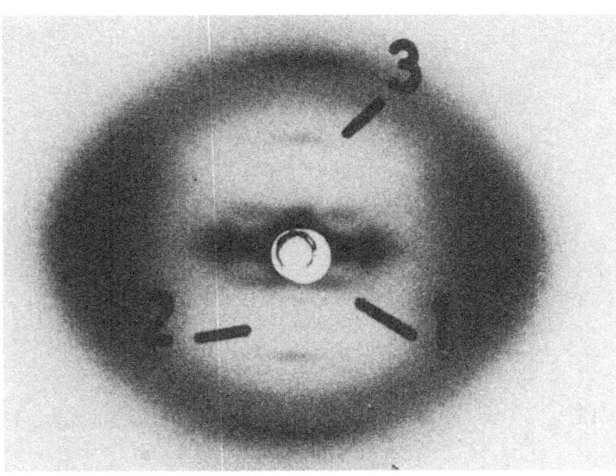

Abb. 7. Röntgenkleinwinkeldiagramm einer nitrierten Mohairprobe, die mit Natrium-dodecylsulfat nachbehandelt wurde. Die 1., 2. und 3. Ordnungen sind markiert

Danksagung

Herrn Prof. Dr.-Ing. Dres. h. c. H. Zahn danke ich für seine stete Diskussionsbereitschaft und der Deutschen Forschungsgemeinschaft für die finanzielle Unterstützung meiner Arbeiten.

Literatur

1. Mac Arthur J (1943) Nature 152:38
2. Bear RS (1943) J Amer Chem Soc 65:1784
3. Spei M, Stein W, Zahn H (1970) Kolloid Z u Z Polymere 238:447
4. Heidemann G, Halboth H (1967) Nature 213:71
5. Spei M, Heidemann G, Halboth H (1968) Nature 217:247
6. Spei M (1970) Kolloid Z u Z Polymere 238:436
7. Spei M (1973) Habilitationsschrift, RWTH Aachen, in „Forschungsberichte des Landes Nordrhein-Westfalen", Westdeutscher Verlag 1975, Nr. 2455
8. Spei M, Proc 6th Int. Wool Text Res Conf, Vol II, p 263, Pretoria 1980
9. Spei M (1983) Colloid and Polymer Sci 261:965
10. Parry DAD, Crewther WG, Fraser RDB, MacRae TP (1977) J Mol Biol 113:449

11. Spei M (1976) Proc 5th Int Wool Text Res Conf, Aachen 1975, Schriftenreihe Deutsches Wollforschungsinstitut an der Technischen Hochschule Aachen, Vol II, p 90
12. Spei M, Zahn H (1971) Monatsh Chem 102:1163
13. Spei M, Heidemann G, Zahn H (1968) Naturwiss 55:346
14. Spei M (1971) Appl Polymer Symposia 18:659
15. Spei M (1972) Kolloid Z u Z Polymere 250:207
16. Bonart R, Spei M (1972) Kolloid Z u Z Polymere 250:385

Anschrift des Verfassers:

Dr. M. Spei
Lehrstuhl für Makromolekulare Chemie
an der RWTH Aachen
D-5100 Aachen

Progress in Colloid & Polymer Science Progr Colloid & Polymer Sci 69:159–166 (1984)

Cosorption von p-Hydroxybenzoesäureestern (Parabene) mit Nonylphenol-Polyglykolen an porösem SiO_2 aus Wasser*)

H. Rupprecht und R. Daniels

Lehrstuhl für Pharmazeutische Technologie, Institut für Pharmazie an der Universität Regensburg, Regensburg

Zusammenfassung: p-Hydroxybenzoesäureester (Parabene) werden aus wässrigen Lösungen von SiO_2-Oberflächen nicht nennenswert adsorbiert. In Gegenwart von Nonylphenol-Polyethylenglykolen findet jedoch eine Cosorption statt, die allerdings bei porösem SiO_2 von engen Poren (< 10 nm) behindert wird. Die Bindung der Parabene nimmt dabei mit der Länge der Alkylkette ihrer Alkoholkomponente zu; kurze PEG-Reste im Tensid führen ebenfalls zu hohen Bindungswerten. Aus den Adsorptionsisothermen und der mittels Tail-Analyse erstellten Desorptionskinetik ergeben sich Hinweise auf den Mechanismus der Cosorption.

Abstract: Although p-hydroxybenzoic acid esters (parabens) were found not to be adsorbed onto silica surfaces from aqueous solutions, in the presence of nonylphenol-polyoxyethylene surfactants co-adsorption was observed. However, this effect was found to be restrained by the narrow pores present in the silica supports (< 10 nm pore diameter). The parabens obviously displaced surfactant molecules in the adsorption layers, giving a ratio of paraben to the surfactant molecules between 1 : 10 and 1 : 36 in the co-adsorbates. The total amount of coadsorbed paraben could be increased by increasing the alkyl-chain length of the alcohol residue in the ester or by combination with nonyl-phenol-surfactants with short PEG residues. The co-adsorption mechanism is discussed here on the basis of the adsorption and desorption kinetics of the co-adsorbates as obtained by tail-analysis.

Key words: Cosorption, SiO_2, Parabene, Nonylphenol-PEG.

1. Einleitung

Wasserhaltige Arzneimittel wie z. B. Suspensionen, Emulsionen oder Hydrogele enthalten neben den therapeutisch aktiven Wirkstoffen eine Reihe von Hilfsstoffen, welche die Arzneimittelqualität hinsichtlich Wirksamkeit, chemischer, physikalischer und antimikrobieller Stabilität garantieren. Hierbei kommen sowohl Konservierungsmittel als auch organische und anorganische Makromoleküle z. B. als viskositätsregulierende Hilfsstoffe oder als Trägermaterialien für Wirkstoffe – zum Einsatz [1, 2]. Da ferner Tenside sehr häufig zur Benetzung hydrophober Feststoffe oder als Emulgatoren verwendet werden, sind in solchen Arzneimitteln vielfältige wechselseitige Reaktionsmöglichkeiten gegeben [3, 4].

Voraussetzung für eine sinnvolle Kombination der oben erwähnten Hilfsstoffe ist die Kenntnis ihrer möglichen gegenseitigen Wechselwirkungen, die physikalisch-chemisch vor allem von Grenzflächenreaktionen bestimmt werden. Vorliegende Untersuchung befaßt sich daher mit Fragen der Adsorption und der Coadsorption von Hilfsstoffkombinationen an Oberflächen von SiO_2-Präparaten in wässrigen Dispersionen, wobei das Sorptionsverhalten der gelösten Substanzen sowohl durch Erstellung von Adsorptionsisothermen als auch durch die Tail-Analyse an entsprechenden Sorbaten während der Desorption ermittelt wird [5].

Die in die Untersuchungen einbezogen SiO_2-Präparate weisen eine definierte Porenstruktur auf (Tab. 1) und kommen daher als Trägermaterialien für Wirkstoffe zur Steuerung der Wirkstoffliberation in Frage

*) Ergänzter Text eines Vortrags auf der 31. Hauptversammlung der Kolloidgesellschaft in Bayreuth 1983

KT-W 802

Tabelle 1. Untersuchte SiO_2-Präparate

Handelsbezeichnung	spez. Oberfläche (BET-N_2) [m^2/g]	mittlerer Porendurchmesser [nm]	mittlerer Teilchengröße [μm]
Spherosil Typ A	200	< 10	100 - 200
Spherosil Typ C	95	20 - 40	100 - 200
Versuchsmuster KR 36 aus Polyethoxysiloxan kondensiert	510	14	200

[6]. Ihr Sorptionsverhalten gibt darüber hinaus Hinweise auf das Bindungsvermögen kolloider SiO_2-Produkte, die vorwiegend zur Viskositätsregulierung herangezogen werden [7]. Bei den Konservierungsmitteln wurden homologe Alkylester der p-Hydroxybenzoesäure ausgewählt, während als Modelle für nichtionische PEG-Tenside Nonylphenol-Polyethylenglykole (Nonylphenol-PEG) mit unterschiedlichen PEG-Kettenlängen eingesetzt wurden (Tab. 2).

2. Experimenteller Teil

2.1 Verwendete Substanzen:

SiO_2-Präparate (Tab. 1):

a) Spherosile: Kugelförmige Teilchen, hergestellt durch Fällung aus Kieselsäure-solen und anschließende Kalzination [9-11]; (Fa. Serva, Heidelberg)

b) Versuchsmuster Kr 36: Kugelförmige Teilchen, hergestellt durch hydrolytische Polykondensation von teilpolymerisiertem Polyethoxysiloxan [12, 13]

Nonylphenol-Polyethylenglykole: Handelsbezeichnung, „Marlophen 8 . .“ (Chemische Werke Hüls, Marl) (Tab. 2).

p-Hydroxybenzoesäureester (Parabene)

Handelsbezeichnung: Nipagin bzw. Nipasol; (Nipa Laboratories Ltd., London GB) (Tab. 2).

Lösungsmittel: Wasser, entspr. Aqua purificata Pharm. Europ. II. [14] Methanol, p. A. (Fa. Merck, Darmstadt).
Alle Substanzen wurden ohne weitere Reinigung verwendet.

2.2 Quantitative Bestimmungsmethoden:

HPLC-Bestimmung:

Geräte: Pumpe – Series 2 Liquid Chromatograph, (Perkin Elmer); Einspritzventil – Rheodyne 7010 mit 10 μl Probenschleife; Detektor – Uvicon LCD mit variabler Wellenlänge (Kontron); Integrator – Chromatography Data Station Sigma 10, (Perkin Elmer)

Bedingungen:

Mobile Phase: Methanol 85 Teile, Wasser 15 Teile; *Fluss:* 1.3 ml/min; *Säule:* RP-18, 5 μm, Länge 25 cm; *Säulenvordruck:* 190 bar; *Detektion:* UV 277 nm.

Die Berechnung der Konzentration erfolgt durch Umrechnung aus den integrierten Peakflächen mittels aus Eichgeraden bestimmten Eichfaktoren.

Spektralphotometrische Bestimmung:

Geräte: Spektralphotometer UV 210 A, (Shimadzu); Wavelength Programmer WP – 1, (Shimadzu); Cellpositioner CP 200 A, (Shimadzu); Küvetten 1 mm Quarz Durchflußküvetten; Referenz: Wasser, (Aqua purificata).

Tabelle 2. Tenside und Parabene

Handelsbezeichnung	chem. Bezeichnung	mittlere Molmasse	CMC (8) [mmol/l]
Marlophen 88	Nonylphenol-PEG-(8,5)	590	0,08
Marlophen 89	Nonylphenol-PEG-(9,4)	630	0,09
Marlophen 812	Nonylphenol-PEG-(12,7)	780	0,10
Marlophen 820	Nonylphenol-PEG-(21,2)	1160	0,17
Marlophen 850	Nonylphenol-PEG-(50)	2470	0,86

Handelsbezeichnung	chem. Bezeichnung	Struktur	R
Nipagin M	Methylparaben	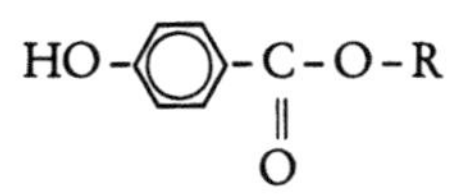	R = $-CH_3$
Nipagin A	Ethylparaben		R = $-CH_2-CH_3$
Nipasol M	Propylparaben		R = $-(CH_2)_2-CH_3$

Messprinzip:

Die simultane Bestimmung von Paraben und Nonylphenol-Polyethylenglykolen beruht auf den unterschiedlichen UV-Spektren der beiden Substanzen. Durch Messung der Extinktionen bei 255 nm und 276 nm lassen sich die Konzentrationen der beiden Substanzen entsprechend dem Lambert-Beer-Gesetz mittels der aus Eichmessungen gewonnenen Umrechnungsfaktoren berechnen [15].

Co-Adsorptionsuntersuchungen:

In 25 ml Rollrandgläschen werden 0.200 g Kieselgel eingewogen, 20.00 ml Tensid-Paraben-Lösung zugegeben und die dichtverschlossenen Gefäße 16 h bei 21 ± 0.5 °C geschüttelt. Anschließend werden im Überstand die Gleichgewichtskonzentrationen von Nonylphenol-PEG und Paraben quantitativ mittels HPLC bestimmt. Aus der Differenz von Ansatz- und Gleichgewichtskonzentration wird die jeweils sorbierte Menge berechnet.

Tail-Analysen (5):

Durch eine mit ca. 0.7 g porösem SiO_2 gepackte Glassäule, Länge 400 mm, Durchmesser (innen) 2,0 mm, werden bis zur Einstellung eines dynamischen Gleichgewichts (d. h. Übereinstimmung der Konzentrationen der gelösten Substanzen in der Lösung vor und nach Passage der Säule) wässrige Tensid-Paraben-Lösungen gepumpt. Anschließend wird die so beladene Säule mit Wasser gespült und die sich dabei ergebenden Elutionskurven quantitativ verfolgt. Die Gehaltsbestimmung im Eluat erfolgt entweder spektralphotometrisch oder hochdruckflüssigchromatographisch.

Bestimmung der Sättigungslöslichkeit von Parabenen

In 25 ml Rollrandgläschen werden jeweils 1 bzw. 2 g Paraben eingewogen, 20 ml Wasser oder Tensidlösung entsprechender Konzentration zugegeben und die dichtverschlossenen Gefäße mindestens 24 h bei konstanter (± 0,2 K) Temperatur geschüttelt. Anschließend wird der ungelöste Anteil durch Filtration, Membranfilter (Zelluloseacetat 0,45 µm Porenweite) abgetrennt und im klaren bzw. opaleszierenden Überstand nach Verdünnen 1 : 10 mit Wasser die Konzentration an gelöstem Paraben mittels HPLC bestimmt.

3. Ergebnisse und Diskussion

3.1 Adsorptionsthermen

Die in die Untersuchungen einbezogenen Paraben-Homologen zeigen selbst aus gesättigten wässrigen Lösungen keinerlei nachweisbare Sorption an die Oberflächen der SiO_2-Produkte[1]).

Im Gegensatz dazu adsorbieren die Nonylphenol-PEG sehr stark an Kieselsäureoberflächen und bilden

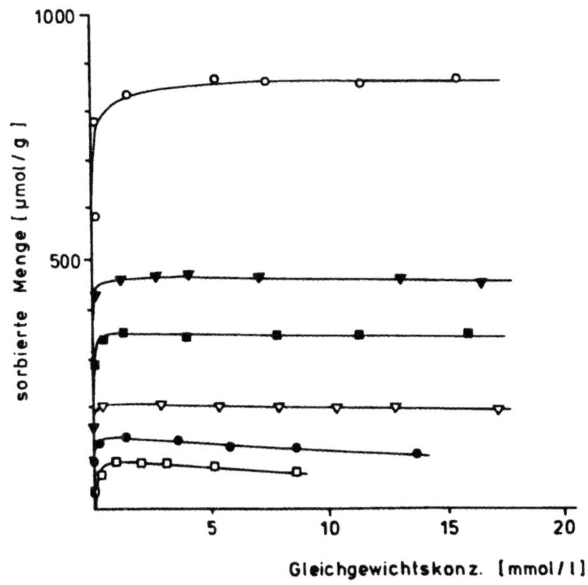

Abb. 1. Adsorption von Nonylphenol-Polyethylenglykolen an SiO_2-Oberflächen aus Wasser, 21 °C. Sorbens: Aerosil 200 (nach 16), PEG-Kettenlänge im Tensid

○——○	8,5	▽——▽	21,5
▼——▼	12,7	●——●	30,0
■——■	14,8	□——□	50,0

dort bereits im Bereich ihrer jeweiligen kritischen Mizellbildungskonzentration dicht gepackte Sorptionsschichten aus [16]. Die sorbierten Mengen in µmol/g nehmen dabei, entsprechend dem Platzbedarf im Sorbat, mit zunehmender PEG-Kettenlänge des Tensides ab (Abb. 1).

Eine deutliche Adsorption der Parabene aus wässriger Lösung an SiO_2-Grenzflächen tritt jedoch dann auf, wenn sie zusammen mit gelösten Nonylphenol-PEG and die Grenzfläche des SiO_2 gelangen (Abb. 2): Welche Bedeutung der Struktur der SiO_2-Träger auf diese Cosorption zukommt, geht aus den Sorptionsisothermen von Methylparaben und Nonylphenol-PEG 9.4 bei unterschiedlichen mittleren Porengrößen der Sorbentien deutlich hervor: Ausgehend von einer Konzentration an Methylparaben von jeweils 1 mmol/l findet bereits bei niedrigen Konzentrationen an Tensid eine deutliche Cosorption statt. Sättigungswerte werden sowohl am engporigen Material (Porendurchmesser < 10 nm) mit 0,068 µmol/m^2 als auch bei größeren mittleren Poren von 30 nm mit 0,189 µmol/m^2 im selben Bereich der Gleichgewichtskonzentration erreicht. Allerdings beobachtet man am weitporigen SiO_2-Träger bei niedrigen Gleichgewichtskonzentrationen an Tensid ein deutliches Maximum der Parabencosorption.

---

[1]) Bei den gewählten Ansatzverhältnissen für die Erstellung der Adsorptionsisothermen lag die Erfassungsgrenze für die Bestimmung der adsorbierten Mengen – bedingt durch das Differenzverfahren bei der Bestimmung und die hohen Ausgangskonzentrationen – bei etwa $1 \cdot 10^{-5}$ mol/l entsprechend ca. $1 \cdot 10^{-6}$ mol/g.

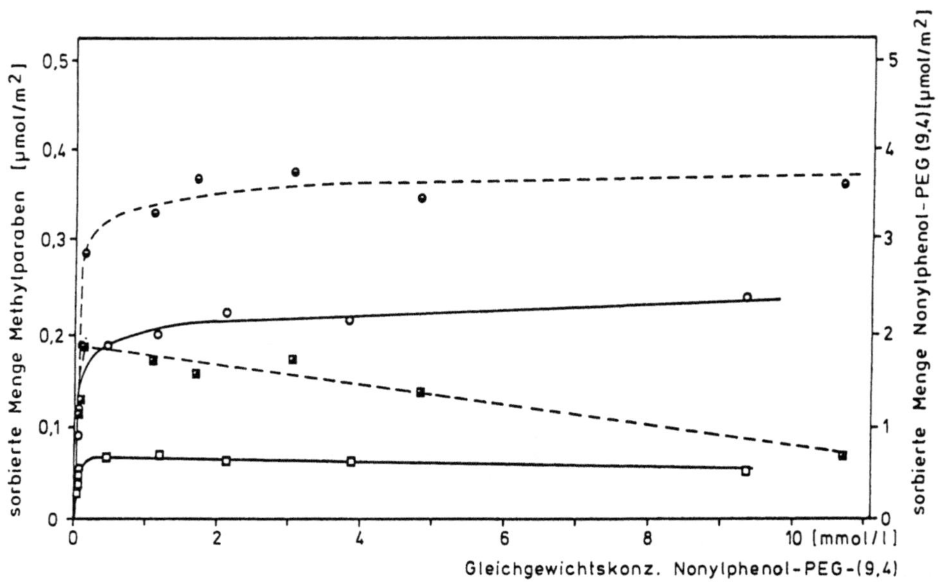

Abb. 2. Cosorption von Methyl-Paraben mit Nonylphenol-PEG 9,4 an poröses SiO_2 aus H_2O (21 °C)
◒—◒ Nonylphenol-PEG an Spherosil C
○—○ Nonylphenol-PEG an Spherosil A
◪—◪ Methylparaben an Spherosil C
□—□ Methylparaben an Spherosil A

Poren mit einer Größe unter 10 nm behindern offensichtlich den Aufbau entsprechender Cosorptionsschichten, da in diesem Material – bei ähnlichen Gleichgewichtsverhältnissen in der Wasserphase – nur etwa die Hälfte der frei zugänglichen Oberfläche[2]) belegt wird und die Ausbildung eines Sorptionsmaximums unterbleibt.

Das Maximum in der Paraben-Cosorption am SiO_2 mit 30 nm mittlerem Porendurchmesser erklären wir damit, daß bei zunehmender Konzentration an Tensid oberhalb der kritischen Mizellbildungskonzentration die Mizellphase in der wässrigen Gleichgewichtslösung zunimmt. Da im gesamten System eine konstante Menge Paraben vorhanden ist, verteilt sich diese zwischen dem Cosorbat und der Solubilisatphase. Zunehmende Menge an Solubilisatphase muß daher konsequenterweise zu einer Abnahme der cosorbierten Menge an der Grenzfläche nach Einstellung der entsprechenden Verteilungsgleichgewichte führen. Beim engporigen Material liegt über dem gesamten Konzentrationsbereich an Tensid eine wesentlich geringere Menge an cosorbiertem Paraben vor, so daß hier eine entsprechende Veränderung der Verteilungsverhältnisse kaum mehr zum Tragen kommt.

[2]) Für N_2 bei der Oberflächenbestimmung zugänglich

Die molare Zusammensetzung der Cosorbate strebt bei höheren Gleichgewichtskonzentrationen an Tensid einem Verhältnis von ca. 36 : 1 zwischen Nonylphenol-PEG und Paraben zu. Eine ähnliche Zusammensetzung findet man in wässrigen Lösungen, wo z. B. in Gegenwart von jeweils 1 mmol/l Methylparaben und Tensid die entsprechenden Mizellen Nonylphenol-PEG Molekel und Parabenmolekel im Verhältnis 33 : 1 enthalten. Danach weisen die Tensidmizellen mit solubilisiertem Paraben und die Cosorptionsschicht an der SiO_2-Grenzfläche etwa die gleiche molare Zusammensetzung auf. Es liegt daher nahe, beim Einbau der Parabene in die Sorbatschichten am SiO_2 einen ähnlichen Mechanismus zu vermuten, wie beim Vorgang der Solubilisation in der Wasserphase [17].

Während für die Solubilisation die Bildung von Mizellen und damit die kritische Mizellbildungskonzentration als Grenzwert eine essentielle Voraussetzung darstellen, findet die Cosorption bereits weit unterhalb der kritischen Mizellbildungskonzentration statt. Danach liegen offenbar bei der Kombination von Nonylphenol-PEG und Parabenen ähnliche Cosorptionsbedingungen vor, wie sie bei Kombinationen von nichtaggregierenden Arzneistoffkationen und PEG-Tensiden an SiO_2Oberflächen beobachtet wurde [18]. Die bevorzugte Bindung wird in diesen Systemen mit gegenseitigen hydrophoben Wechselwirkungen der Sorptionspartner und dem Energiegewinn bei der Sorption an der SiO_2-Grenzfläche erklärt.

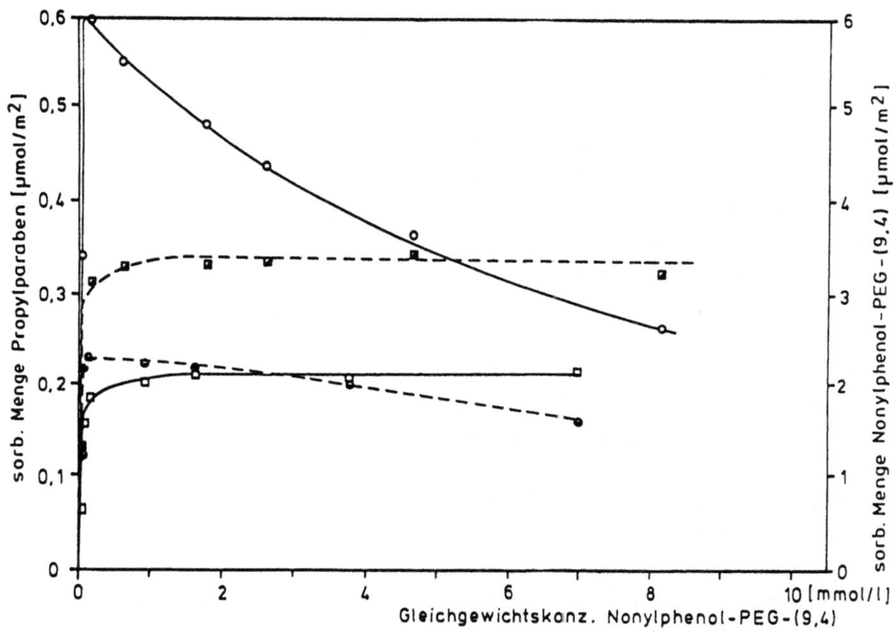

Abb. 3. Cosorption von Propyl-Paraben mit Nonylphenol-PEG 9,4 an poröses SiO_2 aus H_2O (21 °C); □——□ Nonylphenol-PEG an Spherosil Type A; ◪——◪ Nonylphenol-PEG an Spherosil Type C;
◒——◒ Propylparaben an Spherosil Type A } Ausgangskonz. Propylparaben 1 mmol/l
○——○ Propylparben an Spherosil Type C }

Die Cosorption von Propylparaben mit Nonylphenol-PEG 9,4 zeichnet sich gegenüber dem Methylparaben vor allem durch ein wesentlich höheres Maximum der Bindung am weitporigen SiO_2 aus (Abb. 3). Die maximal sorbierte Menge steigt dabei auf 0,6 µmol/m² (Tab. 3 a). Am engporigen Material sind die Sorptionswerte für das Paraben ebenfalls erhöht, wobei gleichzeitig die Sättigungssorptionswerte für das Tensid an SiO_2-Grenzfläche gegenüber dem Methylparaben verringert werden (Tab. 3 b). Entsprechend findet man in den Cosorbaten ein molares Verhältnis von 10 : 1 für Tensid und Paraben.

Aus dem Vergleich der Sättigungswerte in Tabelle 3 folgt, daß beim Einbau von Parabenmolekel in das Cosorbat Tensidmolekel aus der Sorptionsphase annähernd im molekularen Verhältnis verdrängt werden. Zunehmende Lipophilie der Alkylreste im Paraben verstärkt die Tendenz zur Cosorption, verringert damit gleichzeitig die Menge am gebundenen Tensid. Die Sorptionswerte am strukturell unterschiedlichen Kr 36 liegen für beide Komponenten deutlich tiefer, wenn man zum Vergleich vom mittleren Porenradius der verwendeten SiO_2-Träger ausgeht (Tab. 3). Bei Kr 36 handelt es sich um ein SiO_2-Präparat, das an der äußeren Oberfläche vorwiegend kleinere Poren aufweist und somit den Zugang zum Korninneren, vor allem für die Tensidmolekel, behindert [12].

Tabelle 3

a) Abgeschätzte Sättigungssorptions-Werte für Nonylphenol-PEG (9,4)

	Spherosil Type A 10 nm (µmol/m²)	Spherosil Type C 20–40 nm (µmol/m²)	Kr 36 14 nm (µmol/m²)
ohne Paraben	2,48	3,78	1,77
In Gegenwart von			
Methylparaben	2,43	3,63	1,85
Ethylparaben	2,13	3,47	1,81
Propylparaben	2,12	3,25	1,85

b) Maximal sorbierte Menge Paraben in Gegenwart von Nonylphenol-PEG (9,4)

	Spherosil A (µmol/m²)	Spherosil C (µmol/m²)	Kr 36 (µmol/m²)
Methylparaben	0,068	0,189	0,067
Ethylparaben	0,14	0,35	0,103
Propylparaben	0,229	0,6	0,145

Ausgangskonzentration Paraben jeweils 1 mmol/l

3.2 Tail-Analyse

Entsprechend den Grundlagen zur Tail-Analyse [5] wurde das Desorptionsverhalten an den SiO_2-Trägern in einem Säulenbett mittels einer der Flüssigkeits-Chromatographie vergleichbaren Anordnung untersucht. Hierzu wurde das SiO_2-Trägermaterial in der Säule zunächst mit einer Solubilisatlösung der Nonylphenol-PEG und dem jeweiligen Paraben bei konstanter Konzentration beider gelösten Substanzen durchspült, um das Trägermaterial mit der Cosorbatphase zu sättigen. Unmittelbar anschließend an diesen Beladungsvorgang wurde die Desorption eingeleitet, in dem reines Wasser anstelle der Solubilisatlösung durch die Säule gepumpt wurde. Während der Desorption wurde der Konzentrationsverlauf für Tensid und Paraben zeitabhängig verfolgt.

Abb. 4 gibt die Elution von Ethylparaben und Nonylphenol-PEG 50 aus einem Cosorbat an Spherosil A wieder. Beide Substanzen weisen typische Stufen im Konzentrationsverlauf während des Eluierens auf. Während das Tensid jedoch eine kontinuierliche Abnahmekurve zeigt, ist beim Wirkstoff eine Plateauphase deutlich erkennbar. Freisetzung des Parabens und Desorption des Tensides erscheinen danach offensichtlich eng miteinander gekoppelt zu sein.

Die Desorptionskinetik der Parabene ist von der Kettenlänge des Polyglykolrestes im Tensid stark abhängig (Abb. 8). Bei nur 9,4 Ethereinheiten verlängert sich die Elutionszeit gegenüber Nonylphenol

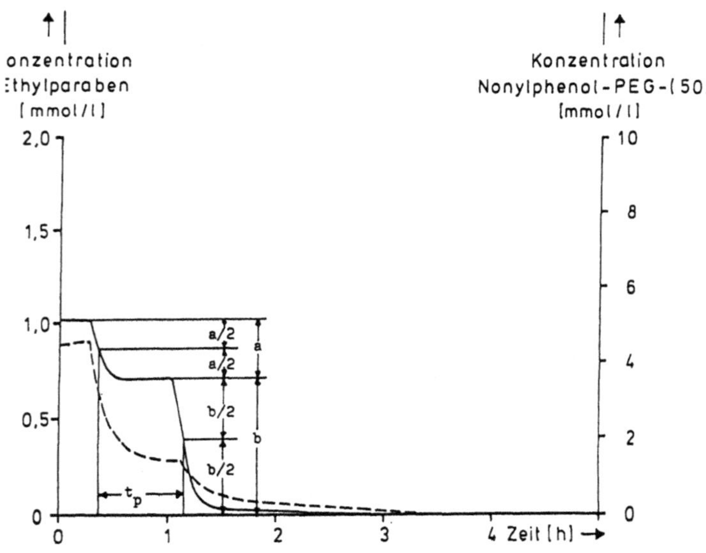

Abb. 4. Desorption von Ethylparaben-Nonylphenol-PEG 50 Cosorbaten aus Spherosil A durch H_2O (21 °C) Tail-Analyse —— Ethylparaben; ---- Nonylphenol-PEG 50; t_p Zeitdauer der Plateau-Phase

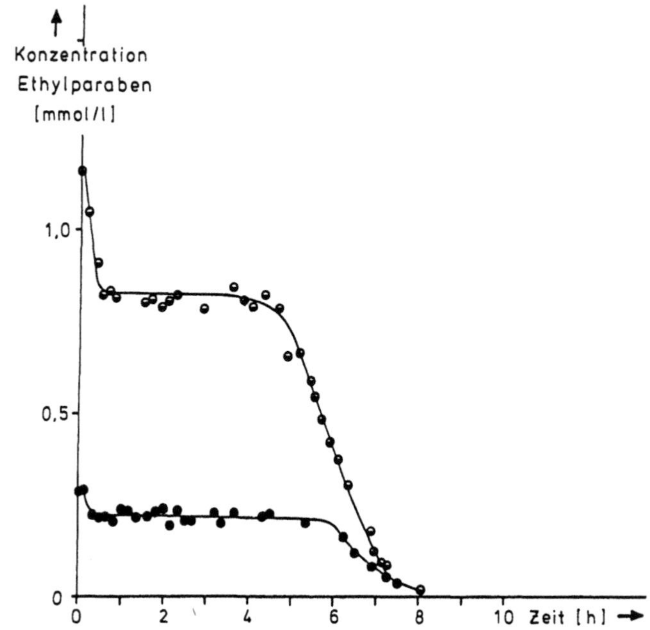

Abb. 5. Desortpion von Ethylparaben aus Cosorbaten mit Nonylphenol-PEG 9,4 aus Spherosil A durch H_2O (21 °C), Tail-Analyse ◒—◒ 1,15 mmol l^{-1} Ethyl-Paraben; ●—● 0,307 mmol l^{-1} Ethyl-Paraben; Konzentration Nonylphenol-PEG 9,4 bei Adsorption 4,018 mmol l^{-1}

PEG mit 50 Ethereinheiten erheblich[3]). Sie wird allerdings auch von der Ausgangskonzentration an Wirkstoff im Solubilisat bestimmt (Abb. 5), wobei höhere Parabenmengen im Solubilisat einen höheren Elutionsspiegel hervorrufen. Innerhalb eines weiten Konzentrationsbereichs ergibt sich dabei eine lineare Abhängigkeit zwischen der Plateaukonzentration bei der Desorption und der Ausgangskonzentration des Arzneistoffes im Solubilisat bei Beladen der Säule (Abb. 6).

Verändert man beim Adsorptionsschritt die Nonylphenol-PEG-Menge bei konstantem Angebot an Ethylparaben, so findet man während der Desorption für die Parabenkonzentration die in Abbildung 7 gezeigte Beziehung: Danach sinkt die Parabenkonzentration beim Eluieren mit Wasser in der Plateauphase, wenn bei der Adsorption eine höhere Tensidkonzentration eingestellt wird.

Variiert man bei gleichen molaren Ansatzverhältnissen zwischen Paraben und Nonylphenol-PEG die Kettenlänge im Tensid, dann verändern sich bei der Desorption die Zeiten konstanter Auswaschraten beträchtlich, während die Konzentrationen für die

[3]) In Abbildung 8 sind zur besseren Übersicht die Desorptionskurven des Tensides nicht wiedergegeben.

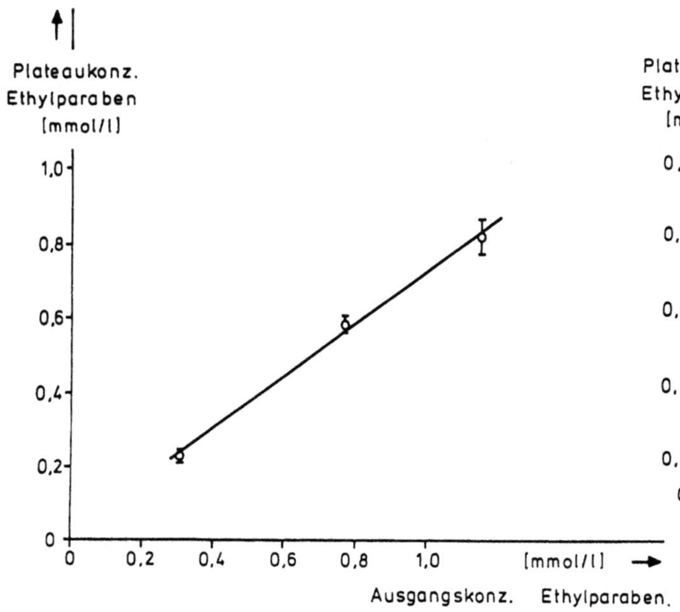

Abb. 6. Plateaukonzentration von Ethylparaben bei der Desorption (Tail-Analyse) aus Cosorbaten mit Nonylphenol-PEG 9,4 in Abhängigkeit von der Parabenkonzentration beim Adsorptionsschritt. Elution mit Wasser, 21 °C.
Ausgangskonzentration Nonylphenol-PEG 9,4 bei der Adsorption jeweils 4,018 mmol/l

Plateauphasen nur geringe Unterschiede aufweisen (Abb. 8). Eine Verkürzung der Polyglykolketten, d. h. zunehmende Lipophilie des Tensides führt zu einer erheblichen Verlängerung der Desorptionszeit. Keinen Einfluß auf die Höhe der Plateaukonzentration zeigt dagegen die Durchflußgeschwindigkeit des Wassers!

Aus der Parabenkonzentration während der

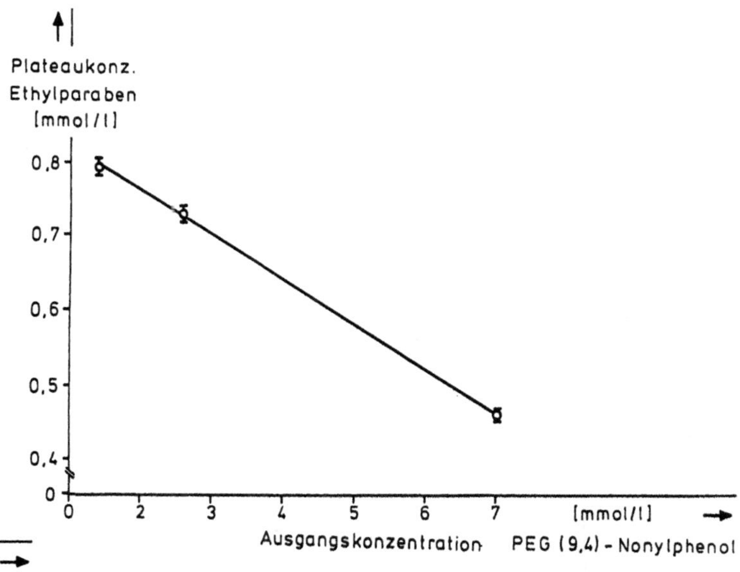

Abb. 7. Plateaukonzentration von Ethylparaben bei der Desorption (Tail-Analyse) aus Cosorbaten mit Nonylphenol-PEG 9,4 in Abhängigkeit von der Tensidkonzentration beim Adsorptionsschritt. Elution mit Wasser, 21 °C.
Ausgangskonzentration Ethyllparaben 0,77 mmol/l

Plateauphase, der Durchflußgeschwindigkeit des Wassers und der Zeit t_p für die Plateauphase kann die Menge an desorbiertem Wirkstoff berechnet werden (die Zeit t_p ist entsprechend Abbildung 4 aus den Elutionskurven entnommen worden). Der Zusammenhang geht dabei aus Abbildung 9 hervor, welche die Abhängigkeit der in der Plateauphase desorbierten Wirkstoffmenge von der Länge der Polyglykolketten im Tensid aufzeigt. Daraus folgt, daß bei vergleichbaren molaren Ansatzverhältnissen Polyethylenglykol-

Abb. 8. Desorption von Ethylparaben (Tail-Analyse) aus Cosorbaten mit Nonylphenol-PEG unterschiedlicher Kettenlänge an SiO_2 (Spherosil A). Elution mit H_2O (21 °C).
Ausgangskonzentration Nonylphenol-PEG 5 mmol/l; Ausgangskonzentration Ethylparaben 1 mmol/l
PEG-Kettenlänge im Tensid
···· 50
·–·– 20
––– 12
—— 8,5

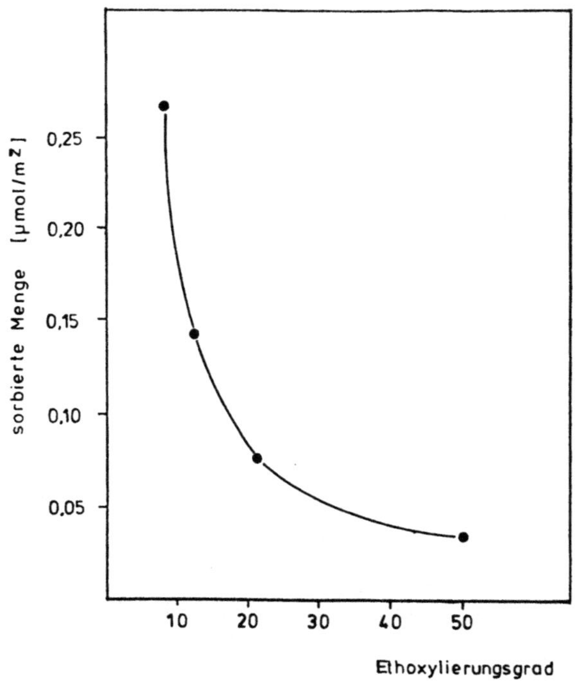

Abb. 9. Cosorption von Ethylparaben mit Nonylphenol-PEG-Ethern an Spherosil A in Abhängigkeit vom Ethoxylierungsgrad der Ether. Sorptionswerte an der Plateauphase der Desorption mit H_2O bei 21 °C berechnet

tenside mit kurzen PEG-Resten eine wesentlich höhere Cosorption des Parabens hervorrufen.

3.3 Diskussion

Aus den Adsorptionsgleichgewichten und dem Verlauf der Desorption ergibt sich zusammenfassend folgendes Bild: Bei der Cosorption von Parabenen mit Nonylphenol-PEG werden – annähernd im molekularen Verhältnis – Parabenmolekel im Austausch gegen Tensidmolekel in die Sorptionsschicht eingebaut. Dieser Einbau nimmt – bei vergleichbaren Ansatzverhältnissen – mit abnehmender Kettenlänge des Polyglykolanteils der Tenside zu. Ebenso erhöht sich die cosorbierte Menge, wenn die Kettenlänge der Alkylkomponente im Paraben zunimmt.

Die Ausbildung der Cosorbatschichten wird von der Dimension des Porengerüstes dann erheblich beeinflußt, wenn der mittlere Porendurchmesser im Bereich der Dimensionen der Cosorbatschicht, d. h. < 10 nm liegt.

Poren in der Größenordnung der Mizellaggregate in wässriger Lösung bestimmen auch die Kinetik der Freisetzung: Die Aufrechterhaltung einer konstanten Elutionskonzentration über längere Zeit und deren Abhängigkeit von der PEG-Komponente im Tensid lassen schließen, daß Transportprozesse bei Ad- und Desorption vorwiegend auf Monomere von Wirkstoff bzw. Tensid beschränkt bleiben [20]. Offenbar behindern die engen Poren im Trägermaterial die Bildung bzw. den Transport frei beweglicher Solubilisatmizellen. Damit bestimmen die Aggregations- und Sorptionsgleichgewichte innerhalb des Porengerüstes die Konzentration an desorbierbaren Monomeren. Erst nach Auswaschen des Tensides am Ende der Plateauphase kommt es zu einem allmählichen Abfall der Konzentration bei Desorption der Restmenge an Paraben (Abb. 8). Die entsprechende Desorptionskinetik ist dabei für die Freisetzung von inkorporierten Wirkstoffen in hydrophilen porösen Matrixsystemen typisch [19].

Danksagung:

Wir danken den Firmen SERVA (Heidelberg), CHEMISCHE WERKE HÜLS (Marl), DEGUSSA (Frankfurt a. M.), NIPA LABORATORIES Ltd. (London) und E. MERCK (Darmstadt) für die großzügige Überlassung von Untersuchungsmaterial, dem Fond der Chemischen Industrie danken wir für die finanzielle Unterstützung der Forschungsarbeiten.

Literatur

1. Sucker H, Fuchs P, Speiser P (1978) (ed) Thieme G, Pharmazeutische Technologie, Stuttgart
2. List PH, Hörhammer L (1971) Handbuch der Pharmazeutischen Praxis, 4. Neuausgabe, Band VII, Teil A, Springer Verlag, Heidelberg
3. Ullmann E (1967) Mitteil Dtsch Pharm Ges 37:89
4. Rupprecht H (1978) Acta Pharm Techn 24:105
5. Suzuki H (1976) Bull Chem Soc Japan 49:375
6. Rupprecht H, Kerstiens B (1981) Pharm Ztg
7. Degussa (1978) Aerosil in Pharmazie und Kosmetik, Schriftenreihe Pigmente Nr. 49
8. Liebl H (1973) Dissertation LMU München
9. Unger K (1972) Angew Chem 84:331
10. De Vries AJV, Le Page M, Bean R, Guillemin CL (1967) Anal Chem 39:935
11. Iler RK (1979) The chemistry of Silica, Cornell University Press, Ithaka
12. Rupprecht H, Unger K, Biersack MJ (1977) Colloid Polymer Sci 255:276
13. Unger K, Rupprecht H, Kircher W (1980) Pharm Ind 42:1027
14. Pharm Europ (1980) 2. Ausgabe, Gori Verlag, Frankfurt, p 8
15. Tinker RB, McBay AJ (1954) J Amer Pharm Assoc 43:315
16. Rupprecht H, Liebl H., Ullmann E (1973) Pharmazie 28:759
17. Shinoda K (1974) Principles of solution and solubility, M. Deccer, New York, Basel
18. Rupprecht H, Liebl H (1975) Pharm Ztg 120:179
19. Desai SJ, Singh P, Simonelli AP, Higuchi WJ (1966) J Pharm Sci 55:1224, 1230, 1235
20. Roy S, Ruckenstein E (1983) J Colloid Interf Sci 92:383

Authors' address:

H. Rupprecht
Lehrstuhl für Pharmazeutische Technologie
Institut für Pharmazie an der Universität Regensburg
8400 Regensburg

Progress in Colloid & Polymer Science Progr Colloid & Polymer Sci 69:167–173 (1984)

Adsorption von Tensiden an Zeolith A*)

M. J. Schwuger, W. v. Rybinski und P. Krings

Henkel KGaA, Düsseldorf

Zusammenfassung: Kationische Tenside werden an der äußeren Oberfläche von Zeolith A adsorbiert. Die Adsorptionsisotherme zeigt einen langmuirartigen Verlauf. Es wird eine deutliche pH-Abhängigkeit der Adsorption beobachtet, wobei mit steigendem pH-Wert die adsorbierten Mengen abnehmen. Nicht-ionische Tenside werden in Abwesenheit anderer Substanzen an Zeolith A nicht adsorbiert. Bei Zusatz geringer Mengen von kationischen Tensiden erfolgt indessen eine Adsorption nichtionischer Tenside an der Zeolithoberfläche. Es bilden sich gemischte Adsorptionsschichten aus, deren Zusammensetzung vom Konzentrationsverhältnis des kationischen zum nichtionischen Tensid abhängt. Mit steigender Konzentration des kationischen Tensids durchläuft die Adsorption des nichtionischen Tensids ein Maximum. Dieses Adsorptionsverhalten kann durch die Veränderung des Ladungszustandes der Zeolithoberfläche bei der Kationtensidadsorption erklärt werden. Die Adsorption kationischer und nichtionischer Tenside an Zeolith A wird durch Temperatur, pH-Wert und Struktur des Kationtensids beeinflußt. Durch die Adsorption kationischer Tenside an Zeolith A trägt der Einsatz zeolithhaltiger Waschmittel dazu bei, kationische Tenside aus Waschprozeß und Abwasser zu entfernen.

Abstract: Cationic surfactants are adsorbed at the external surface of zeolite A. The adsorption isotherm can be described by the Langmuir equation. A marked dependence of the amounts adsorbed of the pH-value exists, and the plateau values of the isotherms decrease with increasing pH. Nonionic surfactants are not adsorbed on zeolite A in the absence of other surfactants. When small amounts of cationic surfactants are added, nonionic surfactants can be adsorbed on zeolite A. Mixed adsorption layers are formed, whose composition depends on the concentration ratio of cationic and nonionic surfactants. With increasing concentration of cationic surfactant the adsorption isotherm of the nonionic surfactant goes through a maximum. This effect can be explained by a change of the surface charge of zeolite A, when cationic surfactants are adsorbed. The adsorption of cationic and nonionic surfactant on zeolite A is influenced by temperature, pH and structure of the cationic surfactant. The use of zeolite A in laundry detergents hence has consequences with regard to the elimination of cationic surfactants in the washing process and in sewage waters.

Key words: Adsorption, surfactant, zeolite.

1. Einleitung

Wasserunlösliches Natriumaluminiumsilikat vom Zeolith-A-Typ wird in zunehmendem Maße als Builder in Waschmitteln eingesetzt [1–4]. Im Waschprozeß und im Abwasser kommen Tenside unterschiedlicher Struktur mit Zeolith A zusammen und können mit der Oberfläche des Natriumaluminiumsilikats Wechselwirkungen eingehen. Auf Grund der negativen Oberflächenladung von Zeolith A werden besonders kationische Tenside an Zeolith A adsorbiert [5]. Kationische Tenside werden als Avivagemittel eingesetzt und wirken wegen ihrer Adsorption an Textilien während des Spülgangs. Bei einer nachfolgenden Wäsche werden die adsorbierten Tenside von der Textilfaser abgelöst und gelangen so in die Waschlauge. Die Kenntnis der Wechselwirkungen zwischen Zeolith A und kationischen Tensiden ist daher von großem anwendungstechnischem Interesse. Die Tensidadsorption aus wäßrigen Lösungen wird von physikalisch-chemischen Parametern wie Temperatur, pH-Wert und Elektrolytgehalt sowie durch die Anwesenheit anderer Tensidtypen beeinflußt. Speziell über die

*) Vortrag, gehalten auf der 31. Hauptversammlung der Kolloid-Gesellschaft, Bayreuth 11. bis 14. Oktober 1983.

KT-W 801

Mischadsorption an Feststoffen ist dabei bisher nur wenig bekannt [6, 7]. Es wurde daher die Adsorption von kationischen und nichtionischen Tensiden an Zeolith A in Abhängigkeit verschiedener Parameter untersucht.

2. Experimentelles

Als Adsorbens wurde synthetischer Zeolith A (SASIL, eingetragenes Warenzeichen der Henkel KGaA) verwendet. Die Teilchengröße war < 10 µm mit einem Maximum der Teilchengrößenverteilung bei 4,2 µm. Die spezifische Oberfläche betrug 2,6 m^2/g und wurde durch Stickstoffadsorption bei 77 K mit der BET-Methode ermittelt.

Ditalgalkyldimethylammoniumchlorid (Präpagen WK, Handelsmarke der Hoechst AG, Frankfurt), Abkürzung DAC, und kettenlängenreines Hexadecyltrimethylammoniumchlorid, Abkürzung HTC, wurden als kationische Adsorptive verwendet. Als nichtionisches Tensid wurde Nonylphenoloctaglycolether (NP 8) eingesetzt.

Die Bestimmung der adsorbierten Mengen des kationischen Tensids erfolgte durch Messung der Lösungskonzentration vor und nach der Adsorption mit der Zweiphasentitration, die Konzentration des nichtionischen Tensids wurde durch UV-Spektroskopie ermittelt.

3. Ergebnisse und Diskussion

Abbildung 1 zeigt die Adsorptionsisothermen von Ditalgalkyldimethylammoniumchlorid an Zeolith A bei pH 8 und 10. Die Isothermen sind gekennzeichnet durch einen steilen Anfangsanstieg, der auf eine präferentielle Adsorption des kationischen Tensids an der Zeolithoberfläche hinweist. Bei höheren Lösungskonzentrationen erreichen die Isothermen einen Plateauwert, der mit steigendem pH-Wert niedriger liegt. Dieses Adsorptionsverhalten von DAC an Zeolith A kann durch elektrostatische Wechselwirkungen zwischen der kationischen Kopfgruppe des Tensids und der negativen geladenen Zeolithoberfläche erklärt werden Zeolith A ist aus einem Aluminiumsilikatgerüst aufgebaut, dessen negative Überschußladung durch Natriumionen ausgeglichen wird. Auf Grund der hohen Beweglichkeit der Natriumionen können diese durch andere positiv geladene Ionen, z. B. Erdalkaliionen, ausgetauscht werden [8].

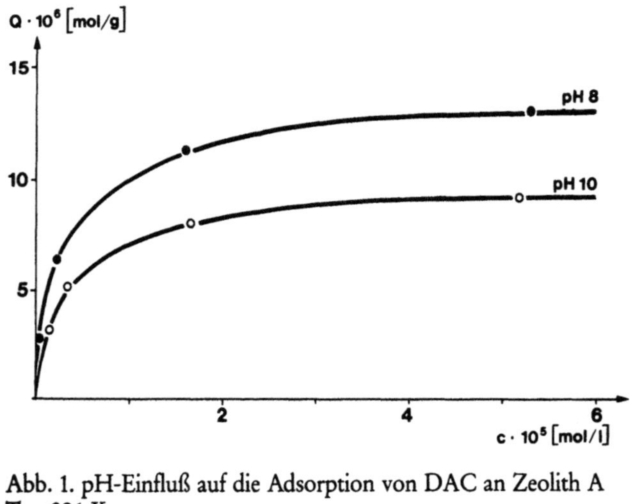

Abb. 1. pH-Einfluß auf die Adsorption von DAC an Zeolith A T = 296 K

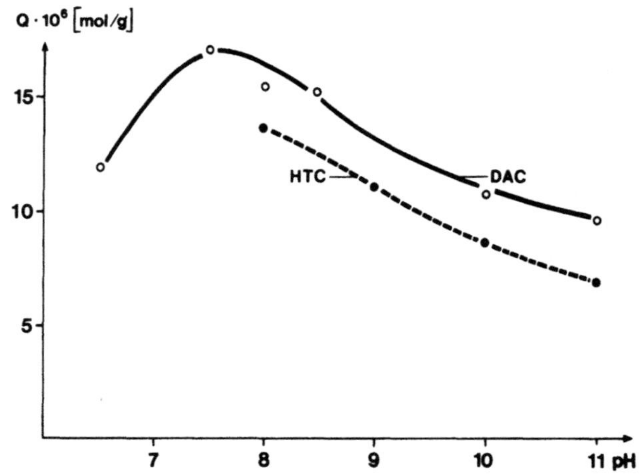

Abb. 2. Plateauwert der Adsorptionsisothermen an Zeolith A $c_{Tensid} = 1 \cdot 10^{-3}$ mol/l, T = 296 K

Da der Ladungszustand der negativ geladenen Zeolithoberfläche einen großen Einfluß auf die Adsorption der positiv geladenen Ammoniumionen hat, besteht eine deutliche Abhängigkeit vom pH-Wert (Abb. 2). Bei gleicher Isothermenform nehmen die Plateauwerte der Isothermen mit steigendem pH-Wert ab. Zeolith A wird im sauren pH-Bereich bereits langsam aufgelöst, die Adsorption bei pH-Werten kleiner als 7 ist daher nicht mehr charakteristisch für Wechselwirkungen zwischen Zeolith und Tensid. Die verringerte Tensidadsorption bei höheren pH-Werten ist jedoch spezifisch für die Oberfläche von Zeolith A. Dieser Effekt ist unabhängig von der Zahl der Alkylketten der quaternären Alkylammoniumverbindung, die adsorbierten Mengen liegen aber bei der Monoalkylverbindung HTC niedriger als bei der Dialkylverbindung DAC. Der Unterschied erklärt sich aus dem unterschiedlichen Platzbedarf der Monoalkyltrimethylverbindung im Vergleich zur Dialkyldimethylverbindung, der aus Messungen der Schichtdicke und der Adsorption kationischer Tenside an Schichtsilikaten abgeschätzt werden kann [9, 10]. Wegen der negativ geladenen Zeolithoberfläche sollte sich die Monoalkyltrimethylverbindung so anordnen, daß das positiv geladene

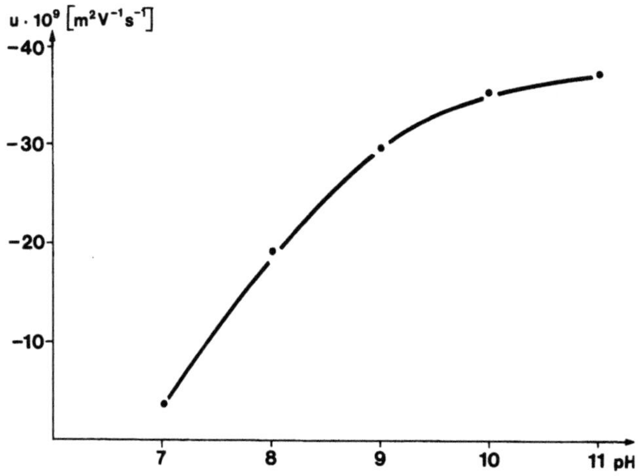

Abb. 3. Elektrophoretische Beweglichkeit von Zeolith A
T = 296 K

quaternäre Ammoniumion einen möglichst geringen Abstand zur Zeolithoberfläche hat. Dies führt zu einer Struktur der Adsorptionsschicht, bei der die drei Methylgruppen möglichst nahe an der Zeolithoberfläche sind. Im Vergleich dazu sind bei der Dialkyldimethylverbindung nur zwei Methylgruppen zur Zeolithoberfläche hin orientiert. Der Platzbedarf der Monoalkyltrimethylverbindung (~ 0,46 nm^2) ist daher größer als der der Dialkylverbindung (~ 0,43 nm^2). Auf Grund des Platzbedarfs der Kationtenside kann der Adsorptionsvorgang nur an der äußeren Oberfläche von Zeolith A stattfinden, da bei einer Porengröße von 0,16 nm^2 die Tensidmoleküle nicht in das Porensystem eindringen können.

Messungen der elektrophoretischen Beweglichkeit von Zeolith A zeigen, daß die negative Ladung mit steigendem pH-Wert zunimmt (Abb. 3). Werden die Adsorptionsergebnisse mit denen der elektrophoretischen Beweglichkeit verglichen (Abb. 2 und 3), kann überraschenderweise ein entgegengesetzter Verlauf beobachtet werden. Die Ergebnisse unterscheiden sich von vielen bisher bekannten [11–13], die auf eine Zunahme der adsorbierten Mengen kationischer Tenside mit steigender Alkalität hinweisen. Dieser charakteristische Unterschied wird wahrscheinlich durch spezielle Wechselwirkungen zwischen der Zeolithoberfläche und dem kationischen Tensid verursacht. An der äußeren Oberfläche einer Elementarzelle von Zeolith A sind bei der idealen Struktur 8 Aluminiumatome, von denen 6 tetraedrisch im Zeolithgitter angeordnet sind. Diese bewirken einen negativen Ladungsüberschuß, der nahezu pH-unabhängig ist und durch Natriumionen ausgeglichen wird. Zwei Aluminiumatome in der äußeren Oberfläche haben nur 3 Bindungen im Zeolithgitter und können daher jeweils eine Hydroxylgruppe bei alkalischen pH-Werten binden. Mit steigendem pH-Wert führt dies zu einer Zunahme der negativen Ladung und der Bindung von zusätzlichen Natriumionen.

Die Natriumionen können durch andere positiv geladene Ionen ausgetauscht werden. Wenn nur der pH-unabhängige Austausch berücksichtigt wird, ergibt sich aus der Zellkonstante des Zeolithen (1,23 nm) ein Platzbedarf von ungefähr 0,25 nm^2 pro ausgetauschtem Kation. Da der Platzbedarf eines Ditalgalkyldimethylammoniumions bei senkrechter Anordnung in einer dicht gepackten Monoschicht ungefähr 0,43 nm^2 ist, können nur 60 % der austauschfähigen Plätze durch das Tensid bedeckt werden. Ein Teil der negativen Ladung der äußeren Zeolithoberfläche kann daher auch nach der Einstellung des Adsorptionsgleichgewichts durch Natriumionen kompensiert werden. Aus der BET-Oberfläche und dem angenommenen Platzbedarf des Tensidmoleküls resultiert ein Wert von $10 \cdot 10^{-6}$ mol/g für die adsorbierte Menge bei einer vollbesetzten Monoschicht. Dieser Wert wird im Plateaubereich der Isothermen bei pH 10 erreicht (Abb. 2). Bei niedrigeren pH-Werten sind die Natriumgegenionen wegen der geringeren negativen Ladung des Zeolithgitters schwächer gebunden (Abb. 3). Im pH-Bereich zwischen 7 und 9 können daher zusätzliche Tensidmoleküle adsorbiert werden, wobei vermutlich die Alkylketten zur Zeolithoberfläche ausgerichtet und über schwächere van der Waals-Wechselwirkungen zwischen den Alkylketten gebunden sind. Dies ist eine mögliche Erklärung für die deutlich erhöhte Adsorption bei niedrigeren pH-Werten, die durch Extraktionsversuche unterstützt wird. Wurde Zeolith A bei pH 10 mit Kationtensid beladen, konnten die adsorbierten Mengen in Wasser nicht desorbiert werden, während die zusätzliche Adsorption bei niedrigeren pH-Werten durch Extraktion mit Wasser abgelöst wurde.

Anwendungstechnisch bedeutsam ist die Adsorption kationischer Tenside in Gegenwart anderer Tensidtypen. Anionische und nichtionische Tenside allein werden an Zeolith A innerhalb der Meßgenauigkeit der verwendeten Verfahren im alkalischen pH-Bereich nicht adsorbiert. Durch die Gegenwart nichtionischer Tenside wird aber die Adsorption kationischer Tenside an Zeolith A deutlich beeinflußt (Abb. 4). Der Zusatz von NP 8 bewirkt eine Abnahme der adsorbierten Mengen von DAC besonders im Plateaubereich der Isothermen. Die Ursache hierfür ist eine Adsorption von NP 8 an Zeolith A aus Mischungen mit DAC (Abb. 5), die mit steigender Kationtensidkonzentra-

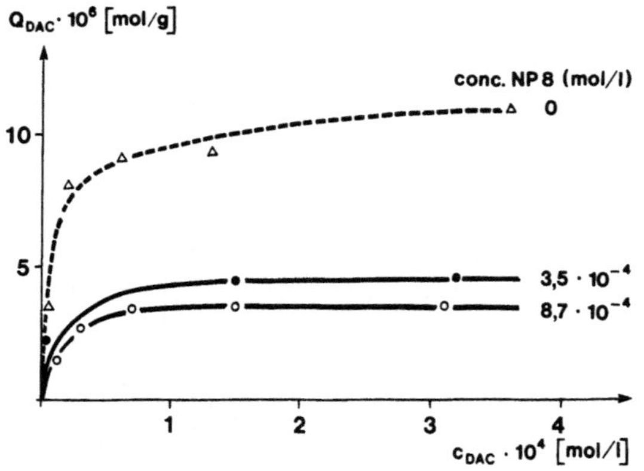

Abb. 4. Adsorption von DAC an Zeolith A nach Zusatz von NP 8
pH = 10, T = 296 K

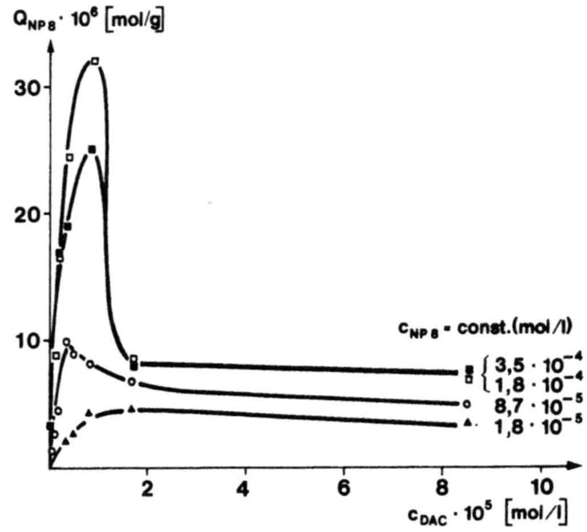

Abb. 6. Adsorption von NP 8 an Zeolith A aus Mischungen mit DAC
pH = 10, T = 296 K

tion abnimmt und bei hohem Kationtensidgehalt der Lösungen einem Plateauwert zustrebt. Höhere Konzentrationen an NP 8 führen zu einer verstärkten Adsorption des nichtionischen Tensids, während gleichzeitig die adsorbierten Mengen des Kationtensids abnehmen.

Bei niedrigen Kationtensidkonzentrationen durchlaufen die adsorbierten Mengen von NP 8 ein Maximum, dessen Höhe von der Konzentration des nichtionischen Tensids abhängt (Abb. 6). Bei höherem Gehalt der Lösungen an nichtionischem Tensid ist dieses Maximum deutlicher ausgeprägt und zu geringfügig höherer Kationtensidkonzentration verschoben. Erklärt werden kann dieses Adsorptionsmaximum durch eine Mischfilmadsorption von nichtionischem und kationischem Tensid an der Zeolithoberfläche. Vergleicht man die elektrophoretische Beweglichkeit von Zeolith A mit den adsorbierten Mengen an DAC in Abhängigkeit der Lösungskonzentration von DAC, so beobachtet man eine Umladung der Zeolithoberfläche mit zunehmender Belegung durch DAC (Abb. 7). Der Umladungspunkt ergibt sich bei einer Oberflächenbelegung mit DAC von ca. 1/3 einer Monoschicht. Ähnliche Werte wurden auch bei der Adsorption von kationischen Tensiden an Mineraloberflächen gefunden [14]. Eine Gegenüberstellung von Abb. 6 und 7 zeigt, daß das Maximum der Adsorption an nichtionischem Tensid aus Gemischen mit DAC im Bereich des Ladungsnullpunkts der Zeolithoberfläche erfolgt. Offensichtlich wird durch die Verringerung der Ladung der Zeolithoberfläche infolge der Kationtensidadsorption eine Adsorption des nichtionischen Tensids begünstigt, die über Wechselwirkungen zwi-

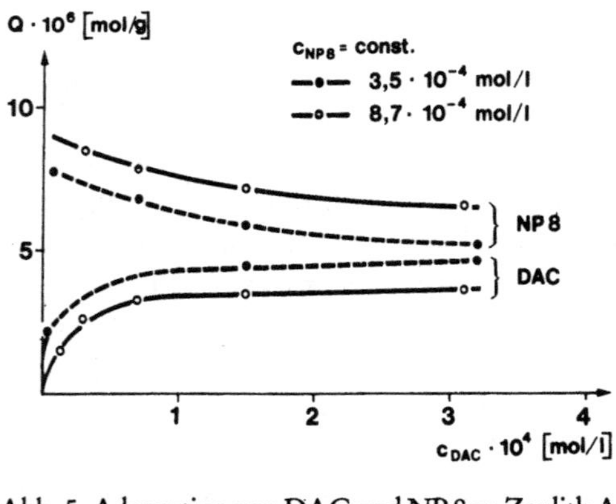

Abb. 5. Adsorption von DAC und NP 8 an Zeolith A aus Mischungen in Abhängigkeit der DAC-Konzentration
pH = 10, T = 296 K

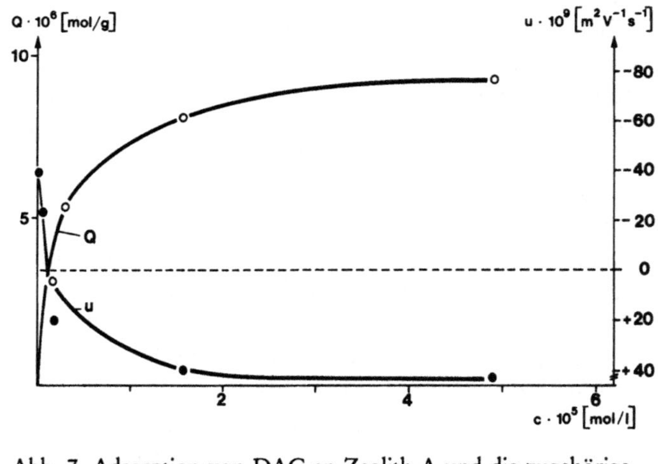

Abb. 7. Adsorption von DAC an Zeolith A und die zugehörige elektrophoretische Beweglichkeit
pH = 10, T = 296 K

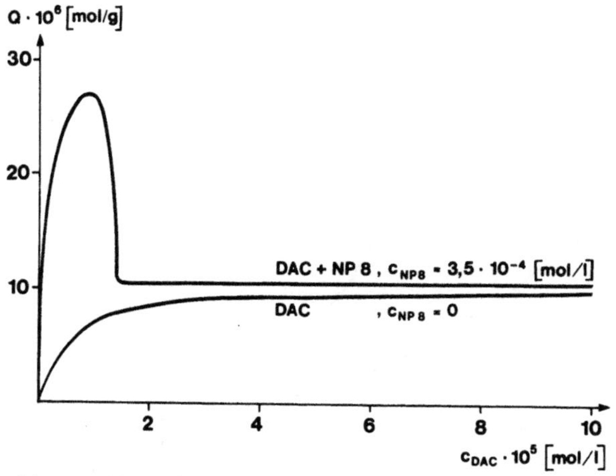

Abb. 8. Gesamtadsorption von DAC und NP 8 an Zeolith A
pH = 10, T = 296 K

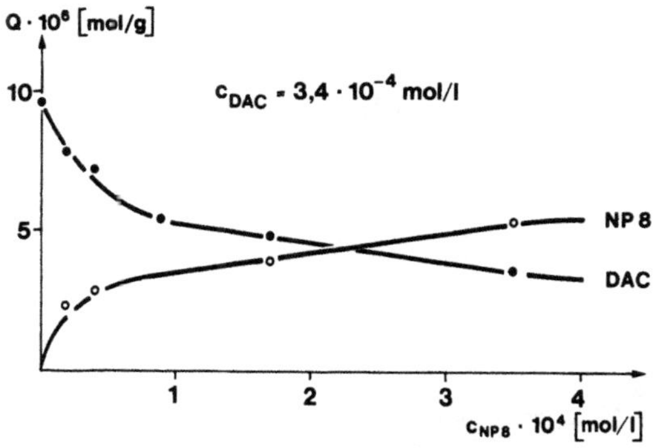

Abb. 9. Adsorption von DAC und NP 8 an Zeolith A aus Mischungen in Abhängigkeit der NP 8-Konzentration
pH = 10, T = 296 K

schen den Alkylketten der beiden Tensidtypen erfolgen kann. Analoge Ergebnisse wurden bereits bei der Mischadsorption von kationischen und nichtionischen Tensiden an Geweben beobachtet [7].

In der Abbildung 8 ist die gesamte adsorbierte Menge von DAC und NP 8 mit der Einzeladsorption von DAC verglichen. Läßt man das Adsorptionsmaximum bei niedrigen Lösungskonzentrationen außer acht, so entspricht der Plateauwert bei der Mischadsorption der Einzeladsorption von DAC. Die adsorbierten Mengen liegen bei der Mischadsorption etwas höher, da die Monoalkylverbindung NP 8 einen geringeren Platzbedarf hat als die Dialkylverbindung DAC und so eine höhere Packungsdichte in der adsorbierten Schicht gestattet. Nimmt man für das nichtionische Tensid einen Platzbedarf von 0,25 nm^2 und für das kationische Tensid einen Wert von 0,43 nm^2 an und berücksichtigt die adsorbierten Mengen aus Abb. 5, so ergibt sich für das Isothermenplateau eine Belegung der Zeolithoberfläche, die geringfügig unter der einer theoretischen Monoschicht liegt. Für das Adsorptionsmaximum berechnet sich ein Wert, der dem einer theoretischen Doppelschicht entspricht. Diese Ergebnisse unterstützen den vorgeschlagenen Adsorptionsmechanismus mit einer zusätzlichen Bindung des nichtionischen Tensids über Wechselwirkungen mit den Alkylketten des kationischen Tensids. Abbildung 9 veranschaulicht die Adsorption von DAC und NP 8 aus Mischungen an Zeolith A in Abhängigkeit der Konzentration des nichtionischen Tensids bei konstanter Konzentration des Kationtensids. Die Isotherme von NP 8 ist durch einen langmuirartigen Verlauf gekennzeichnet, während die adsorbierten Mengen von DAC mit steigender Konzentration des nichtionischen Tensids bis auf einen Wert von ca. 30 % einer theoretischen Monoschicht abnehmen (vgl. Abb. 7). Auch bei sehr hohen Konzentrationen an NP 8 kann das kationische Tensid DAC nicht vollständig von der Zeolithoberfläche verdrängt werden. Dies ist ein weiterer Hinweis dafür, daß für eine Adsorption des nichtionischen Tensids eine bestimmte Oberflächenbelegung mit DAC nötig ist.

Die Temperaturabhängigkeit der Adsorption von kationischen und nichtionischen Tensiden an Zeolith A zeigen die Abbildungen 10 und 11. Im untersuchten Bereich von 296 K bis 333 K wird die Adsorption des kationischen Tensids DAC von der Temperatur nur sehr wenig beeinflußt. Die adsorbierten Mengen des nichtionischen Tensids NP 8 liegen dagegen bei höherer Temperatur deutlich niedriger. Erklärt werden kann dieses unterschiedliche Verhalten wiederum

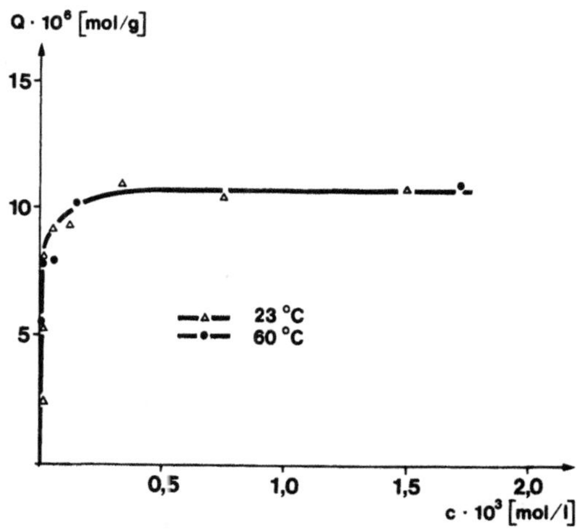

Abb. 10. Einfluß der Temperatur auf die Adsorption von DAC an Zeolith A
pH = 10

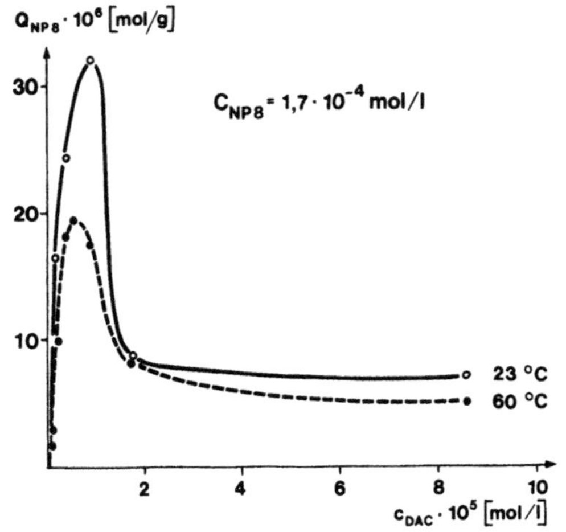

Abb. 11. Einfluß der Temperatur auf die Adsorption von NP 8 aus Mischungen mit DAC
pH = 10

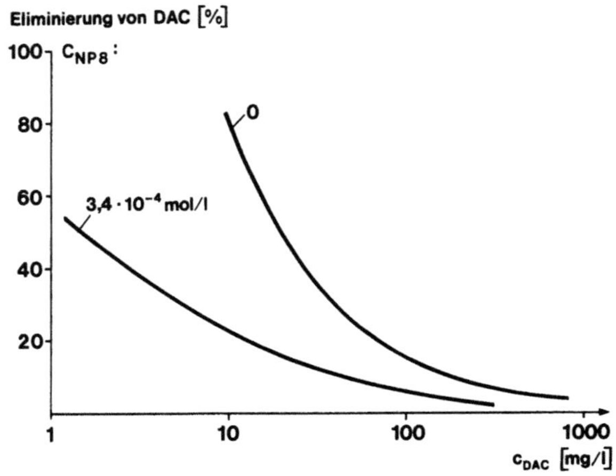

Abb. 13. Eliminierung von DAC durch Zeolith A in Abwesenheit und Gegenwart von NP 8
$c_{Zeolith} = 2$ g/l, pH = 10, T = 296 K

durch den Adsorptionsmechanismus der Tenside an Zeolith A.

Die Bindung des kationischen Tensids zur Zeolithoberfläche durch elektrostatische Wechselwirkungen nach einem Ionenaustausch gegen Natriumionen ist wenig temperaturabhängig, während die Adsorption des nichtionischen Tensids über van der Waals-Wechselwirkungskräfte eine größere Temperaturabhängigkeit aufweist.

Die Struktur des kationischen Tensids hat auf die Adsorption des nichtionischen Tensids an Zeolith A nur geringen Einfluß (Abb. 12). Das Adsorptionsverhalten von NP 8 in Gegenwart der Monoalkyltrimethylammoniumverbindung HTC entspricht dem bei

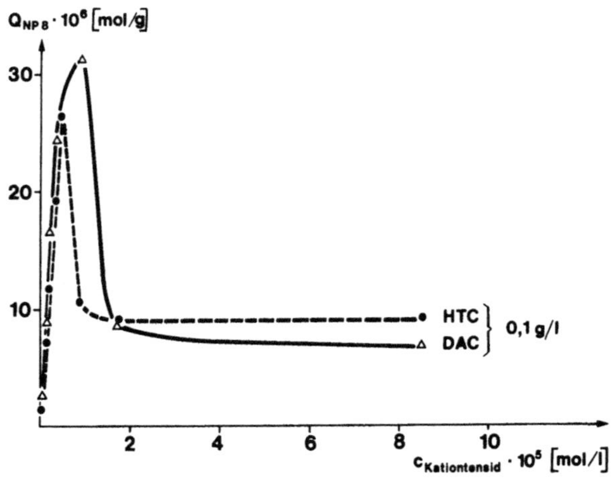

Abb. 12. Adsorption von NP 8 an Zeolith A aus Mischungen mit HTC bzw. DAC

Anwesenheit der Dialkyldimethylammoniumverbindung DAC. Geringe Unterschiede in der Höhe des Maximums und des Plateauwertes können durch unterschiedliche molare Konzentrationen bei gewichtsgleichem Zusatz der beiden Kationtenside erklärt werden.

Die anwendungstechnische Bedeutung der Adsorption von Tensiden an Zeolith A liegt in der Eliminierung der kationischen Tenside aus Waschgang und Abwasser durch Zeolith A. Abbildung 13 veranschaulicht, daß besonders bei niedrigen Konzentrationen ein hoher Prozentsatz des Kationtensids an Zeolith A adsorbiert wird. Die Gegenwart von nichtionischem Tensid verringert zwar die Adsorptionskapazität von Zeolith A gegenüber DAC, ändert aber nur wenig am prinzipiellen Adsorptionsverhalten. Die Ursache hierfür ist der steile Anfangsanstieg der Adsorptionsisotherme von DAC an Zeolith A (siehe Abb. 4 und 10). Dies führt dazu, daß speziell unter Abwasserbedingungen und pH-Werten größer als 7 Kationtenside durch Zeolith A aus wäßriger Lösung entfernt werden. Die Verwendung zeolithhaltiger Waschmittel unterstützt daher die Eliminierung kationischer Tenside aus Waschprozeß und Abwasser.

Literatur

1. Schwuger MJ, Smolka HG, Rostek HM (1974) DOS 24 12 837, Henkel
2. Schwuger MJ, Smolka HG (1976) Colloid Polymer Sci 254:1062
3. Berth P (1978) J Amer Oil Chemists Soc 55:52
4. Krings P, Verbeek H (1981) Tenside Detergents 18:260

5. Schwuger MJ, von Rybinski W, Krings P (1983) (ed) Ottewill RH, in: Adsorption from Solution, Academic Press, London
6. Schwuger MJ, Smolka HG (1977) Colloid Polymer Sci 255:589
7. Rubingh DN, Jones T (1982) Ind Eng Chem Prod Res Dev 21:176
8. Schwuger MJ, Smolka HG (1978) Colloid Polymer Sci 256:1014
9. Lagaly G, Witter R (1982) Ber Bunsenges Phys Chem 86:14
10. Weiß A (1958) Kolloid-Zeitschrift 158:22
11. Giles CH, D'Silva AP, Easton JA (1974) J Colloid Interface Sci 47:766
12. Connor P, Ottewill RH (1971) J Colloid Interf Sci 37:642
13. Ginn ME (1970) (ed) Jungermann E, in: Cationic Surfactants, Surfactants Science Series Vol 4, Marcel Dekker Inc, New York
14. Jaycock MJ, Ottewill RH (1963) Bull Inst Mining Met 677:497

Authors' address:

M. J. Schwuger
Henkel KGaA
4000 Düsseldorf

Progress in Colloid & Polymer Science Progr Colloid & Polymer Sci 69:174–180 (1984)

Massen- und Selbstdiffusion in Systemen wechselwirkender Brownscher Teilchen*

W. Heß und R. Klein

Fakultät für Physik, Universität Konstanz, Konstanz, F. R. G.

Zusammenfassung: Dynamische Eigenschaften konzentrierter kolloidaler Systeme werden mit Hilfe verallgemeinerter hydrodynamischer Gleichungen behandelt. Für das Beispiel eines Polystyrolkugel-Latex zeigt sich, daß, ausgehend von einer Paar-Wechselwirkung, die experimentellen Ergebnisse für dieses System quantitativ berechnet werden können.

Abstract: Dynamical properties of concentrated colloidal systems are treated by means of generalized hydrodynamic equations. For the example of a latex consisting of polystyrene spheres we show that, starting from a pair-potential, the experimental results for this system can be calculated quantitatively.

Key words: Kolloidale Lösungen, verallgemeinerte hydrodynamische Gleichungen, wellenvektor- und frequenzabhängige Transportkoeffizienten, viskoelastische Eigenschaften, Diffusion.

1. Einleitung

Die Einführung moderner Streumethoden, wie z. B. der kohärenten und inkohärenten Neutronenstreuung [1], der Photonenkorrelations-Spektroskopie [2–6] oder der erzwungenen Rayleighstreuung [7], zur Untersuchung kolloidaler Systeme hat die Möglichkeit eröffnet, neben den üblichen Transportkoeffizienten auch die vollständige Zeitabhängigkeit von Korrelationsfunktionen zu bestimmen. Diese Korrelationsfunktionen, z. B. die Geschwindigkeits-Autokorrelationsfunktion oder die Konzentrations-Autokorrelationsfunktion, die häufig auch als „intermediate scattering function" bezeichnet wird, können eine wesentlich genauere Information über die Dynamik des untersuchten Systems liefern, als es die Transportkoeffizienten tun. Betrachten wir als Beispiel das mittlere Verschiebungsquadrat eines diffundierenden Teilchens,

$$W(t) = \frac{1}{6} < (\underline{r}(t) - \underline{r}(0))^2 >, \tag{1.1}$$

so hat dieses für ein wechselwirkendes System das in Abbildung 5 gezeigte typische Aussehen. (Die in Abbildung 5 dargestellten Ergebnisse ergeben sich theoretisch für geladene kugelförmige Teilchen und sind in qualitativer Übereinstimmung mit Computersimulationen für ein ähnliches System.)

Für kleine Zeiten wächst $W(t)$ linear an, wird dann jedoch flacher und verhält sich für große Zeiten wieder linear mit der Zeit. Das Abflachen der Kurve wird durch die Wechselwirkung oder, man kann auch sagen, durch die Stöße mit den anderen Kolloidteilchen verursacht. Asymptotisch für große Zeiten sieht man dann die Verschiebung des Teilchens, gemittelt über viele Stoßprozesse hinweg, und in diesem Zeitbereich ist $W(t)$ dann durch den Selbst- oder Tracer-Diffusionskoeffizienten bestimmt. Wir sehen also, daß die vollständige Funktion $W(t)$ wesentlich mehr Information enthält, als nur diesen asymptotischen Anteil.

Das Gleiche gilt auch für die räumliche Struktur und den räumlichen Ablauf der Dynamik. Wenn man ein Streuexperiment unter einem bestimmten Winkel durchführt, untersucht man das System mit einem Streuvektor $\underline{k} = \underline{k}_{aus} - \underline{k}_{ein}$, und beobachtet gerade die $\underline{k}$-ten Fourierkomponenten der räumlichen Fluktuationen des Systems. Nehmen wir z. B. ein Lichtstreu-

* Vortrag, gehalten auf der 31. Hauptversammlung der Kolloid-Gesellschaft, Bayreuth 11. bis 14. Oktober 1983.

KT-W 790

experiment, dann sind typische Wellenlängen der untersuchten Fluktuationen von der Größenordnung einiger 1000 Å, und damit durchaus von der gleichen Größenordnung wie der Abstand nächster Nachbarn in üblichen kolloidalen Systemen. Das bedeutet aber, daß man unter diesen Umständen das System nicht als ein Kontinuum ansehen darf, sondern daß man im Streuexperiment die diskrete kurzreichweitige Struktur und die zeitliche Entwicklung dieser Struktur sieht.

Die Dynamik kann aber nun nicht mehr durch die wohlbekannten phenomenologischen Transportgleichungen wie die Diffusions-Gleichung oder die Navier-Stokes-Gleichung beschrieben werden, denn diese sind typische Kontinuumsgleichungen. Zum vollständigen Verständnis solcher Streuexperimente ist es deshalb notwendig, Transportgleichungen zu entwickeln, die die Dynamik kolloidaler Systeme auf einer beliebigen Zeit- und Längenskala beschreiben und die erst im langwelligen Limes und für große Zeiten in die üblichen hydrodynamischen Gleichungen übergehen. In diesem Bereich wurden in den letzten Jahren erhebliche Fortschritte erzielt, eine zusammenfassende Darstellung dieser Arbeit ist kürzlich erschienen [8], und eine kurze Darstellung der wesentlichen Ergebnisse soll hier gegeben werden.

2. Verallgemeinerte Transportgleichungen für Massendiffusion

Die zeitliche Entwicklung der Konzentrationsfluktuationen ist durch die Kontinuitätsgleichung bestimmt,

$$\frac{\partial}{\partial t} c(\underline{r}, t) = -\underline{\nabla} \underline{j}(\underline{r}, t), \tag{2.1}$$

$\underline{j}(\underline{r}, t)$ ist der Teilchenstrom. Um nun zu einer geschlossenen Gleichung für $c(\underline{r}, t)$ zu kommen, würde man für die langwelligen Fluktuationen den Fickschen Ansatz

$$\underline{j}(\underline{r}, t) = -D_m \underline{\nabla} c(\underline{r}, t) \tag{2.2}$$

machen. Dabei ist D_m der Massen-Diffusionskoeffizient. Für die kurzwelligen Fluktuationen in einem wechselwirkenden System muß man aber berücksichtigen, daß der Teilchenstrom am Ort $\underline{r}$ auch noch durch einen Konzentrationsgradienten an einem benachbarten Ort $\underline{r}'$ beeinflußt wird, und daß andererseits der Strom zur Zeit t auch noch von Konzentrationsgradienten, die zu früherer Zeit existiert haben, bewirkt wird. Man verallgemeinert den Fickschen Ansatz (2.2) deshalb in der Form

$$\underline{j}(\underline{r}, t) = -\int_0^t dt' \int dV' D(\underline{r}-\underline{r}', t-t') \underline{\nabla}' c(\underline{r}', t), t \geq 0, \tag{2.3}$$

wobei die verallgemeinerte Diffusionsfunktion $D(\underline{r}, t)$ die hydrodynamische Größe D_m ersetzt.

Die wesentliche Funktion zur Interpretation von Photonen-Korrelationsspektroskopie- oder kohärenten Neutronenstreuexperimenten ist der dynamische Strukturfaktor

$$S(\underline{k}, t) = \langle c(\underline{k}, t)\, c(-\underline{k}, 0) \rangle; \tag{2.4}$$

dabei sind die $c(\underline{k}, t)$ die Fourierkomponenten der Konzentrationsfluktuationen, und die eckigen Klammern bedeuten eine Mitteilung über die Gleichgewichtsverteilung. Gehen wir zur Laplace-Transformierten über, dann geben die Gleichungen (2.1) und (2.3) einen geschlossenen Ausdruck

$$\tilde{S}(\underline{k}, z) = \frac{\tilde{S}(\underline{k})}{z + \tilde{D}(\underline{k}, z) k^2}, \tag{2.5}$$

wobei $S(\underline{k}) = S(\underline{k}, t=0)$ der statische Strukturfaktor und $\tilde{D}(\underline{k}, z)$ die Fourier-Laplace-Transformierte der verallgemeinerten Diffusionsfunktion ist. Im hydrodynamischen Grenzfall geht diese Funktion in den üblichen Massen-Diffusionskoeffizienten über,

$$\lim_{\substack{k, z \to 0 \\ \underline{k}^2/z = \text{const}}} \tilde{D}(\underline{k}, z) = \tilde{D}(0,0) = D_m, \tag{2.6}$$

und der dynamische Strukturfaktor ist dann eine Exponentialfunktion,

$$\lim_{\substack{k, z \to 0 \\ \underline{k}^2/z = \text{const}}} S(\underline{k}, t) = S(0)\, e^{-D_m \underline{k}^2 t}, t \geq 0 \tag{2.7}$$

so wie wir es auch unter Verwendung des Fickschen Gesetzes (2.2) erhalten hätten. Die Verallgemeinerung der hydrodynamischen Gleichung besteht also darin, wellenvektor- und zeitabhängige, beziehungsweise wellenvektor- und frequenzabhängige Transportkoeffizienten einzuführen. In gleicher Weise kann man bei der Bewegungsgleichung für die Stromfluktuationen vorgehen. Diese lautet für die Fourierkomponenten

$$m \frac{\partial}{\partial t} \underline{j}(\underline{k}, t) = \underline{f}(\underline{k}, t), \tag{2.8}$$

wobei $f(\underline{k}, t)$ die Fourierkomponenten der Kraftdichte bezeichnet. Die Kraftdichte zerlegt man in einen Gleichgewichtsanteil, der als Gradient eines lokalen osmotischen Drucks π aufgefaßt werden kann, und einen Reibungsanteil, der durch die Bewegung der Teilchen verursacht wird.

$$\underline{f}(\underline{k}, t) = -i\underline{k}\pi(\underline{k}, t) - \int_0^t dt' \underline{\underline{\zeta}}(\underline{k}, t-t') \cdot \underline{j}(\underline{k}, t'), t \geq 0 . \quad (2.9)$$

Der lokale osmotische Druck wird wieder durch die Konzentrationsfluktuationen ausgedrückt,

$$\pi(\underline{k}, t) = \frac{\partial \pi(\underline{k})}{\partial c(\underline{k})} c(\underline{k}, t) = \frac{k_B T}{S(\underline{k})} c(\underline{k}, t) . \quad (2.10)$$

Dann bilden die Gleichungen (2.1), (2.8)–(2.10) ein geschlossenes Gleichungssystem, mit dessen Hilfe wir eine verallgemeinerte Stokes-Einstein-Beziehung herleiten können in der Laplace-transformierten Form,

$$\tilde{D}(\underline{k}, z) \frac{S(\underline{k})}{z + \frac{k_B T / S(\underline{k})}{mz + \tilde{\zeta}_{\|}(\underline{k}, z)}} ; \quad (2.11)$$

$\tilde{\zeta}_{\|}(\underline{k}, z)$ ist die longitudinale Komponente des verallgemeinerten Reibungstensors $(\tilde{\zeta}_{\|}(\underline{k}, z) = \underline{k} \cdot \tilde{\underline{\underline{\zeta}}}(\underline{k}, z) \cdot \underline{k}/\underline{k}^2)$. Auch hier erhalten wir im hydrodynamischen Grenzfall das wohlbekannte Ergebnis

$$\lim_{\substack{k, z \to 0 \\ \underline{k}^2/z = \mathrm{const}}} \tilde{D}(\underline{k}, z) = D_m = \frac{k_B T / S(0)}{\tilde{\zeta}(0, 0)} = \mu \frac{\partial \pi}{\partial c} \Big|_T , \quad (2.12)$$

wobei $\mu = \tilde{\zeta}(0, 0)^{-1}$ die Beweglichkeit ist und für ein verdünntes System $\frac{\partial \pi}{\partial c} \Big|_T$ gerade gleich $k_B T$ ist.

Mit (2.11) wird die Laplace-Transformierte des dynamischen Strukturfaktors

$$\tilde{S}(\underline{k}, z) \frac{S(\underline{k})}{z + \frac{k_B T / S(\underline{k})}{mz + \tilde{\zeta}_{\|}(\underline{k}, z)}} . \quad (2.13)$$

Die verallgemeinerte Reibungsfunktion steht für eine Vielzahl von dissipativen Prozessen, sie enthält die direkte Reibung durch das Lösungsmittel, die hydrodynamische Wechselwirkung und die innere Reibung auf Grund der Wechselwirkung der Kolloidteilchen untereinander. Vernachlässigen wir jedoch die hydrodynamische Wechselwirkung, die die Wechselwirkung der Kolloidteichen über Strömungsfluktuationen im Lösungsmittel berücksichtigt, dann kann man die Reibungsfunktion in zwei Anteile zerlegen,

$$\tilde{\zeta}_{\|}(\underline{k}, z) = \zeta_0 + \underline{k}^2 \tilde{\eta}_{\|}(\underline{k}, z)/c , \quad (2.14)$$

den Stokesschen Reibungskoeffizienten ζ_0, der durch die Reibungskraft jedes einzelnen Teilchens am Lösungsmittel verursacht wird, und eine longitudinale Viskositätsfunktion $\tilde{\eta}_{\|}(\underline{k}, z)$, die von der Wechselwirkung zwischen Kolloidteilchen herrührt. Durch die starke Reibung mit dem Lösungsmittel, die durch die interne Reibung noch verstärkt wird, wird das System überdämpft sein, und der Ausdruck für den Strukturfaktor vereinfacht sich zu

$$\tilde{S}(\underline{k}, z) = \frac{S(\underline{k})}{z + \frac{k_B T / S(\underline{k})}{\zeta_0 + \underline{k}^2 \tilde{\eta}_{\|}(\underline{k}, z)/c}} . \quad (2.15)$$

Mit Hilfe eines „Linear Response“-Formalismus kann ein mikroskopischer Ausdruck in Form einer Kraft-Autokorrelationsfunktion für die Viskositätsfuntion abgeleitet werden. Betrachten wir kurz den Fall eines wechselwirkungsfreien Systems. Dann ist $S(\underline{k}) = 1$ und vor allem $\tilde{\eta}_{\|}(\underline{k}, z) = 0$. Wir erhalten also das wohlbekannte Einsteinsche Ergebnis

$$S(\underline{k}, t) = e^{-\frac{k_B T}{\zeta_0} \underline{k}^2 t} , t \geq 0 . \quad (2.16)$$

Für endliche Konzentrationen finden wir jedoch, daß sich der Einfluß der Wechselwirkung einmal in Form einer wellenvektorabhängigen osmotischen Kompressibilität, und zum anderen in der wellenvektor- und frequenzabhängigen Viskosität äußert. Diese Frequenzabhängigkeit bringt es mit sich, daß die Korrelationsfunktionen im allgemeinen keine einfache Zeitabhängigkeit, wie etwa in Form von Exponentialfunktionen in (2.7) und (2.16) besitzen. In einer Reihe von Photonkorrelations-Experimenten [6, 9] an Polystyrol-Latices, deren Teilchen stark geladen waren, wurde gezeigt, daß $S(\underline{k}, t)$ keine einfache Exponentialfunktion ist und daß die Abweichung von der Exponentialfunktion stark mit dem Streuvektor variiert (Abb. 1).

Berechnet man nun die Viskositätsfunktion in (2.15), so ist dies im allgemeinen nur numerisch möglich, und ein geschlossener Ausdruck für $S(\underline{k}, t)$ ist nicht erhältlich. Man kann jedoch gewisse Parameter einführen, die die zeitliche Korrelationsfunktion charakterisieren. Zur Beschreibung des Kurzzeitverhal-

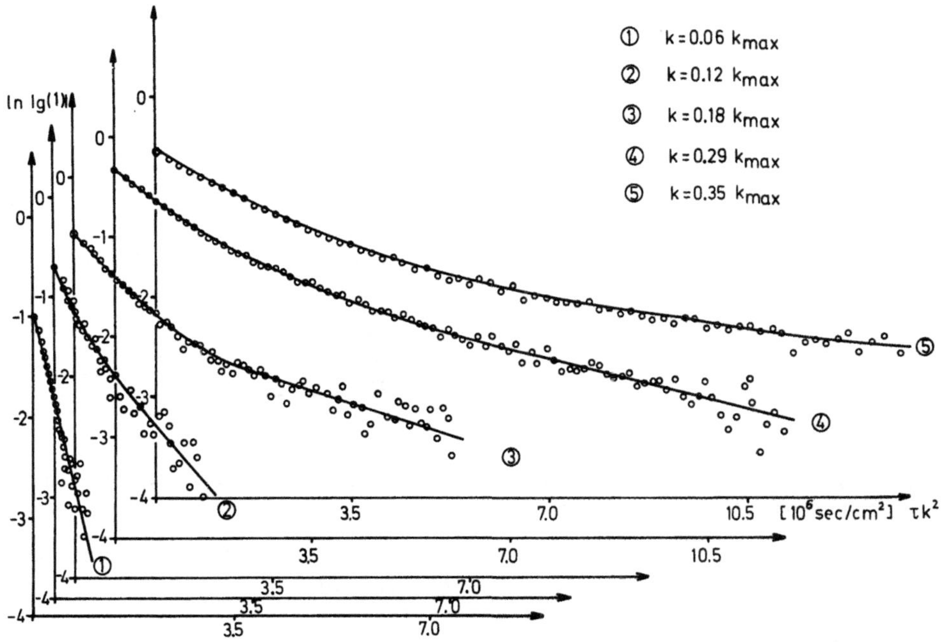

Abb. 1. Experimentelle Ergebnisse für den dynamischen Strukturfaktor eines Polystyrol-Latex bei verschiedenen Streuvektoren [9]. k_{max} ist der Streuvektor bei dem der statische Strukturfaktor sein Maximum hat

tens benutzt man häufig eine Kumulantenentwicklung,

$$S(\underline{k}, t) = S(\underline{k}) \exp\left[-\sum_{n=1}^{\infty} \frac{1}{n!} \mu_n(\underline{k})\, t^n\right], \; t \geq 0 \quad (2.17)$$

Die ersten beiden Kumulanten sind dann

$$\mu_1(\underline{k}) = \frac{k_B T k^2}{\zeta_0 S(\underline{k})} \quad (2.18\,\mathrm{a})$$

$$\mu_2(\underline{k}) = -\frac{k_B T k^2}{\zeta_0 S(\underline{k})} \frac{\eta_{\parallel}(\underline{k}, t=0)\, k^2}{c\,\zeta_0}. \quad (2.18\,\mathrm{b})$$

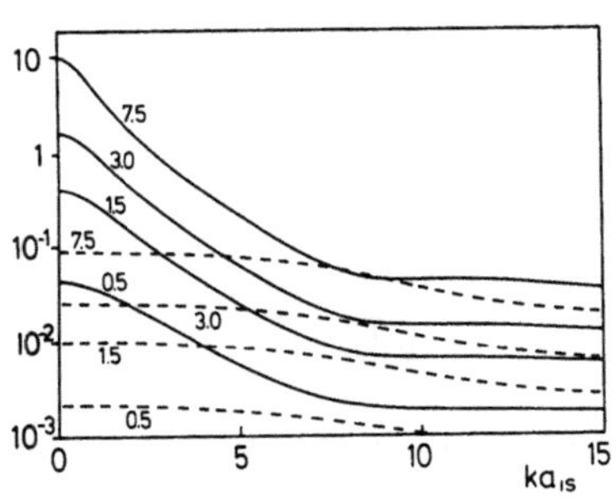

Abb. 2. Hochfrequente elastische Moduln als Funktion von $ka_{is} = k\,d/2\,\phi^{-1/3}$; d ist der Durchmesser eines Teilchens bei verschiedenen Volumenkonzentrationen ϕ zwischen $5.0 \cdot 10^{-3}$ und $7.5 \cdot 10^{-3}$. Volle Linien: $e^{\infty}(\underline{k}) \equiv G^{\infty}(\underline{k})\, 4\pi\,(d/2)^3/(3\,k_B T)$; unterbrochene Linie: $g^{\infty}(\underline{k}) \equiv G^{\infty}(\underline{k})\, 4\pi\,(d/2)^3/(3\,k_B T)$

Dieser zeitliche Anfangswert der dynamischen Viskosität, $\eta_{\parallel}(\underline{k}, t=0)$, ist gerade das, was in den Neutronenstreu-Untersuchungen einfacher Flüssigkeiten als hochfrequente elastische Konstanten bezeichnet wird

$$\eta_{\parallel}(\underline{k}, t=0) = E^{\infty}(\underline{k}) - c\frac{k_B T}{S(\underline{k})} \quad (2.19)$$

$$E^{\infty}(\underline{k}) = c\left[3\,k_B T + c\int d^3 r\, g(r) \frac{\partial^2 U}{\partial z^2} \frac{1 - \cos kz}{k^2}\right] \quad (2.20)$$

und was man in der Rheologie den Speicher-Modul nennen würde. Wir haben diese Parameter für ein System berechnet, das gerade dem von Grüner und Lehmann [10] untersuchten Polystyrolkugel-Latex entspricht. Die Ergebnisse für verschiedene Konzentrationen sind in Abbildung 2 gezeigt. Die gebrochenen Kurven stellen den entsprechenden Schermodul dar. Man sieht, daß der longitudinale Modul bei kleinen Streuvektoren sehr viel größer ist als der Schermodul. Der k-Wert, bei dem beide sich annähern, entspricht etwa dem reziproken mittleren Abstand zweier benachbarter Teilchen. Untersucht man die Konzentrationsabhängigkeit der Moduln bei $k = 0$, dann erhält man Abbildung 3. Der longitudinale Modul wächst quadratisch mit der Konzentration an,

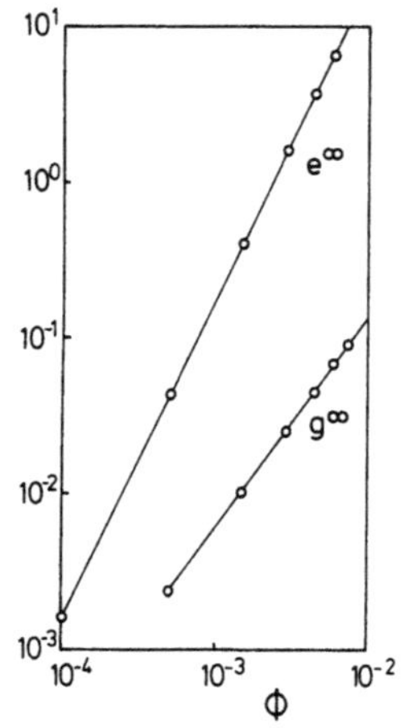

Abb. 3. Langwelliger Limes der hochfrequenten elastischen Moduls $e^{\infty}(0)$ und $g^{\infty}(0)$ als Funktion der Volumenkonzentration ϕ

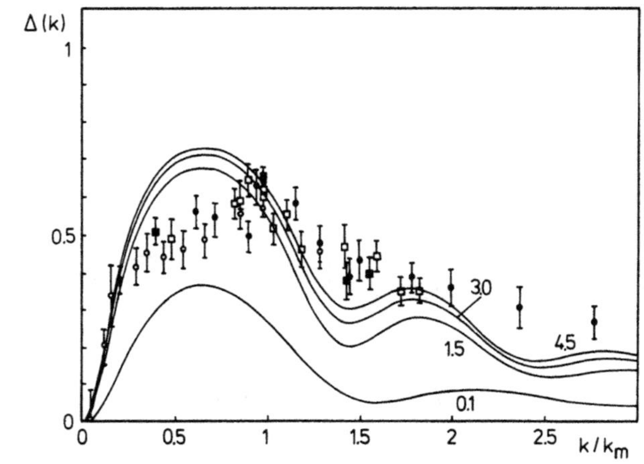

Abb. 4. $\Delta(\underline{k})$, Gleichung (2.24). Die Kurven sind theoretische Ergebnisse für $\phi = 0.1; 1.5; 3.0; 4.5\,10^{-3}$. Datenpunkte sind experimentelle Ergebnisse von Grüner und Lehmann [9] zwischen $\phi = 1.5\,10^{-3}$ und $7.5\,10^{-3}$. k_m ist der Streuvektor bei dem $S(\underline{k})$ sein Maximum hat

während der Schermodul überraschenderweise wie $c^{4/3}$ variiert. Diese quadratische Konzentrationsabhängigkeit des longitudinalen Moduls ist auch von Grüner und Lehmann experimentell festgestellt worden [10].

Es scheint hier besonders wichtig, darauf hinzuweisen, daß, unter bestimmten Umständen, die Bestimmung der Streuvektorabhängigkeit des longitudinalen elastischen Moduls eine direkte Möglichkeit eröffnet, das Zweiteilchen-Wechselwirkungspotential zu messen. Man kann den mikroskopischen Ausdruck (2.20) für den Modul in folgender Weise umschreiben, wobei jetzt $U(\underline{k})$ die räumliche Fourier-Transformierte des Zweiteilchen-Wechselwirkungspotentials ist.

$$E^{\infty}(\underline{k}) = c\,k_B T \left[\frac{c\,U(\underline{k})}{k_B T} + 3 - \frac{1}{k_B T} \frac{1}{(\underline{k}^2)^2} \frac{1}{(2\pi)^3} \int d^3k'\, h(\underline{k}') \left[(\underline{k}\cdot\underline{k}')^2\, U(\underline{k}') - (\underline{k}\cdot(\underline{k}-\underline{k}'))^2\, U(\underline{k}-\underline{k}') \right] \right] \tag{2.21}$$

$h(\underline{k})$ ist die Fouriertransformierte der statischen radialen Korrelationsfunktion. Die numerische Berechnung ergibt nun, daß für das stark geladene Polystyrolkugel-System, für genügend kleine Wellenvektoren, die restlichen Terme der rechten Seite in (2.21) gegenüber dem ersten Term vernachlässigt werden können, und wir in sehr guter Näherung schreiben können

$$\frac{\mu_2(\underline{k})}{\mu_1(\underline{k})} - \mu_1(\underline{k}) = -\frac{c}{\zeta_0} \underline{k}^2\, U(\underline{k})\,. \tag{2.22}$$

Durch die Messung der Wellenvektorabhängigkeit der ersten beiden Kumulanten könnte also die Fouriertransformierte des Wechselwirkungspotentials bestimmt werden. Diese Aussage gilt unter der Voraussetzung, die für das betrachtete System gegeben ist, daß die Wechselwirkung stark und langreichweitig ist.

Betrachten wir wieder die vollständige Zeitabhängigkeit von $S(\underline{k}, t)$, so ist das charakteristische Ergebnis, daß die Relaxation für große Zeiten langsamer ist als am Anfang. Diese Verlangsamung der Dynamik der Konzentrationsfluktuationen ist zurückzuführen auf die zusätzliche interne Reibung aufgrund der Wechselwirkung der Kolloidteilchen untereinander, die sich in der dynamischen Viskosität ausdrückt. Definieren wir eine mittlere Relaxationszeit des dynamischen Strukturfaktors als

$$\tau(\underline{k}) = \frac{1}{S(\underline{k})} \int_0^{\infty} dt\, S(\underline{k}, t) = \frac{\tilde{S}(\underline{k}, z=0)}{S(\underline{K})}, \tag{2.23}$$

und als Maß für die Abweichung von $S(\underline{k}, t)$ von einer Exponentialfunktion die Größe

$$\Delta(\underline{k}) = \frac{\mu_1(\underline{k}) - \tau(\underline{k})^{-1}}{\mu_1(\underline{k})} = \frac{k^2\, \tilde{\eta}_{\|}(\underline{k}, z=0)}{c\,\zeta_0 + k^2\, \tilde{\eta}_{\|}(\underline{k}, z=0)}, \tag{2.24}$$

so wird diese gerade durch den Wert der frequenzabhängigen Viskosität bei der Frequenz Null bestimmt. Mit Hilfe einer Modenkopplungsapproximation wurde $\tilde{\eta}_{\parallel}(\underline{k}, z=0)$ berechnet, das Ergebnis für 4 verschiedene Konzentrationen ist in Abbildung 4 gezeigt. Die experimentellen Werte sind der Arbeit von Grüner und Lehmann entnommen [9]. Man sieht eine ausgezeichnete Übereinstimmung bei den 3 höheren Konzentrationswerten, die den experimentellen entsprechen. Die kleinste gerechnete Volumen-Konzentration von $\phi = 10^{-4}$ ist experimentell nicht untersucht worden.

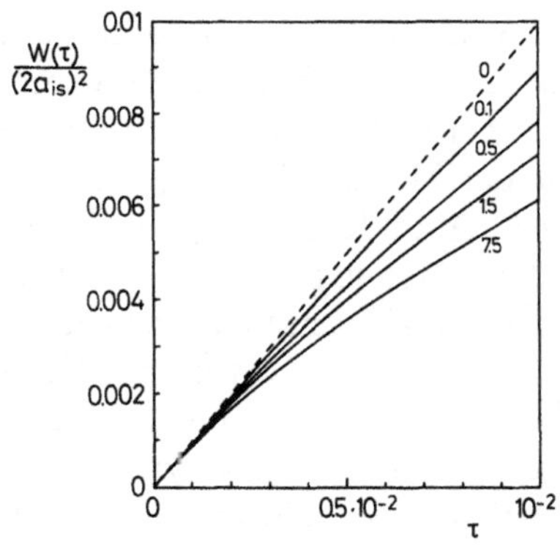

Abb. 5. Mittleres Verschiebungsquadrat als Funktion der reduzierten Zeit $\tau = t\,(k_BT/\zeta_0)\,(2a_{is})^{-2}$, bei 4 verschiedenen Konzentrationen zwischen $0.1 \cdot 10^{-3}$ und $7.5 \cdot 10^{-3}$. Die gebrochene Linie entspricht freier Diffusion ($\phi = 0$)

3. Selbstdiffusion

Die wesentliche Größe zur statistischen Beschreibung der Bewegung eines einzelnen Teilchens ist der Selbstdiffusionspropagator $G(\underline{r}, t)$. Dieser gibt die Wahrscheinlichkeit dafür an, daß ein Teilchen, das zur Zeit 0 am Ort $\underline{r}'$ war, zur Zeit t am Ort $\underline{r}$ sein wird. Um nun im Rahmen der verallgemeinerten Hydrodynamik einen Ausdruck für den Propagator zu erhalten, geht man ganz analog vor, wie es im letzten Kapitel für den dynamischen Strukturfaktor gezeigt wurde. Wieder erhält man geschlossene Ausdrücke für die Fourier-Laplace-transformierten Größen,

$$\tilde{G}(\underline{k}, z) = \frac{1}{z + \tilde{D}_s(\underline{k}, z)\,\underline{k}^2} \tag{3.1}$$

$$\tilde{D}_s(\underline{k}, z) = \frac{k_B T}{mz + \tilde{\zeta}_s(\underline{k}, z)} \tag{3.2}$$

$\tilde{D}_s(\underline{k}, z)$ ist die verallgemeinerte Diffusionsfunktion für den Selbstdiffusionsprozeß, $\tilde{\zeta}_s(\underline{k}, z)$ die entsprechende Reibungsfunktion. Beide Größen dürfen nicht mit den jeweils entsprechenden Funktionen, die die Dynamik der Konzentrationsfluktuationen beschreiben, gleichgesetzt werden. Die Diffusionsfunktion $\tilde{D}_s(\underline{k}, z)$ kann mit zwei weiteren, für die Charakterisierung der Einteilchendynamik wichtigen Korrelationsfunktionen, in Zusammenhang gebracht werden, mit der Geschwindigkeits-Autokorrelationsfunktion

$$V(t) = \frac{1}{3} < \underline{v}(t) \cdot \underline{v}(0) >, \tag{3.3}$$

wobei $\underline{v}(t)$ die Geschwindigkeit des betrachteten Teilchens ist und mit dem mittleren Verschiebungsquadrat $W(t)$, Gleichung (1.1). Denn es gilt

$$V(t) = D_s(k=0, t) \tag{3.4}$$

und andererseits

$$W(t) = \int_0^t dt'\,(t - t')\,V(t'). \tag{3.5}$$

Im hydrodynamischen Limes geht $\tilde{D}_s(\underline{k}, z)$ in den Selbstdiffusionskoeffizienten über,

$$D_s = \lim_{\substack{k, z \to 0 \\ k^2/z = \text{const}}} \tilde{D}_s(\underline{k}, z) = \tilde{D}_s(0, 0), \tag{3.6}$$

und aus den Gleichungen (3.4) und (3.5) folgen dann die wohlbekannten Ergebnisse

$$D_s = \int_0^\infty dt\,V(t) \tag{3.7}$$

und

$$W(t) \to D_s t \quad \text{für } t \to \infty. \tag{3.8}$$

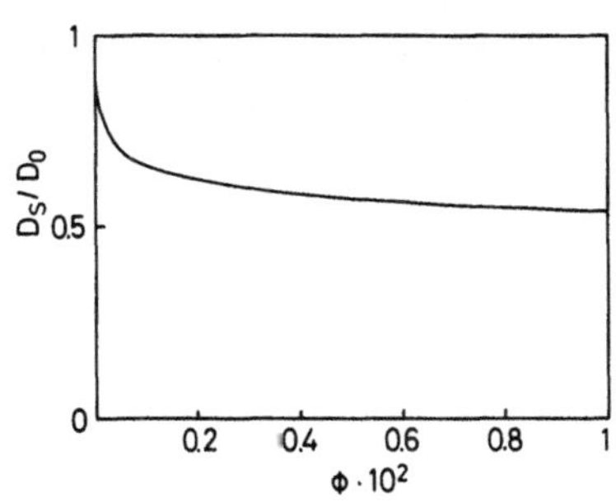

Abb. 6. Selbstdiffusionskoeffizient als Funktion der Volumenkonzentration für Polystyrolkugeln mit $d = 900$ Å, Oberflächen-Potential $\psi_0 = 73$ mV und einer Abschirmlänge von 5000 Å. D_0 ist der Wert bei $\phi = 0$, $D_0 = k_BT/\zeta_0$

Auch für die Selbstdiffusions-Reibungsfunktion $\tilde{\zeta}_s(\underline{k}, z)$ kann mit Hilfe des „Linear Response"-Formalismus ein mikroskopischer Ausdruck angegeben werden [8], der mit Hilfe der Modenkopplungsapproximation berechnet wurde. Abbildung 5 zeigt das Ergebnis für das mittlere Verschiebungsquadrat, die verschiedenen Kurven zeigen das bereits in der Einleitung diskutierte Verhalten. In Abbildung 6 wird die Konzentrationsabhängigkeit des Selbstdiffusionskoeffizienten für das System aus geladenen Polystyrolkugeln gezeigt. Man sieht, daß für kleine Konzentrationen eine scharfe Abnahme von D_s erfolgt, während danach eine relativ geringe Konzentrationsabhängigkeit einsetzt. Interessanterweise liegt der Übergang in diesen „Plateau-Bereich" bei einer Konzentration, bei der der Nächste-Nachbar-Abstand gerade von der Größenordnung der Debye-Länge der Gegenionen ist! Man kann das Ergebnis so interpretieren, daß bei höheren Konzentrationen die Kolloidteilchen selbst zur Abschirmung beitragen und daß deshalb ein Sättigungseffekt einsetzt. Ein ähnliches Verhalten kann man auch in der Funktion $\Delta(\underline{k})$, Abbildung 4, sehen. Diese zeigt ebenfalls eine relativ schwache Konzentrationsabhängigkeit im Bereich $1.5 \leq \phi \cdot 10^3 \leq 4.5$, in Übereinstimmung mit dem experimentellen Ergebnis, während die Kurve für $\phi = 10^{-4}$ deutlich darunter liegt.

4. Schluß

Wir haben hier versucht, an Hand einiger Beispiele zu zeigen, daß man mit Hilfe verallgemeinerter hydrodynamischer Gleichungen die dynamischen Eigenschaften konzentrierter kolloidaler Systeme qualitativ und quantitativ verstehen kann. Dabei richtet sich die quantitative numerische Berechnung vor allem auf ein System stark geladener, langreichweitig wechselwirkender Polystyrolkügelchen, das in den letzten Jahren eingehend experimentell untersucht wurde. Das Ergebnis zeigt, daß man ein konzentriertes kolloidales System als eine überdämpfte Flüssigkeit verstehen kann, bei der auf Grund der kurzreichweitigen Ordnung lokale elastische Spannungsfluktuationen auftreten, die die Relaxation der Konzentrationsfluktuationen und die Teilchendiffusion verlangsamen. Obwohl unsere Arbeit sich soweit auf kugelförmige Teilchen beschränkt, lassen sich doch die allgemeinen Ergebnisse, wie sie hier präsentiert wurden, ohne weiteres auf Systeme aus stäbchenförmigen Teilchen oder solche mit inneren Oszillationsfreiheitsgraden übertragen. Lediglich die mikroskopischen Definitionen der verallgemeinerten Transportkoeffizienten, die in Referenz [8] gegeben werden, müßten für diese Systeme neu abgeleitet werden und diese zusätzlichen Freiheitsgrade berücksichtigen. Selbstverständlich müßten in diesem Fall auch die detaillierten numerischen Berechnungen neu durchgeführt werden.

Referenzen

1. Hayter JB (1978) In: Neutron Diffraction, Dachs H (ed) Springer, Berlin
2. Cummins HZ, Pike ER (eds) (1974) Photon Correlation and Light Beating Spectroscopy, Plenum Press, New York
3. Cummins HZ, Pike ER (eds) (1977) Photon Correlation Spectroscopy and Velocimetry, Plenum Press, New York
4. Berne BJ, Pecora R (1976) Dynamic Light Scattering, Wiley, New York
5. DeGiorgio V, Corti M, Giglio M (eds) (1980) Light Scattering in Liquids and Macromolecular Solutions, Plenum Press, New York
6. Pusey PN, Tough RJ (1982) In: Dynamic Light Scattering and Velocimetry: Applications of Photon Correlation Spectroscopy, Pecora R (ed) Plenum Press, New York
7. Rondelez F, in Referenz 5
8. Hess W, Klein R (1983) Advances in Physics 32:173
9. Grüner F, Lehmann W (1979) J Phys A 12:L 303
10. Grüner F, Lehmann W (1982) J Phys A 15:2847

Anschrift der Verfasser:

W. Heß
Fakultät für Physik
Universität Konstanz
7750 Konstanz

Subject Index

GPSR Compliance

The European Union's (EU) General Product Safety Regulation (GPSR) is a set of rules that requires consumer products to be safe and our obligations to ensure this.

If you have any concerns about our products, you can contact us on ProductSafety@springernature.com

In case Publisher is established outside the EU, the EU authorized representative is:

Springer Nature Customer Service Center GmbH
Europaplatz 3
69115 Heidelberg, Germany

Batch number: 10578403

Printed by Printforce, the Netherlands